NORTH AMERICA
A Geographical Mosaic

Edited by

FREDERICK W. BOAL

Professor Emeritus of Human Geography, The Queen's University of Belfast

and

STEPHEN A. ROYLE

School of Geography, The Queen's University of Belfast

A member of the Hodder Headline Group
LONDON

First published in Great Britain in 1999 by
Arnold, a member of the Hodder Headline Group
338 Euston Road, London NW1 3BH

http://www.arnoldpublishers.com

Co-published in the United States of America by
Oxford University Press Inc.
198 Madison Avenue, New York, NY 10016

British Library Cataloguing in Publication Data
A catalogue entry for this book is available from the British Library

Library of Congress Cataloging-in-Publication Data
A catalog record for this book is available from the Library of Congress

Production Editor: Julie Delf
Production Controller: Iain McWilliams
Cover Design: T. Griffiths

ISBN 0 340 69262 6 (hb)
ISBN 0 340 69261 8 (pb)

Typeset by J&L Composition Ltd, Filey, North Yorkshire
Printed and bound in Great Britain by MPG Books Ltd, Bodmin, Cornwall

What do you think about this book? Or any other Arnold title?
Please send your comments to feedback.arnold@hodder.co.uk

To the US citizens in my family (FWB)
To Lisa (SAR)

CONTENTS

SECTION E THE URBAN SCENE

SECTION F REGIONAL VIGNETTES

SECTION G THE CONTINENT AND THE WORLD

CONTRIBUTORS

Henk Aay is Professor and Chair, Department of Geology, Geography and Environmental Studies, Calvin College, USA

John A. Agnew is Professor and Chair of Geography, University of California at Los Angeles, USA

James P. Allen is Professor of Geography, California State University, Northridge, USA

William B. Beyers is Professor of Geography, University of Washington, USA

Frederick W. Boal is Professor Emeritus of Human Geography, The Queen's University of Belfast, UK

Larry S. Bourne is Professor of Geography, University of Toronto, Canada

William J. Carlyle is Professor of Geography, University of Winnipeg, Canada

Donald G. Cartwright is Professor of Geography, University of Western Ontario, Canada

Kenneth S. Coates is Dean of Arts, University of New Brunswick at Saint John, Canada

David Demeritt is Lecturer in Geography, King's College London, UK

Owen J. Furuseth is Professor of Geography, University of North Carolina at Charlotte, USA

Stephen J. Hornsby is Director, Canadian–American Center and Associate Professor of Anthropology, University of Maine, USA

Loretta Lees is Lecturer in Geography, King's College London, UK

John Mercer is Professor and Chair of Geography, Syracuse University, USA

William R. Morrison is Professor of History, University of Northern British Columbia, Canada

Jeffrey C. Rogers is Professor of Geography, Ohio State University, USA

Stephen A. Royle is Senior Lecturer in Geography, The Queen's University of Belfast, UK and President of the Geographical Society of Ireland

Houston C. Saunderson is Professor of Geography, Wilfrid Laurier University, Canada

Donald J. Savoie holds the Clément-Cornier Chair in Economic Development, Institut Canadien de Recherche sur le Développement Régional, Université de Moncton, Canada

Richard H. Schein is Associate Professor of Geography, University of Kentucky, USA

David M. Smith is Professor of Geography, Queen Mary and Westfield College, University of London, UK

Iain Wallace is Professor of Geography, Carleton University, Canada and President of the Canadian Association of Geographers

James O. Wheeler is the Merle Prunty Jr Professor of Geography, University of Georgia, USA

William C. Wonders is University Professor Emeritus of Geography, University of Alberta, Canada

Maurice Yeates is Professor of Distinction, Centre for the Study of Commercial Activity, Ryerson Polytechnic University, Ontario, Canada

ACKNOWLEDGEMENTS

The editors are from Queen's University, not that of Kingston, Ontario, but the Queen's University of Belfast in Northern Ireland, although we both have much experience of travelling to, living and working in Canada and the United States. We would like to thank former colleagues at the University of Michigan, the University of Calgary, the University of Toronto, Carleton University, the University of Iowa and the University of Prince Edward Island for helping to stimulate our interest in North America. Thanks, too, to colleagues at Queen's University Belfast with whom we have presented a course on the regional geography of North America from which emanated the idea of editing this book. We are both involved, too, in the Centre of Canadian Studies at the Queen's University of Belfast, which has helped us to maintain our interest in the larger part of the North American continent.

The staff at Arnold have been ever courteous and patient with us in our efforts and thanks are especially due to Laura McKelvie, former Publisher, Geography and Environmental Science. The maps were prepared for us by the cartographic staff of the School of Geography at the Queen's University of Belfast and we are grateful indeed to Gill Alexander and Maura Pringle for their work. We would also like to thank Larry Bourne, Jenitha Orr and Nick Tate for help of various kinds.

The photographs used to open Chapters 2, 22 and 23 were taken by Noel Mitchel of the Queen's University of Belfast; those to open Chapters 3–7, 9–13, 15, 16, 19–21 and 24 by Stephen Royle; that to open Chapter 8 by Frederick Boal; that to open Chapter 14 by David Smith; that to open Chapter 17 by Stephen Hornsby; and that to open Chapter 18 by Julie Delf of Arnold.

Finally we must thank the contributors for their efforts. Almost all North Americans, they none the less agreed to write for a couple of foreigners who would turn their miles to kilometres, acres to hectares, beat their plows into ploughs and add seemingly redundant 'u's to all their 'labors'.

Section A

Introduction

1

INTRODUCING THE CONTINENTAL MOSAIC

FREDERICK W. BOAL AND STEPHEN A. ROYLE

CONTINENTS IN COLLISION

North America, geologically speaking, has been formed by separation and accretion. Separation entailed the break-up of the former super continent of Pangaea. This break-up created what became the North Atlantic. Accretion geologically constructed the North American continent by adding rock to the core (craton) now known as the Canadian Shield (see Chapter 2). Indeed, land mass is still being 'docked' on to the western edge of the continent.

Very recently, in geological terms, the Eurasian and North American continents have been brought together again socially. Two writers have forcefully dramatized this process. First, Jared Diamond, in his 1997 volume *Guns, germs and steel*, entitles one of his chapters 'Hemispheres colliding':

> After at least 13 000 years of separate developments, advanced American and Eurasian societies finally collided within the last 1000 years. Until then, the sole contacts between human societies of the Old and the New Worlds had involved the hunter-gatherers on opposite sides of the Bering Strait (Diamond, 1997, p. 370).

Second, Alfred Crosby, in *Ecological imperialism* (1986, p. 131) strikingly summarizes this process when he declares that 'the seams of Pangaea were closing, drawn together by the sailmaker's needle.' The closing of the seams was an overwhelmingly European project, with profound effects on the peopling and cultural characteristics of Canada and the United States. Jared Diamond asks the question 'why did Europeans reach and conquer the lands of Native Americans, instead of vice versa?' His answer is that Europeans had the advantage of a greater range of domesticable plants and animals, that they had a more advanced technology and, sadly, that they were the carriers of a much greater range of harmful germs. These impacted on the Native American populations with a terrible severity, and did much more to undermine native resistance to European intrusion than any force of arms.

CONTINENTAL ACCRETION

Just as the Canadian Shield provided the foundation on to which the geological accretion of North America has taken place, so the St Lawrence and the eastern seaboard have provided a similar base for the 'accretion' of the human geography – the former predominantly French, the latter British. This accretion process can be viewed at two geographical scales and at two different time periods. Initially, beyond the eastern attachment points, vast and somewhat vague claims were being made – the French up the St Lawrence, the Great Lakes and the Mississippi, and the Spanish to the southern continental fringes. The British, initially, were rather hemmed in east of the Appalachians. Consequent upon the outcomes of various wars between the competing European powers, the British succeeded in taking over almost all the French claims (the St Lawrence, the Great Lakes and the eastern part of the Mississippi Basin). Spain hung on tenuously to the southern and southwestern territories.

Perhaps one of the great ironies of North American history is the fact that very soon after the British had succeeded in extinguishing the French claims, the separatist sentiments of the 'British' American colonists on the eastern seaboard were converted into reality by the American Revolution (1776). The so-called '13 colonies' became the 13 founding states of the United States of America. Additionally, in 1783 the British trans-Appalachian claims were ceded to the new country. At the same time Britain retained the area to the north of the Great Lakes and also the lower St Lawrence Valley and the lands in close proximity to the Gulf of St Lawrence (Newfoundland, Nova Scotia, New Brunswick and Prince Edward Island). This territory subsequently became the core for the accretion of present-day Canada.

Several things are noteworthy with regard to the human accretion of the United States and Canada. First, territory was acquired ahead of what Walter Prescott Webb has called 'the migratory horde that was moving west' (1953, p. 5). The United States obtained a huge addition to its land area by the purchase of Louisiana from France in 1803 (France having obtained it from Spain three years earlier). Louisiana, as then defined, extended from the Mississippi west to the foothills of the Rockies. Subsequently Florida was acquired from Spain (1819), Texas was annexed in 1845 (having previously seceded from Mexico), the Oregon Territory south of the 49th parallel was ceded by Britain in 1846, while a huge increment of land was gained in the southwest at the expense of Mexico in 1848. Other additions followed, such as the purchase of Alaska from Russia in 1867.

Lying behind this expansion was what a New York newspaper editor, John L. O'Sullivan, in 1845, called 'manifest destiny': 'Our manifest destiny to overspread and to possess the whole of the continent which Providence has given us for the development of the great experiment of liberty and federated self-government.'

Canada, perhaps, never had such an imperative to motivate continental acquisition. None the less a process not unlike that observed in the case of the United States can be seen north of what became the Canada–US border as well. Beyond the core settlement area of the St Lawrence and the Lower Great Lakes, a huge expanse known as Rupert's Land (basically the entire Hudson Bay drainage basin) was granted by Charles II to the fur trading Hudson's Bay Company in 1670. Rupert's Land was transferred to Canada 200 years later. In 1871 the transcontinental extent of the country was completed when the then British colony of British Columbia entered the Canadian Federation.

A second dimension of the accretion process occurred when sections of the 'unorganized' lands became states or provinces – that is when governmental functions below that of the federal are transferred to the lower level political unit (the 'state' [US] or the 'province' [Canada]). Fig. 1.1 maps this accretion process. Building on the core areas (St Lawrence–Lower Great Lakes in the case of Canada, the 13 original states in the case of the US), a westward expansion process can readily be detected. However there is one major exception to this – the jump to the Pacific, followed by a later back-filling of the Prairies and

the Great Plains. In addition, in the case of Canada there is evidence of a northward expansion, seen not in the formation of new provinces out of former territories, but in the northward extension of existing provincial units – specifically Québec, Ontario and Manitoba. This northward accretion at the level of provinces is not complete, with the Yukon Territory, the Northwest Territories and the newly proclaimed Nunavut (1 April 1999) not yet having acquired provincial status. In all this Newfoundland and Labrador is somewhat of an anomaly. According to the continental east–west dynamic, Newfoundland should have been part of the Canadian core, or at least an early addition. However Newfoundland only joined the Canadian Federation in 1949. Such a delay may reflect, at least in part and somewhat ironically, a prolonged attachment to the 'mother country' (the United Kingdom).[1]

TURNING THINGS AROUND

The accretion of the human geography of the North American continent began in the east and subsequently extended westwards (with the exception of the prehistoric entry of Native peoples and limited Russian exploration southwards along the Pacific coast). Our whole perception of North American geography is shaped by this east–west sequence. But what if the early colonization had not been European but Asian, with the settlement core lying along the Pacific seaboard rather than the Atlantic? Donald Meinig, in Volume II of his *magnum opus*, *The shaping of America*, gives us some 'what if' geography[2] when he suggests that under different historical circumstances the United States might have been more extensive than it actually is, taking in northern Mexico and possibly Cuba (with major implications for the linguistic and cultural composition of the country). Alternatively, Meinig suggests a smaller US, with part of the present northwest being a component of 'British North America', and the southwest occupied by a more extensive Mexico, by a 'Republic of California', a 'Republic of Texas', and a Mormon polity of 'Deseret' (Meinig, 1993, p. 215).

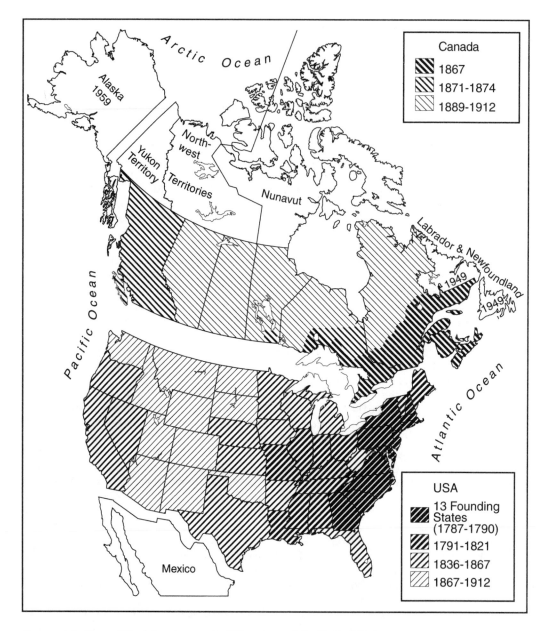

FIGURE 1.1 The political development of Canada and the United States
Note: Hawaii became a State in 1959

However, here we want to suggest a counterfactual geography where the core early non-Native American was Chinese, with a smaller but still significant Japanese presence in the Puget Sound–lower Fraser Valley area. A transcontinental accretion process starting thus in the West could have created a North America dominated by Chinese culture and language, with a Japanese enclave remaining on the lower Fraser River (the Native Americans, in these circumstances, certainly would not have been referred to as 'Indians'!). Of course environmental conditions for such an eastward expansion would have been less favourable than those encountered by the westward marching Europeans. The Rockies would have presented a more formidable barrier than the Appalachians, and the land to the east of the Rockies (the dry High Plains and the western Prairies) would have been less welcoming territory than the land between the Appalachians and the Mississippi. No matter how fanciful such a geographical reconstitution may be, by reorientating our perspective we can, as Meinig suggests, jar ourselves 'out of habits of mind, loosen images so familiar, so constantly put before us, so deeply imprinted on our national consciousness, that they are assumed to be fixed and inevitable' (Meinig, 1993, p. 217).

Of course there is an unlimited supply of possible 'what ifs' – what if the French had won on the Heights of Abraham in 1759 rather than the British? Would North America have become a French-speaking continent, with a still-persistent English speaking enclave on the Atlantic seaboard? If French had become the dominant language of the 'United States', would today's world *lingua franca* be French rather than English? Think about it!

ON THE FRONTIER

'Collision' and 'accretion' may be major background themes in the geography of North America. The idea of the 'frontier' is much more up-front, so to speak. The widespread use of the term undoubtedly owes its prominence to the writings of the American historian, Frederick Jackson Turner. In a famous paper presented in 1893 he noted two things. First, as observed by the Superintendent of the United States Census, while the country had had a frontier up to and including 1880 (that is an expanding edge of settlement beyond which human occupancy dropped below two persons per square mile), after that date such a situation could no longer be recognized. Second, Turner made the grand claim that 'the existence of an area of free land, its continuous recession and the advance of American settlement westward, explain American development' (Turner, 1893). This claim has been subjected to subsequent criticism, but the pervasiveness of the idea of the 'frontier' itself has been sustained. Turner's frontier was an edge of European settlement that pushed westwards for 100 years. There were many frontiers over this period, each marking a specific stage in the westward expansion. The frontier zone was one of perceived opportunity, where, as William Cronon puts it, 'unexploited natural abundance was the central meaning' (Cronon, 1991, p. 150) or, as Walter Prescott Webb wrote (1953, p. 2), 'the American thinks of the frontier as lying within and not at the edge of the country. It is not a line to stop at but *an area inviting entrance*' (our emphasis). Of course, in all this any pre-existing population becomes well-nigh invisible. There is little concern for what things may look like from 'the other side of the frontier' (Bohannan, 1967; Reynolds, 1982).

Providing an alternative perspective, however, Patricia Limerick has drawn our attention to another North American frontier – '*la frontera*.' *La frontera* is the borderland between Mexico and the United States. Here, according to Limerick, 'there is no illusion of vacancy, of triumphal conclusions' (Limerick, 1994, p. 90), though in a curious twist, it still can be seen 'as an area inviting entrance' – people moving from Mexico into the US in search of work, and US entrepreneurs moving south of the Rio Grande to set up manufacturing plants sustained by cheap labour.

The Turnerian frontier is seen as an expanding edge of rural settlement. However Cronon argues that the frontier and the metropolis are not two separate phenomena, but two parts of the same thing:

> Turner's frontier, far from being an isolated rural society was in fact the expanding edge of the [city] boosters' urban empire. ... Frontier and metropolis turned out to be two sides of the same coin. ... As village became metropolis, so frontier became hinterland (Cronon, 1991, pp. 51, 53).

As we will now see, the frontier is not just the other side of the same coin – it has actually appeared as a metropolitan phenomenon in its own right.

While the frontier has been seen as the expanding edge of the accretion process that contributed to the shaping of the human geography of North America, it has also made an appearance in many other guises. As Limerick has pointed out 'as a mental artefact, the frontier has demonstrated an astonishing stickiness and persistence. It is virtually the flypaper of our mental world; it attaches itself to everything' (Limerick, 1994, p. 94). As early as 1953 Webb expressed concern that 'new frontiers are being conjured up in every direction' (1953, p. 280). Nevertheless, let us look briefly at three instances of geographical applications – to the suburbs, to the inner city, and to the night.

The spreading edges of North American cities, particularly those in the United States, are seen as frontier environments. This notion is contained in the title of Kenneth Jackson's book *Crabgrass frontier: the suburbanization of the United States* (1985), though he only indirectly brings on board the frontier concept in his text when he suggests a land use sequence at the urban edge from 'nomadic Indians' through agriculture, suburbanization, urban decay and then renewal (p. 286). Grady Clay more directly addresses the frontier when he claims that

> I think it is useful to ask what realm of American life is having a similar formative effect [as the earlier western frontier] upon the American character. What is today's frontier? ... What we are seeing today is a shift in the old frontier from an east–west polarity to the zone of metropolitan influence of urban fronts [edges] (Clay, 1973, p. 76).

Clay sees suburbanites as being like earlier frontiersmen in that they have moved out to the edge 'to get away from those bad influences' (in this case from racial change in the older, central cities). Finally the journalist, Joel Garreau (see Chapter 18 and Fig. 18.2) in his discussion of 'edge cities' (see Chapter 13) claims that they represent 'the third wave in our lives pushing into new frontiers in this half century. Today we have moved our means of creating wealth, the essence of urbanism – our jobs – out to where most of us have lived and shopped for two generations.

This has led to the rise of edge city' (Garreau, 1991, p. 4).

The final urban frontier entails an interpretation, not of the burgeoning suburban edge, but of the economic disaster areas of the black ghetto. Here we find David Ley proposing *The black inner city as frontier outpost* (1974). This is not a land of opportunity but of threat, an attribute which would not have been absent on the western frontiers – 'the solution becomes encasement, the construction of a stockade as a defense measure' (p. 97). However in his exploration of the inner city Ley failed to uncover the frontier solidarities that he expected to find, particularly at community level.

Perhaps one of the most original applications of the frontier model is that proposed by Murray Melbin, where he sees the 'night as frontier' (Melbin, 1987). Basically, as the land frontier closed so there was a changeover from space to time as the realm for the most vigorous expansion in the United States, 'as the flow across the continent swerved into the night rather than spilling into the sea' (p. 30). Thus the night – and particularly the urban night – has become colonized by increasing activity involving a growing number of people. Of course there is nothing particularly North American in this, but the enlisting of the word 'frontier' to describe it certainly is.

CONNECTIONS

The expanding human occupance of North America relied fundamentally on good connections – trans-Atlantic navigation, movement by river and road, railroads, airways and telecommunications. Of these the railroad is historically the most significant, functionally and symbolically.

Early movement by Europeans into the heart of the continent, particularly into its northern, forested expanses relied heavily on the birch bark canoe, technology initially deployed by Native peoples and subsequently adopted by European fur traders. Most of inland Canada was first explored using such canoes, and they continued to be employed until the early nineteenth century.

Also redolent of the pioneering era was the covered wagon – the 'prairie schooner' – which was a key instrument in the trans-Appalachian and the trans-Mississippi migrations.

However effective the canoe and the covered wagon may have been, they were none the less slow means of transport of limited capacity. It was the railroad that had by far the greatest transformative effect on North America's human geography. There is a powerful sense evident amongst many historians and historical geographers that the railroad was the key device, binding east to west – 'from sea to shining sea.' The need to provide linkages was forcefully expressed even in the pre-rail area. As early as 1817 Senator Albert Leacock of Pennsylvania proclaimed 'the imperious necessity in a government ... of tying together the whole community by the strongest ligatures.' He was calling for roads and canals, but his request was really answered by the railroad. Rail networks, and particularly transcontinental lines are a dominant theme in the integration of east and centre with west in nineteenth-century North America:

> Bonds of steel as well as of sentiment were needed to hold the new confederation together. Without railways there would be and could be no Canada.... The development of steam powered railways in the nineteenth century ... was integral to the very act of nation building (Marsh, 1988, pp. 1821–2; see also Wynn, 1987; Meyer, 1987).

The early concentration of US railroad construction lay from just west of Chicago to the Atlantic seaboard, and north of the Ohio River (see the map in Meyer, 1987, p. 322). But the great continental integrating developments were the transcontinental lines – lassoing the west and tying it to the east. The first of these connections was consummated at Promontory, Utah in 1869, when the westward-extending Union Pacific met the eastward-extending Central Pacific. A second US transcontinental line was opened in 1881 (the Atchison, Topeka and Santa Fe, and the Southern Pacific). Canada lagged somewhat behind, but a condition for the entry of British Columbia into the Con-federation was that a rail link to that province from the east be established – thus the Canadian Pacific project, with the eastern and the western limbs of the line being joined at Craigellachie, BC in 1885.[3] The Canadian Pacific was seen as important not just because it provided an east–west Canadian link but because it forestalled possible northward railroad penetration from the United States, a situation that might not only have directed traffic flows on to a north–south axis but which might also have raised concerns about loss of Canadian control of the Prairies themselves.

Other technologies have had integrating effects – the telegraph, the telephone and the highway system (which, in the United States following the Second World War became the Interstate Freeway System). Airlines, radio and television and computer networks have refined the earlier contributions of the railroads. Indeed they are now involved in an integration that is no longer just transcontinental but global.

ORDER UPON THE LAND

There is one feature of the human spatial organization of the North American continent that has a pervasive, if taken-for-granted quality – the grid, that 'famous American template' (Meinig, 1993, p. 241). Entry into new land required some means of organizing the allocation and subsequent occupation of space. The French in the St Lawrence area employed the distinctive 'rang' or long-lot system of narrow strips fronting on to river banks, coasts or lake shores. This system was later used in the interior – near Winnipeg, in the vicinity of Detroit, and in Louisiana, Wisconsin, and Illinois. The early English colonies on the eastern seaboard were characterized by an apparently much less orderly system (or at least one lacking in rectangularity). Here 'metes and bounds' were used, where pre-existing natural or human created features served as markers for establishing property limits.

It was the expansion into the interior (in both Canada and the United States) that led to the creation of the almost ubiquitous grid as we know it today. In the US, the Land Ordinance of 1785 established a set of procedures, producing a federal system that was at once simple and comprehensive. The fundamental building block was the township, each township being six square miles (9.65 km^2), which in turn was subdivided into 36 rectangular sections of one square mile (259 ha) each. These could be further subdivided into half and quarter sections, and even into smaller units (see Fig. 9.5 this volume; also Meinig, 1993, p. 243; Hilliard, 1987, pp. 155–62). A similar scheme was applied in Ontario, though here the township size varied from the American model, being 9 by 12 miles (14.48 by 19.3 km) or 10 by 10 miles (16.09 by 16.09 km) (Harris and Warkentin, 1974, pp. 123–4). Prior survey was a prerequisite to any land grant – therefore survey preceded settlement, as was the case in the US, except where squatting took place in advance of official survey and allocation. Thus the appropriateness of Meinig's description of the rectangular survey as 'template'. Later, when Canada began to open up the Prairies, the Dominion government adopted the American pattern of the 640 acre (259 hectare) section and the 36 section township. In this way much of the settled lands of North America west of the Appalachians was imprinted indelibly with the grid. With reference to Ontario: 'Each of these surveys imparted its own geography to the landscape, affecting the layout of roads, the shape of fields, the location of woodlots and, to a substantial degree, the location of farmsteads' (Harris and Warkentin, 1974, p. 124).

However Meinig strikes a cautionary note when he writes that

> Only as the woods were cleared, fields were laid out, and, especially, the local road system neared completion, was this repetitive rectilinear system manifest (and only much later, when it could be viewed from above – first from balloons, then from airplanes – did it become famous as a distinctive human imprint on the earth) (Meinig, 1993, p. 243).

The land survey grid was a great device for placing 'order upon the land', to use the evocative title of Hildegard Binder Johnson's book (1976). On the western, unforested plains the invention of barbed wire provided a tool for imposing this order. Jonathan Rabin, writing of the settlement of eastern Montana, claims that the fences were not merely functional:

> They were a statement of the belief that this unruly land could be subdued. Rectangles rule.... This was no longer mere land, it was landscape, and it was an American classic. It was American in its newness, its hard angularity, its generous spaces and solitudes, as in the mix of its people and their individual architectures (Rabin, 1996, p. 129).

While the gridded survey fundamentally shaped the rural landscape, it also became a dominant element in the urban scene. Arguably, some of the early towns of the eastern seaboard adopted a grid-iron street pattern as an import from a European classical tradition, and arguably such a scheme may have itself been suggestive for the development of the rural survey grid. Be that as it may, the rural grid became a major shaper of the North American city. As the urban areas expanded into the surrounding rural spaces, so they became locked on to the land survey template itself. One has only to examine present-day street maps of, for instance, Chicago or Toronto to see this effect. The overwhelming presence of the grid in the city – in the form of the city 'block' – not only provides a distinctly North American dimension to the sense of the city, it also provides the basis for what must be the most dramatic expression of landscape angularity – the skyscraper. It may not be too fanciful to suggest that the skyscraping cores of the continent's cities are no less than the expression of the rectangular land survey template into a third – vertical – dimension. Here the engineering skills of William Le Baron Jenny (iron and steel frame construction) and the vertical transportation innovations of Elisha G. Otis (the elevator) provided the necessary technological underpinnings.

We have seen how the rectangular 'gridded' survey provided a foundational framework for the North American mosaic, whether this be visualized in land use terms or in the multicultural dimensions of human occupancy. *North America: a geographical mosaic* now goes on to build upon this mosaic theme.

THE GEOGRAPHICAL MOSAIC

This book is edited by two geographers from Northern Ireland, albeit with considerable experience of having lived and worked in, and lectured upon, both Canada and the United States. This gives us a distance and perspective from which to view North America. For the other chapters we requisitioned the expertise of geographers and other scholars practising in North America. They range in age from the relatively young to senior academics – in one case, a retired scholar, William Wonders. Professor Wonders, who tells us his chapter on the Canadian North is likely to be his last publication, does our book signal honour. The authors are drawn from both Canada and the US; only David Smith, who writes on Atlanta, was not working in North America when invitations to write were issued, although Loretta Lees and David Demeritt subsequently moved to England. Each chapter is prefaced by a photograph chosen to represent its theme or region as a whole – thus no captions are provided.

The authors were given a fairly detailed brief and a word limit. Most stuck to the former, if not the latter. Some chapters are thematic and are fundamental to any understanding of North America. These include those in the first section of the book: Houston Saunderson on the physical background, Jeffrey Rogers on climate, and David Demeritt on the forest, both for its own sake and as a metaphor for resource use in North America. On to this physical background the book then introduces people: aboriginal inhabitants (Kenneth Coates and William Morrison) and the waves of immigration that gave North America its diverse ethnic base (Richard Schein). The cultural regions that emerged as a result are discussed by Henk Aay, followed by a chapter on political and other divisions by Donald Cartwright. The book moves on to the economy: agriculture, which has shaped much of the North American landscape (Owen Furuseth); manufacturing

(Iain Wallace) and the ever-ramifying service economy (William Beyers). The next section recognizes that North America in terms of its peoples' residences is now largely urban and suburban with chapters on urban systems (Larry Bourne) and on the geography of the city (John Mercer). Then the book moves on to a run of ten shorter chapters forming a series of urban, rural and regional vignettes.

The use of vignettes as a tool to the understanding of large geographical areas is hardly new: Preston James did it for his *Latin America* in 1941; more recently *Geographical snapshots of North America* appeared (Janelle, 1992), but Janelle's book had no thematic chapters on which to base the vignettes. The use of vignettes in this volume is an acknowledgement that we have not sought to produce a continental regional geography in the manner of the many editions of John Paterson's classic *North America* (1960–94) with its relatively restricted thematic sections followed by a comprehensive regional survey of the geography of the continent. Once the editors of this book, in consultation with the publisher, had decided to commission a dozen substantial thematic chapters, it was inevitable that there would not be room for adequate coverage of all this huge continent's diverse areas. Instead, we offer carefully selected parts of the mosaic that is North America, using 'mosaic' in the book's title to signify this approach. Interestingly, three authors also used this word in their chapters – before they were aware of the final title of the volume which was commissioned under an earlier working title.

What this approach permits is a detailed look at issues or places representative of things that are happening in other areas too, or which are intrinsically important in their own right. Thus Los Angeles was commissioned from James Allen because it is perceived as the quintessential American city, which everybody thinks they know if only through the movies; Atlanta (David Smith) represents a number of cities in the newly-growing South and Southwest; and Boston (Stephen Hornsby) is one of the more established northeastern cities facing different challenges to those of the Sunbelt. Frederick Boal looks at the two largest cities in Canada:

Toronto and Montréal – great rivals but also key contributors to the essence of the country. The regional vignettes in Section F include an area that has, perhaps, been left behind by North American progress: Atlantic Canada (Donald Savoie). Another, on the Great Plains and Prairies (William Carlyle) is about an area struggling to adapt to modern life. By contrast, the once-derided South (James Wheeler) is studied to appreciate why it is now more successful, whilst the Pacific Northwest (Loretta Lees) marches to a different tune as the environmentally-sensitive 'Ecotopia'. Maurice Yeates looks at the St Lawrence and Great Lakes cross-border area, historically one of the engines of the North American economy, whilst William Wonders considers the Canadian North, the last great frontier of a continent that epitomized the concept of the frontier itself.

These depictions of parts of the mosaic of North American geography when embedded upon the knowledge imparted through the thematic chapters should give the reader a sense of North America. Finally, in the last chapter John Agnew goes beyond the continental borders to consider North America's dominant influence in the wider world. For no other continent today – with the end of colonialism and, later, the Cold War – would it be necessary to commission such a chapter. This is a measure of how important is the continent that is North America.

ENDNOTES

1. London, England is nearer St John's, Newfoundland than is the case with one of Canada's own major cities, Vancouver.
2. A fascinating collection of 'what if' history is to be found in Niall Ferguson's edited collection *Virtual history* (1998).
3. The Canadian spike was made of iron; that used at Promontory, Utah was made of gold!

KEY READINGS

Meinig, D.W. 1986: *The shaping of America: a geographical perspective on 500 years of history, Vol. I: Atlantic America 1492–1800.* New Haven: Yale University Press.

Meinig, D.W. 1993: *The shaping of America: a geographical perspective on 500 years of history, Vol. II: continental America 1800–1867.* New Haven: Yale University Press.

Sale, R.D. and Karn, E.D. 1962: *American expansion: a book of maps.* Homewood, Illinois: The Dorsey Press.

The historical atlas of Canada 1987–1993, Vols I, II and III. Toronto: University of Toronto Press.

Vance, J.E. 1990: *Capturing the horizon.* Baltimore: Johns Hopkins University Press.

Section B

The physical and biotic milieux

2

PHYSIOGRAPHIC STRUCTURE AND EARTH SURFACE PROCESSES

HOUSTON C. SAUNDERSON

The landmass of North America extends over more than 50 degrees of latitude, 3000 nautical miles (5556 km), from the Arctic Ocean to the Gulf of Mexico. From the Strait of Belle Isle it stretches westwards about 2600 nautical miles (4815 km) to Queen Charlotte Sound and maintains this width southwards to the parallel from the Carolinas to southern California. Marine invasion of the northern and southern interior has reduced the distance between the polar and tropical seas to about 1200 nautical miles (2224 km). Oceans to the west and east have been kept more distant from one another by a concentration of rockmass into mountain barriers elongated parallel to their coasts.

This skeletal framework of north–south mountain chains and central lowlands has important effects on the motion of air in the lower atmosphere, permitting polar temperatures in winter to penetrate as far south as the Gulf Coast states and tropical humidity to reach north to the Great Lakes in summer. The pathways of European and post-colonial settlement are also related to the orientation of the main elements of the terrain. But the relation between terrain, the movement of air, exploration and settlement are large topics in themselves and are addressed by others elsewhere in this volume. The focus here is to outline the geology and surface processes of North America.

We begin with a mapping of surface elevations, for three reasons. First, the spatial distribution of elevations correlates with the ages of rocks and is the product of the evolution of North America. Second, the surface generated, when combined with gravity and rock density, transforms to a surface of potential energy derivatives of which (energy gradients) drive the downwasting of all continental interiors. Third, a numerical mapping of relief leads to an estimation of the rock mass stored in the terrain. One can then calculate the ratio of rock mass stored to rock mass being eroded by contemporary processes.

MAPPING THE TERRAIN: A COMPUTER MODEL

A numerical mapping of elevations is an exercise in surface fitting, a topic with many applications in geography. Hardy's (1971) multiquadric method of fitting is an exact, global method of particular relevance to terrain science because of its discovery as a method for describing topography. It also has a similarity to the vector sum rule, giving the equations a natural, physical basis. To compute elevations for the whole continental surface using multiquadrics, one needs only a sample of spot heights and their co-ordinates. The surface generated in Fig. 2.1 is a mapping of 3.6×10^5 interpolated elevations obtained from a sample of 334 heights.

THE CONTINENTAL SURFACE

A relief map of North America (Fig. 2.1) shows not only the geographical pattern of elevations. It also is the surface expression of the main structural divisions of the continent.

Surface of the Laurentian Shield

The surface of the Precambrian Shield consists of the worn-down remnants of the most ancient continent-forming processes. Though absolute elevations are generally lower than 600 m, the surface is not flat and relative relief is often tens of metres. As some of the surface undulations continue beneath the Palaeozoic cover, at least part of this extensive surface has been exhumed. Many of the depressions, including the Great Lakes basins, have been deepened much more recently by glacial abrasion and plucking during the multiple glaciations of the Quaternary period. Denudation during and since the Precambrian era has not removed all traces of the Precambrian surface. Some of the rock basins, including the Sudbury Basin in Ontario, are believed to be remnant structures from Precambrian meteorite impacts. The impact energy required for the formation of the Sudbury structure has been estimated at 8.6×10^{23} joules from a projectile 14 km in

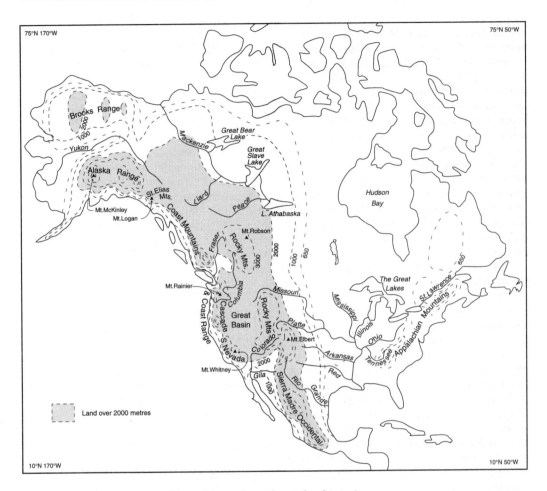

FIG. 2.1 North America: multiquadric surface of mainland terrain

diameter moving at a speed of 20 km sec^{-1} (Stöffler *et al.*, 1994). Fault planes, master joints and depressions in the crystalline rocks control much of the modern drainage, giving it a rectangular to deranged pattern. Closer to sea level there is a regional trend in drainage towards Hudson Bay. Drainage paths are not always permanent, as it may take several weeks at the end of winter before rivers flowing over melting snow and ice reoccupy the previous year's channels.

Younger, marginal surfaces

Along the eastern and western margins of this ancient cratonic core lie the younger Appalachian (Palaeozoic) and Cordilleran (Mesozoic-Cenozoic) mountain concentra-

tions which converge to the south (Fig. 2.1). If the 2000-metre contour were used as the lowest contour on a relief map, only a few peaks would be plotted from the Appalachians: Mt. Mitchell (2037 m), Clingman's Dome (2025 m) and Mt. Guyot (2018 m), all in the southern part of the range. In contrast, much of the mountain mass stands well above 2000 m in the Cordilleran ranges. Between the Appalachian and Cordilleran mountains lies the broad Mississippi Basin extending from south of the Great Lakes to the Gulf of Mexico and from the western flanks of the Appalachians to the foothills of the southern Rocky Mountains. In this central plains area elevations are rarely above 600 m, the Mesozoic and Cenozoic rocks being relatively undisturbed by the orogenesis to the west. The contour pattern

(Fig. 2.1) shows a parallel between the modern coast of western North America and the western edge of high relief. The highest features are the Alaskan Range and St Elias Mountains, including Mt. McKinley (6194 m) and Mt. Logan (5951 m), respectively the highest mountains in the United States and Canada along the northwestern edge of the Western Cordillera. The western edge consists of the Coast Mountains of British Columbia, the Cascades of Washington and Oregon, the Coast Ranges and Sierra Nevada of California and Sierra Madre Occidental of Mexico. Though not so high as the Alaskan Mountains, these ranges reach elevations above 4000 m on volcanic peaks (Mt. Rainier, 4392 m) and batholiths (Mt. Whitney, 4418 m). Along the eastern margin of the cordillera stretch the Rocky Mountains from northern British Columbia to Wyoming, Colorado and New Mexico, including Mt. Robson (3954 m), Wind River Peak (4021 m), Pike's Peak (4301 m) and Mt. Elbert (4399 m). Natural gaps – Yellowhead Pass, Kicking Horse Pass and Crowsnest Pass – have provided paths for lines of transport through this eastern wall of mountains. Between the western and eastern mountain ranges are several interior basins and plateaux of lower elevation. Included are the interior plateaux of southern British Columbia, Columbia Plateau, Basin and Range Province and Colorado Plateau, the last two separated by the Aquarius–Wasatch–Uinta Mountains. Striking parallel to the southwestern coast lie the long Central Valley of California, the peninsula of Baja California and Gulf of California.

The gravimetric surface

On Earth, the value of g, the acceleration due to gravity, varies with respect to latitude, altitude and rocks of different densities. Typically, an observed value of g is reduced to sea level and compared to the theoretical value. The difference between the reduced and theoretical values is called a Bouguer anomaly. Marked differences in rock density explain the negative anomalies of continents and the positive anomalies of ocean basins (Goodacre et al., 1987). The negative anomalies are attributed to the concentration of less dense (granodioritic) rocks in continents and positive anomalies to the concentration of denser (basaltic) rocks in ocean basins. Strongly negative anomalies in regions of high terrain are usually accepted as evidence that deep roots of less dense rock underlie most mountain ranges. Along the California–Nevada boundary Bouguer anomalies reach as low as −250 mgals and near the source of the Arkansas River in Colorado as low as −330 mgals (Woollard and Joesting, 1965). A number of large rivers, the Colorado, North and South Platte, Arkansas and Rio Grande, all rise in this region of negative anomalies in the mountains of Colorado. The high potential energy for river degradation correlates in this region with the most negative anomalies.

EVOLUTION OF NORTH AMERICA

North America, like the other continents, consists of a central craton of Precambrian rocks, tectonically stable, to which rocks of younger ages have been added along continental margins (Goddard et al., 1965). Discussion of the general geology can be conveniently divided into sections dealing with the ancient and younger rocks.

Precambrian craton

The central craton (Fig. 2.2) shows evidence of having enlarged geographically by successive episodes of continental welding or accretion. The broad term 'Canadian Shield' or 'Laurentian Shield' includes a number of geological provinces, not just one, though all are of Precambrian age (> 570 Ma). The oldest rocks are of Archean age (> 2.5 Ga). Fold mountains, as high as those today in the Western Cordillera, formed during continental collisions but have been eroded to their roots in the lower crust and upper mantle. Precious metals formed in these roots are more accessible as a consequence of the long, extensive erosion.

Palaeozoic geology: Appalachian orogeny

By the end of the Precambrian era (~570 Ma), the margins of the Laurentian Shield were

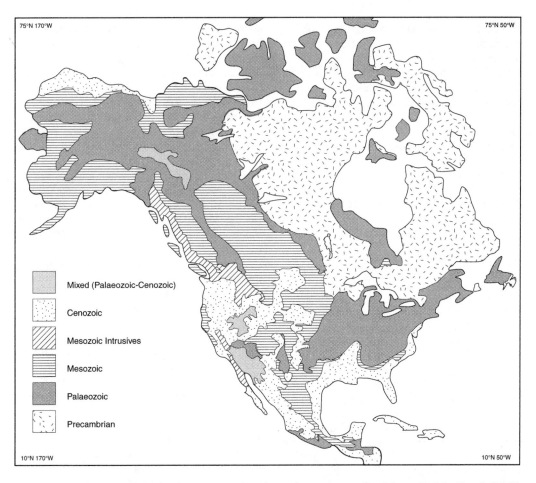

FIG. 2.2 North America: bedrock outcrops of geological eras, generalized from Goddard *et al.* (1965)

inundated by seas containing the first evidence of marine life in any abundance. A marginal apron of nearshore, lagoonal and deep-water sedimentary rocks now overlaps the margins of the Canadian Shield (Fig. 2.2). Early Palaeozoic sandstones originated as beaches in these seas; marine limestones and dolomites, often containing abundant corals and intercalated with shales, accumulated in thick sequences by the end of the Silurian period. During the Mississippian and Pennsylvanian (Carboniferous) extensive vegetation accumulated to form coal layers. The thickness of Palaeozoic rocks led to the proposal that they had accumulated in a geosyncline at the margin of the ancient continent and were then uplifted to form the Appalachian mountain range (Duff, 1993,

p. 733). By the end of the Palaeozoic era, these mountains stood above the general terrain of Pangaea (Kearey and Vine, 1990; Duff, 1993), the supercontinent which subsequently broke up to form Earth's modern continents. The Appalachian orogeny has been divided into two distinct belts, the Older Appalachians corresponding to the Caledonian orogeny of Europe, and the Newer Appalachians to the Hercynian (Duff, 1993). The Older belt contains folds and reverse faults of Taconic (Ordovician) and Acadian (Devonian) age, whereas the Newer belt is Alleghenian (Carboniferous-Permian). Rocks comprising the mountain chain are Precambrian to Palaeozoic in age. Hibbard *et al.* (1995) distinguish folding and thrusting to the west over a Laurentian foreland and to the east over a foreland that is probably now part of Africa. Orogenesis and igneous intrusion have produced distinctive regional terrain ranging from broad plateaux (Cumberland–Allegheny Plateau) to narrow ridge-and-valley (Pennsylvania) types, the latter best viewed from the air and showing clearly a structural control on rivers such as the Shenandoah, Susquehanna and Delaware, except where they turn abruptly across the strike of ridges.

Mesozoic and Cenozoic: cordilleran orogeny

Just as research in the Appalachians gave rise to the geosynclinal theory in the second half of the nineteenth century, research 100 years later in the Western Cordillera and Pacific Ocean basin has produced the mobilistic theory of accretionary tectonics to explain the observed geology. Rather than the product of subsiding geosynclinal troughs and subsequent uplift, most of the Western Cordillera is now considered to consist of a collage of 'suspect terranes' (Kearey and Vine, 1990; Duff, 1993; Howell, 1995). The word 'terrane' connotes a distinctive tectonostratigraphic unit (Howell, 1995), not to be confused with the topographic 'terrain'. Each 'terrane' consists of a suite of rocks having a common tectonic and stratigraphic history. After welding to continental margins, these 'terranes' contrast sharply with one another, though geographically adjacent. The terranes comprising the cordillera did not form where they are now found. They were docked on to the western margin from other source regions. Some are the result of right-slip faulting, such as along the San Andreas fault (Fig. 2.3), and total displacement of hundreds of kilometres. Other terranes are products of the accretion of offshore volcanic island arcs and compression of back-arc basinal sediments. The Alaska Range consists of the highest terrain in North America, comprised of volcanic peaks and marine sedimentaries like the Aleutian island arc and its back-arc basins.

Accretion of suspect terranes probably began at the beginning of the Mesozoic era when subduction developed along the western margin of the continental craton (Kearey and Vine, 1990). Palaeozoic and Mesozoic sedimentary rocks were thrust to the east over a disrupted, Precambrian basement complex. The buried edge of this basement may be marked on the surface by the trace of the Rocky Mountain Trench northward of latitude 50°N (Berry et al., 1971). The eastern, overthrust mountain ranges comprise the Rocky Mountains – in Canada the Selkirk, Purcell and Kootenay ranges, and in the United States the Teton, Wind River, Medicine Bow, Front, Sawatch and Sangre de Cristo mountains. In Late Mesozoic times, the Laramide Orogeny began in the southern Rocky Mountains (Tweto, 1975), uplifting Palaeozoic and Precambrian rocks that had underlain the Cretaceous seas.

Along the western margin of the cordillera overthrusting has been to the west, in contrast to the eastern margin. Subduction of the Juan de Fuca plate is currently marked by a cluster of earthquakes in coastal Oregon, Washington, and Vancouver Island (Fig. 2.3). The presence of volcanic cones (Mt. Rainier, Mt. Hood, and Mt. St Helens) that extrude basaltic and andesitic lavas are further evidence of deep-seated subduction and upwelling. Details of the tectonics along the West Coast are complicated by the presence of all three types of plate boundary which meet in at least one location as a triple junction. The East Pacific Rise (a spreading boundary) approaches the west coast of Mexico and meets the Middle America Trench (a subducting boundary) to the right while to the left it undergoes multiple offsets in the Gulf of California and transforms to a strike-slip boundary (Fig. 2.3).

The Basin and Range Province, between the eastern and western fringing mountains, has distinctive fault-block mountains and valleys that are surface expressions of tensional stresses in the underlying lithosphere. Tension has extended and thinned the continental crust, producing multiple gravity faults orientated north–south. In two areas, both along edges of the province, there is a preferred strike in directions different to the typical north–south strike of faults. Along the California–Nevada border a major fault trends NW–SE at the junction with the Sierra Nevada and the downthrown side is to the southwest. At the southeastern edge of the

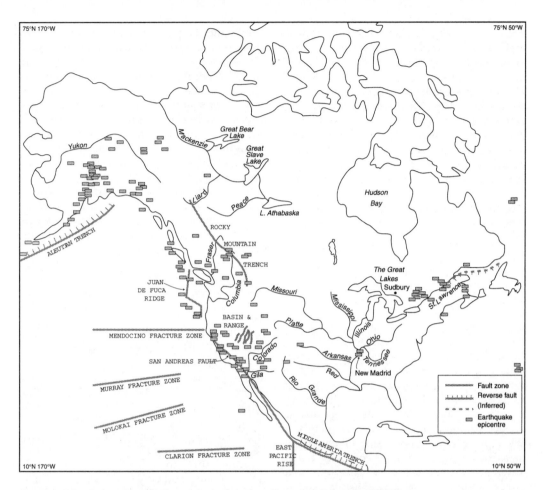

Fig. 2.3 North America: faults and earthquakes (January–February 1998)

province, normal faulting trends NE–SW and the downthrown side is to the southeast, down to the Colorado Plateau. Stewart (1978) has reviewed theories of the extensional mechanism, concluding that upwelling behind a subduction zone was the most satisfactory way to explain the presence of thin crust, high geothermal heat and regional uplift of the whole province.

CONTEMPORARY SURFACE PROCESSES

Tectonic processes

Earthquakes are manifestations of surface-forming faults and volcanic activity. The strain energy released from a magnitude-8 earthquake, like the one in San Francisco in 1906, is about 10^{18} joules, equivalent to a large nuclear explosion. Large quantities of stored energy are contained inside rocks and this energy is released during all scales of fracturing, not just at the scale of the San Andreas fault. When a fracture reaches a critical length it becomes self-propagating, drawing on the strain energy stored inside the rock. Many rivers flow along the traces of fractures, sites of strain energy release. As yet unquantified are the relative effects of this release, gravitational acceleration, and river and glacier mechanics on the initiation and development of slope forms in mountainous regions. Though undisturbed by mountain-building processes, the central lowlands are not without seismic risk. The New Madrid earthquakes of 1811–12 occurred in the heart of the Mississippi River basin. The largest magnitude was >7.0 and the areas affected by shaking were larger than either the 1964 Alaskan or 1906 San Francisco earthquakes (see Stover and Coffman, 1993, pp. 59, 67, 119 for isoseismal maps of these earthquakes). An elongated, positive gravity anomaly (Duff, 1993, p. 691) extends from the Lake Superior Basin southwards to Kansas across the Mississippi Basin. This anomaly has been interpreted as a remnant Proterozoic volcanic belt, possibly an ancient rift valley. Seismic activity is common at the sites of other rifts in Precambrian rock, such as in the Ottawa Valley. The New Madrid location lies just south of a positive anomaly outlier (Woollard and Joesting, 1965). It is also at the southwestern end of a linear track of epicentres, historical and recent (Fig. 2.3), stretching along the western margin of the Appalachians up into the St Lawrence Valley. It is a common misconception that the only region with significant earthquake risk is in the Western Cordillera. Though earthquakes of high magnitude are more frequent there because of active tectonics, tremors occur frequently in the Appalachian region and neighbouring areas of Ontario and Québec, along the St Lawrence and its tributary valleys, and in Arctic locations (Fig. 2.3). Seismic risk, of course, is determined by the density of population as much as by magnitude of earthquakes and susceptibility of substrate to failure. In California, seismic risk is compounded by heavy winter rainfall which concentrates in ravines and saturates soils. These soils are more likely to liquefy when tremors occur simultaneously.

Slope processes

It is not yet possible to estimate the total mass of earth material involved in mass wasting at a continental scale and over a geological time frame. Studies that have been done are reduced to a spatial scale compatible with observations on particular types of movement. In alpine areas there may be a focus on types such as snow avalanches, rock glaciers or on individual events like the

Frank and Hope slides in Canada or the Sherman and Blackhawk slides in the United States. Large rockslides have received much attention because of their size and momentum, a momentum that is often sufficient to carry debris across a valley floor and up the opposite valley side. As many of these events are episodic at intervals of tens, hundreds or thousands of years it is difficult to evaluate their net effect on the evolution of terrain. Distinctive types of motion are found in alpine, lowland and submarine environments. Those in alpine regions are controlled by such factors as the frequency with which rock temperatures cross freezing point, the orientation of rock structures, the local energy gradient, the degree of weathering, microclimatic regime and tectonic activity. Higher slopes are typically dissected from avalanching of rock and snow which leave linear tracks where they have cut into the underlying slope. The lower slopes have aprons of debris formed from the coalescence of debris fans at the exits of avalanche chutes and from talus cones resulting from rockfalls on free-face slopes. The motion of large volumes of material is not confined to alpine regions. During heavy rainfall, channels of small streams in areas of lower relief may become enlarged and the solid load so concentrated that streamflow turns into debris flow. Williams and Guy (1973) estimated that individual tributary flows in Virginia during Hurricane Camille removed up to $\sim 0.06 \times 10^6 \ \mathrm{m}^3 \mathrm{km}^{-2}$. Large slope failures can be frequent even where regional slopes are nearly horizontal. At the close of the Wisconsinan glaciation, marine invasion of the lower St Lawrence deposited clays on the floor of the Champlain Sea. These clays were subsequently uplifted above sea level during the isostatic rebound which followed glaciation. During heavy rainfall, or when saturated by groundwater, they mobilize on slopes as low as two or three degrees. Mass movements are even competent agents of erosion below sea level, as deltas prograde by failure of delta-front sediments.

Fluvial processes

The floods of 1993 on the Mississippi and 1997 on the Red River (North Dakota and southern Manitoba) are recent reminders that rivers are not only powerful agents of erosion, they are capable of making life difficult for people living on their floodplains. Large rivers, the Mississippi and St Lawrence, have received the attention of scholars for some time, partly because of their historical importance as paths of entry for European explorers but also because they were navigable and facilitated commercial penetration of the continent. Big rivers, like the Colorado and its Grand Canyon, are monuments to the power of erosion. Smaller rivers have been given less attention and where records exist, they rarely exceed 40 years. These records consist of observations, often by government agencies in Canada and the United States, from a network of gauging stations on trunk streams and tributaries. They give useful insights into seasonal and annual variations in water and sediment discharges from which sediment yields may be estimated. Such estimates show that small rivers frequently have yields as high as large rivers. Data compiled by Milliman and Meade (1983) show a range of yields from about $60 \, t \, km^{-2} \, a^{-1}$ (Mississippi River) to $0.2 \, t \, km^{-2} \, a^{-1}$ (Colorado River). The values from both of these rivers have been markedly reduced from those of a century ago. Reservoir construction on the Missouri in the 1950s accounts for much of the reduction in the Mississippi's sediment discharge. Construction of the Hoover Dam and the withdrawal of water for irrigation and water supply explain the extremely small amounts of water and sediment in the lower Colorado.

The total tonnage of sediment discharged by rivers can be used to estimate the mass (m_e) of rock and soil eroded. The multiquadric elevations used for surface fitting (Fig. 2.1) make it possible to compute the mass (m_t) of rock stored in terrain. The mass ratio, m_t/m_e, is then an estimate of how long it would take to erode the continental surface. The mass removed per annum by the North American rivers tabulated by Milliman and Meade (1983) is $\sim 500 \times 10^6 \, t \, a^{-1}$ (0.5 Gt). The numerical integration of rock mass stored in North America above sea level (Fig. 2.1) is about 1.0×10^8 Gt, giving m_t/m_e equal to 2×10^8 years. To erode the total mass of North

America to sea level would require about 200 million years.

Contaminant transport

Rivers transport contaminants as well as sediment. Some of these contaminants are part of the dissolved load, but many of the most toxic are insoluble in water and are adsorbed to the fine-grained solids carried as suspended load. They are then transported downstream or become buried in alluvium, only to be exhumed later. Residence time in alluvium may be tens, hundreds or thousands of years and consequently the contaminants may persist in the environment for a long time, long after their manufacture and use have been banned by legislation. Several detailed articles in Meade (1996) document clearly the effects of agricultural use of pesticides, industrial use of toxic chemicals and their delivery to the Mississippi from its tributaries, particularly the Ohio River.

Coastal processes

The continent of North America is bounded by four large oceans which erode huge quantities of rock and sediment by direct wave attack. In polar latitudes the wave attack is replaced in winter by ice-push. Sediment from river mouths is picked up in longshore currents and redistributed along the coast. Large bodies of inland water, the Great Lakes, also show the effects of coastal erosion and redistribution by wave-generated currents. Cliff erosion rates exceeding $1 \, m \, a^{-1}$ are in soft, unconsolidated deposits (in the northern part of the continent, frequently glacial). Individual storms, including storm surges from hurricanes in the Gulf Coast area, are capable of effecting tens of metres of recession. Coastal flooding may result from subsidence of the sea floor, as in the Galveston–Houston area of Texas, or from rising water levels, a common occurrence in the Great Lakes. Where there is a combination of low elevations, soft materials and high-density human population, coastal erosion and flooding become serious problems because of the threat to property and life. Protective structures, such as the seawalls of Galveston

Island and coastal New Jersey, are attempts to reduce the hazard.

Beach erosion is really a study of sediment budgets. A beach is simply temporary storage of sediment. Solid particles are added and removed continually on tidal and seasonal time-scales. If the volume of sediment stored is reduced, a beach shrinks. This shrinkage, though theoretically related to changes in wave climate, is more likely the consequence of a reduction in the amount of sediment being delivered to the beach. Less sand will be added to a beach if the source of sand changes. Along the Great Lakes coasts, if the coastal source changes from glaciofluvial sands to clay-rich tills, sand-sized sediment is no longer available and beach shrinkage is the consequence. Rivers also deliver huge quantities of sand to world coasts, but the amount delivered has been severely reduced by dams and reservoirs along the major channels. Prograding deltas release sediment farther offshore and a longer time elapses before it is brought back onshore. The Mississippi is such a prograding delta and may soon reach the edge of the continental shelf where delta-front sediments will slide into deep water and be unavailable for beach replenishment in Texas, Louisiana, Mississippi and Alabama.

Glacial and periglacial processes

Extensive areas in the northern archipelago and alpine areas of North America are either currently glaciated or show the effects of ground ice. In the north, extensive ice caps such as Barnes Ice Cap, on Baffin Island, scour broad areas and feed narrower outlet glaciers. These ice caps are but very small remnants of the much more widespread Laurentide ice sheet which, in the most recent glaciation (Wisconsinan), covered all of Canada except for a Yukon re-entrant and small pockets of southern Alberta and Saskatchewan (Prest et al., 1967). Alpine glaciers and their ice-field sources are concentrated in the Alaska, St Elias, Coast and Rocky Mountain ranges. They too are small by comparison with the Cordilleran ice sheet, a contemporary of the Laurentide and com-

petitor with it along the eastern foothills of the Western Cordillera.

The Laurentide ice sheet had two major centres of dispersal, one in the District of Keewatin to the west of Hudson Bay and the other centred on the Labrador–Ungava peninsula. From these centres the ice flowed radially, including to the north, as shown by the radial orientation of eskers and glacial lineations at right angles to end moraines. In and around the centres of ice dispersal the surface of the Laurentian Shield shows extensive evidence of glacial abrasion, accentuating a preglacial surface already uneven from the exposure of folds in the Precambrian rock. The smoothness of rôches moutonnées is in some instances the result of glacial abrasion, but in others a consequence of the shearing and folding of rock strata in Precambrian times with much younger trimming by abrasion. Around this inner zone of abrasion is a zone of glaciogenic deposits closer to the margins of the ice sheet. These deposits include thick, extensive outwash and clays of ice-marginal lakes. One of these lakes, Lake Agassiz, covered almost all of Manitoba and part of Ontario north of the Superior Basin. The Great Lakes – Superior (183 m), Michigan (176 m), Huron (176 m), Erie (174 m) and Ontario (75 m) – currently connected like a staircase and emptying into the Gulf of St Lawrence, were not always so connected. In late-glacial times, when lower lake outlets were blocked by ice, the upper lakes discharged via the Mattawa–Ottawa valley into the lower St Lawrence. At the ice-front lakes overflowed, feeding the Mississippi drainage network in the case of the Chicago outlet or along the Mohawk Valley from the Ontario Basin into the Hudson River and out to the Atlantic. In the United States, overflow from ice-dammed Lake Missoula and pluvial Lake Bonneville produced extensive flood deposits. A recent estimate of the peak discharge during the Bonneville flood is about $1.0 \times 10^6\,\mathrm{m}^3\,\mathrm{sec}^{-1}$ (O'Connor, 1993).

A legacy of continental glaciation is the swath of soils which have developed on the ice-marginal deposits spread across the central lowlands. On some areas of the Canadian

Shield, a 'clay-belt' of glaciolacustrine deposits may be the only cultivable soils available. Alpine glaciation contributes meltwater to rivers for use in hydroelectric power generation, but also poses considerable risk in the potential rupture of ice dams and disruption of communications. Beyond the margins of current glaciers and ice caps, the ground is either frozen solid for most of the year or frozen in sporadic patches. These are the areas of continuous and discontinuous permafrost. Where ground temperature fluctuates around freezing, shattering of rock and heaving and melting of ground ice produce distinctive rock-strewn surfaces and thermokarstic features. Patterned ground, stone stripes and polygons, result from the sorting of different particle sizes in the soil. The constant shifting and heaving of soil and rock fragments makes pipeline and foundation engineering a challenge in periglacial areas.

SUMMARY

The physiographic structure of North America reflects an underlying pattern of rock types, rock structures and geophysical processes that range in age from the earliest chapters of Earth's history to the closing moments of the twentieth century. The primordial core, now stable, is exposed over much of the northern half of the continent. Wrapped around this central mass of crystalline rocks are younger rocks, many of marine origin, which have been folded and thrust up episodically into mountain ranges. The vigour of tectonics since Mesozoic times has produced the highest terrain and largest concentration of rock mass along the Pacific margin, now considered to be a collage of terranes added to the continent by accretion and strike-slip faulting. Mountains predating the Mesozoic have been reduced to lower elevations by isostatic adjustment and subaerial erosion. The work done by surface processes, though driven by gravitational energy gradients and regional climates, has been influenced by human occupance of drainage basins. Technological attempts to reduce

environmental hazards such as flooding and erosion have made human use of the land as significant as Newtonian mechanics in understanding the effects of earth surface processes. The association of contaminant transport, burial and re-entrainment with fluvial processes and sediment budgets is but one example of where an understanding of linkages among environmental elements, human and non-human, is essential in reducing environmental degradation.

ACKNOWLEDGEMENT

The author thanks Pam Schaus for producing the figures for this chapter from colour slides of original computer plots. Epicentres on Fig. 2.3 are courtesy of the United States Geological Survey, National Earthquake Information Website at http://wwwneic.cr.usgs.gov/ and of Natural Resources Canada, Geological Survey of Canada Website at http://www.seismo.nrcan.ge.ca/. Co-ordinates of faults/fracture zones are from Reader's Digest (1997) and from Geological Survey of Canada (1969).

KEY READINGS

Goddard, E.N., Billings, M.P., Levorsen, A.I. et al. 1965: Geologic map of North America. Washington, DC: United States Geological Survey.

Hardy, R.L. 1971: Multiquadric equations of topography and other irregular surfaces. Journal of Geophysical Research 76, 1905–15.

Howell, D.G. 1995: Principles of terrane analysis. Topics in the Earth Sciences 8. 2nd edn, New York: Chapman & Hall.

Kearey, P. and Vine, F.J. 1990: Global tectonics. Oxford: Blackwell Scientific Publications.

Milliman, J.D. and Meade, R.H. 1983: Worldwide delivery of river sediment to the oceans. Journal of Geology 91, 1–21.

Prest, V.K., Grant, D.R. and Rampton, V.N. 1967. Glacial map of Canada, Map 1253A. Ottawa: Geological Survey of Canada.

3

WEATHER AND CLIMATE

JEFFREY C. ROGERS

CAUSES OF REGIONAL CLIMATES

Climate is defined as the average weather conditions occurring at a place or region over a period of time. It consists of several elements including air temperature, precipitation, atmospheric pressure, humidity, cloud cover, wind direction and speed. This chapter focuses on aspects of North American climate including the regional distributions of precipitation and air temperature and the role of human activities in altering climate. First however, it is important to understand the physical factors that produce the distribution of climates observed in North America and other parts of the world. These include:

- the latitudinal distribution of solar radiation;
- heating contrasts between water and land;
- ocean currents;
- the distribution of atmospheric pressure systems;
- topography and orography.

Latitudinal distribution of solar radiation

Temperature variability in climate is primarily controlled by the amount of solar radiation falling upon Earth's surface. This in turn is dependent upon the elevation angle at which the sun's radiation strikes the Earth's surface and the length of daylight, both of which vary seasonally and with latitude. In summer, when Earth's Northern Hemisphere is tilted towards the sun, solar elevation is relatively high, radiation strikes Earth more directly over northern middle latitudes and length of daylight increases towards the North Pole. By the start of winter, the Northern Hemisphere is tilted away from the sun, producing a much lower solar elevation across the hemisphere and length of daylight decreases towards the North Pole. The combination of solar elevation above the horizon and length of day work together to enhance solar radiation receipt in summer but decrease it in winter. This produces large seasonal variations in solar radiation receipt, especially towards higher latitudes. Aside from these seasonal radiation receipt differ-

ences, solar elevation over North America always decreases by one degree towards the southern horizon for each degree of latitude one travels further north, producing decreases in temperature with increasing latitude. Temperature is always higher at lower latitudes, where the sun angle is consistently high and, while it decreases toward high latitudes, it also becomes seasonally more variable in keeping with the large seasonal radiation changes.

Heating contrasts between water and land

Heating contrasts between land and any adjacent large water body (an ocean, or the North American Great Lakes) produce the first major modification to solar radiation's latitudinal control of air temperature. Continents heat faster than large water bodies but lose the acquired heat within weeks once radiation input decreases or ceases. In contrast, sunlight penetrates water bodies to a considerable depth and absorbed energy from the sun can be mixed and stored to a depth of 100–200 m rather than immediately heating the air. The stored heat is, furthermore, slowly released back to the atmosphere over several months into the winter. Due to these land/water heat capacity differences, continents are relatively warm in the summer compared to oceans at similar latitudes whilst in winter oceans are comparatively warmer than land. Also affected is the average annual temperature range, defined as the mean temperature difference between July and January. The temperature range over higher latitudes of continents is very large due to both sizeable seasonal solar radiation differences at those locations and strong heating and cooling characteristics of land. Coastal areas have smaller annual temperature ranges with milder winters and cooler summers than interior locales at similar latitudes.

Ocean currents

North America is influenced by three major ocean currents that further amplify climatic effects along coastal areas. The comparatively warm northward-flowing Gulf Stream (Fig. 3.1) serves as a source of moisture and heat along coastal areas of the eastern United States, helping to maintain relatively mild winters at east coast cities such as Boston and New York. The cold southward-flowing California Current creates a very cool climate along the west coast of the US. The adage, attributed to American writer Mark Twain, 'the coldest winter I ever spent was a summer in San Francisco' refers to the ability of this ocean current to keep summer coastal air temperatures below normal for their latitude. The cold Labrador Current flows southwards from the Davis Strait along eastern Canada and Newfoundland and produces similar coastal climate effects.

The distribution of atmospheric pressure systems

Semi-permanent subtropical anticyclones, located over the Atlantic and eastern Pacific

Oceans at latitudes 30–35°N (Fig. 3.1), are important atmospheric high pressure features affecting North American climate. Clockwise circulation of air around the large Atlantic high pressure cell produces northward flow along its western flank, bringing relatively warm and humid air across eastern North America. Clockwise flow across the eastern flank of the Pacific anticyclone produces a southward flow of relatively cool dry air over western North America. Northward-flowing mild air over eastern North America also rises and tends to produce clouds and rain, while southward-flowing cool air over western North America generally descends and is comparatively clear and dry.

Low pressure systems, or cyclones, continually traverse North America and critically affect the spatial distribution of climatic elements. Cyclones form along the polar front, a boundary that lies across Canada in summer and the United States in winter, separating cold dry Canadian air masses from warmer

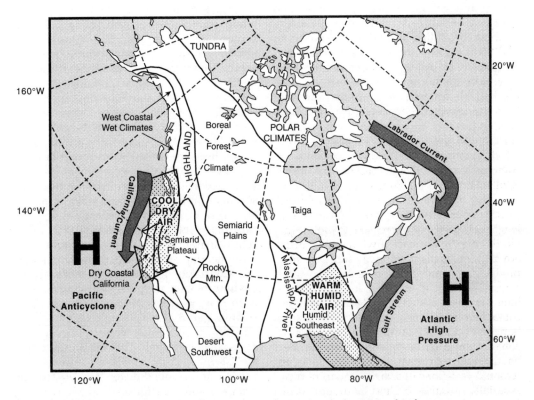

FIG. 3.1 The climatic regions of North America and associated physiographic features

humid air to the south. Cyclones originating near the Gulf of Mexico, Colorado or in Alberta during autumn and winter move across the Great Lakes and leave the continent either towards Baffin Bay or along the St Lawrence Seaway (Whittaker and Horn, 1984). Their attached cold and warm fronts produce widespread areas of poor weather, cloud cover and precipitation. Frequent cyclone passages across the eastern Great Lakes and St Lawrence Seaway make it one of the cloudiest places in the world, often having direct sunlight on fewer than 30 per cent of winter days.

Topography and orography

Air temperature rapidly decreases with elevation over the lowest 10 km of the atmosphere, producing comparatively cold climates in mountainous areas. Highland areas of North America, outlined in Fig. 3.1, are remarkable for climatic and vegetation diversity occurring across vertical zones along different elevations. Vegetation and forest cover change dramatically with height, ultimately yielding to tundra and snow over the highest peaks. Mountains generate precipitation on their windward (western) slopes by forcing moist air to rise and cool, leading to condensation, clouds and precipitation. Eastern, or leeward, slopes have descending air and are generally dry and clear with low precipitation, although descending air helps produce cyclones in Colorado and Alberta that affect weather further east.

CLIMATES OF NORTH AMERICA

The United States east of the Rocky Mountains

Several classifications of world climates have been developed. Most widely used is that of Wladimir Köppen (1936), a late nineteenth- and early twentieth-century German climatologist. Köppen identifies regional climates based on spatial distributions of mean monthly precipitation and air temperatures measured at weather stations and takes into account the regional vegetation cover in determining climatic boundaries. This section presents a general overview of North American climates, drawing on Köppen's classification scheme, details of which are available in many textbooks (e.g., Henderson-Sellers and Robinson, 1986).

One of the most distinctive climates of North America is that of the southeastern United States (Fig. 3.1). Warm and humid, with high annual precipitation and a wide variety of lush vegetation, the southeastern US is one of the wettest areas of the world outside the tropics. High precipitation is caused by proximity to moisture evaporated from the Gulf of Mexico and drawn northwards and westwards into spring and summer thunderstorms and winter cyclones in the prevailing southerly flow around the west side of the Atlantic subtropical anticyclone. Moisture flow and total annual precipitation over eastern North America gradually diminish to the north, away from the Gulf. On the Gulf Coast, New Orleans receives an average 157 cm of rainfall per year (Fig. 3.2h) while Atlanta receives 129 cm (Fig. 3.2i) and Boston only 105 cm (Fig. 3.2f). Warm humid climates occur over southeastern corners of most continents including southeastern South America (Uruguay and adjacent parts of Argentina and southern Brazil), eastern Australia and southeastern China, regions also near warm ocean currents and warm water bodies.

Southeastern United States precipitation is distributed evenly across the months of the year (see Figs 3.2h and 3.2i). Spring and summertime rainfall is convective in nature and Gulf coastal states, particularly Florida, experience the highest number of thunderstorms in North America due to proximity to unstable, northward-moving, moist air. The tropical variety thunderstorms are caused by daytime heating, onshore winds and passage of frontal systems extending from cyclones. Winter months also have abundant precipitation due to continual passage of cyclones across the US. Humid Gulf air is drawn into the storm's large circulation, generating either rain or snow, depending on latitude and elevation.

Among the unique regions in the South-

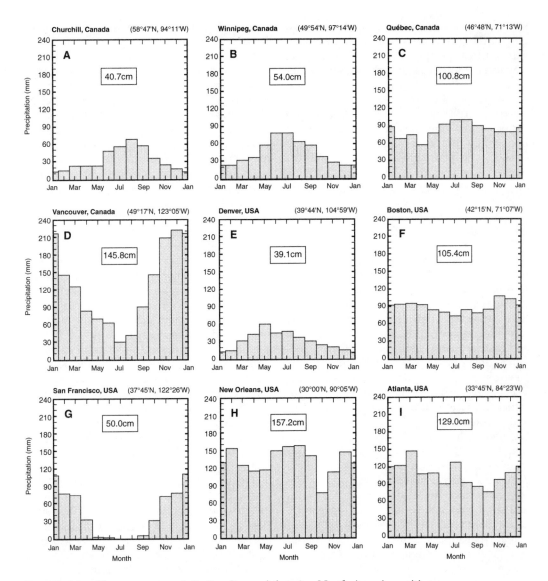

Fig. 3.2 Monthly average precipitation (in mm) for nine North American cities

east is the climate of Florida, which becomes very tropical as one goes south along the peninsula. Southern Florida's climate permits production of citrus crops including oranges, lemons and grapefruit. Efforts to grow these crops in extreme southern Texas and northern Florida met with some success until the late 1970s when sub-freezing air masses became more frequent in winters over the Southeast and led to destruction of citrus trees in these areas (Rogers and Rohli, 1991).

States and provinces surrounding the Great Lakes are a transition zone from broadleaf deciduous trees (to the south) to evergreen needleleaf trees that especially dominate Canadian forests (Fig. 3.1). Climate across the Great Lakes transition zone is marked by increasingly colder winters and cooler and shorter summers. Summers are still comparatively warm and humid around the southern lakes, characterized by northward intrusions of humid maritime tropical air from the southern United States, and convective activity may still lead to thunderstorm rainfall. Precipitation continues to be evenly distributed throughout the year and annual values remain quite large because of frequent cyclone activity across southeastern Canada, the St Lawrence Seaway and the northeastern US. For example, about 100 cm of precipitation occurs at Québec City (Fig. 3.2c).

The Great Lakes exert a substantial regional influence on climate (Eichenlaub, 1979). Areas within 50–100 km of shore, which include many large cities, are milder in winter but cooler than inland areas in summer due to the heating characteristics of the lakes and associated lake-induced winds. Cold season precipitation is enhanced by lake-effect snows occurring along southern and eastern coasts of individual lakes. Snowfall in some areas may exceed 400 cm per year. Lake-effect snow occurs as very cold Canadian air masses pass over open water, evaporating moisture that turns readily into cloud and precipitation as air traverses the lake. Snowfall is then precipitated as it encounters rougher coastal terrain and as it rises over inland hills 100–400 m higher than the lake.

Annual precipitation dramatically decreases towards the West in the United States and southern Canada. Abundant vegetation of the Southeast is replaced by a semi-arid climate that supports few trees and primarily consists of grassland and prairie except near streams where cottonwood and willow trees grow. This Great Plains region, generally west of the Mississippi River but extending into Canada (Fig. 3.1), also rises steadily in surface elevation to between 1.5–2 km above sea level at the base of the Rocky Mountains (the highlands in Fig. 3.1). Denver, Colorado, on the High Plains just east of the Rocky Mountains receives 39 cm annual precipitation (Fig. 3.2e). Near the eastern limit of the plains, Winnipeg, Manitoba, receives 54 cm per year (Fig. 3.2b) and, as with Denver, precipitation favours late spring and early summer months. Köppen labelled the mid-latitude prairie regions a steppe climate,

named after the vast semi-arid grasslands of Russia.

Semi-arid plains are created by descent of eastward-moving air crossing the Rocky Mountains. Air descending the mountains is generally dry and warm, favouring clear skies and low precipitation, and, typically, spreads across the high plains east of the mountains. Over the southern Plains, including Texas, Oklahoma and Kansas, the descending dry air collides with humid moist air moving westwards from the Gulf of Mexico on many spring and early summer days, forming a distinct boundary known as the dryline. Thunderstorms form along the dryline and subsequently move eastwards into the moist air. These Plains thunderstorms often become supercells (Bluestein, 1993), large, long-lasting storms having air flow dynamics that create high precipitation, hail, and tornadoes: strong rotating vortices with winds sometimes in excess of 300 km per hour. Tornadoes are nowhere more abundant than in Oklahoma, Texas, and Kansas; the last being the site of the famous motion picture tornado that transported Dorothy and her dog, Toto, in *The Wizard of Oz*. Thunderstorms over all of North America typically produce about 1000 tornadoes each year and cause roughly 100 fatalities and considerable property damage

The semi-arid Plains of the United States and Canada constitute much of America's breadbasket (see Chapters 9 and 20), supporting vast areas of wheat, corn and soya beans, and requires irrigation due to low precipitation. Much of the irrigation water comes from the Ogallala aquifer, a vast underground reservoir spreading across the plains from the Dakotas to Texas with subterranean water deeper than 300 m in some locales, rivalling depths found in the Great Lakes. Heavy aquifer usage is however causing water shortages to become commonplace along the 'shores' of this underground sea in places such as Texas.

The boreal and polar climates

The boreal climate dominates a vast area of Canada and small parts of the northern United States (Fig. 3.1), extending westwards through the Yukon, Northwest Territories and into southern Alaska. This climate is characterized by long, cold, snowy winters and short summers. Mean summer air temperatures steadily decrease northwards across the boreal forest zone but at least one summer month has a mean temperature above 10°C. In general, precipitation is substantially lower than occurs further south over eastern North America. Although precipitation occurs throughout the year, it is higher in summer (Figs 3.2a and 3.2b) when cyclone tracks are farthest north. Vegetation is dominated by taiga, a coniferous pine forest cover that thrives in cold and snow and whose northern boundary is the treeline. The region leeward of the Rocky Mountains in Alberta generates cyclones that move quickly across southern Canada or the Great Lakes before recurving northeastward across the St Lawrence Seaway and Canadian Maritime provinces. These storms often bring some precipitation across the taiga on at least a few days of each month.

Precipitation over the western Canadian boreal forest and southern Alaska comes from cyclones entering the continent after passage across the Pacific Ocean. The high latitude West Coast climates along the immediate Pacific coast of British Columbia and Alaska (noted in Fig. 3.1) are milder than those of the Canadian boreal forest interior. They are similar, in Köppen's classification, to the climate of Western Europe; unusually wet but quite mild for their latitudinal location and affected by cyclone passages throughout the year. Vancouver (Fig. 3.2d) is most representative, having very high precipitation in winter. Vancouver summers are however somewhat drier than those occurring further north along the West Coast.

Polar desert climates dominate the northernmost regions of North America, poleward of the treeline. They are characterized in the Köppen system by at least one summer month with mean air temperature above 0°C but lower than 10°C. The ground cover of polar climates is tundra – a treeless plain covered by stones, moss, lichens and some dwarf shrubs. Tundra is also interspersed in the boreal forest of southern Alaska and the Yukon and is often found on mountain peaks

above 3000 m. The polar desert lies outside the main band of cyclone activity and the low precipitation is due to the fact that cold Arctic air only holds small amounts of water to precipitate from weather systems that do, occasionally, traverse the region. The distinctive forest/tundra boundary can be traced through time in the records of tree pollens stored in soils and lake sediments and is an indicator of past climates.

Climates of western North America, including mountain highlands

Climates of the United States west of the Rocky Mountains are dominated by gradual descent of air around the eastern flank of the Pacific subtropical anticyclone (Fig. 3.1). The subtropical anticyclone moves northwards in summer and spreads descending air over all of coastal California, creating a notable dry season. In winter, the anticyclone retreats southwards, permitting passage of occasional Pacific storms across California and bringing occasional rain and snow to inland mountains. San Francisco, for example (Fig. 3.2g), receives a majority of its annual 50 cm rainfall between November and March and summer months are extremely dry. This coastal climate helps make California a popular place to live in the US. Summers are dry, contrasting with the tropical heat and humidity of the east, while winters are cool but pleasant. The Köppen classification scheme groups the coastal California climate with that of the expansive northern Mediterranean Basin, characterized also by dry, warm summers but mild and wetter winters. Natural vegetation of California consists of poor grasslands interspersed with evergreen scrub vegetation and evergreen oaks, a type referred to as chaparral. Further north along the West Coast in Washington, Oregon and southern British Columbia (see Vancouver in Fig. 3.2d), summers remain comparatively dry, but the subtropical anticyclone is less influential and increasingly frequent storms bring more precipitation than occurs in California.

The dry summer/wet winter pattern of coastal areas extends over the Sierra Nevada Mountains of eastern California as well as along the Cascade Range in Oregon and the Olympic Mountains of Washington State. These mountains, as well as those along coastal areas in Canada, are among the first to intercept moisture from Pacific storms moving onto the continent during the wet season. They receive some of the highest snow accumulations in the world, with the snow pack around each mountain range serving as a regional water source.

Highlands of the western ranges and those of the Rocky Mountains lying to the east, extending from Colorado to Alberta (Fig. 3.1), are characterized both by rapid climate variation with elevation and by a variety of mountain-slope orientations with respect to the sun and prevailing winds. Solar heating is dependent upon mountain slope orientation with respect to the sun's path across the sky, with sunny slopes experiencing more substantial warming and variation in diurnal range of temperature. The amount of radiation reaching exposed highland surfaces is greater than at lowlands at similar latitudes due to increased penetration of sunlight brought on by the steady decrease in atmospheric density with height. Exposed skin on inhabitants and visitors in mountainous areas is prone to sunburn and windburn. The humidity of the air also decreases with height, further adding to the rapid heating of the ground during the day and rapid cooling at night on sun-exposed slopes. Despite increased exchanges of radiant energy at ever higher elevations, diurnal surface air temperature changes are superimposed upon steadily decreasing mean air temperatures with increasing elevation. The slope orientations with respect to the prevailing wind determine where large precipitation amounts will occur. With westerly prevailing winds and a general north–south orientation of the Rocky Mountains and Ranges of the Far West, precipitation is abundant on westward-facing (windward) slopes and decreases along leeward slopes facing east.

The high elevation Colorado Plateau lies between the Sierra Nevada and the Colorado Rockies (Fig. 3.1). Moisture flow across this region is weak, creating a semi-arid climate. Air descends mountain slopes or is associated with stationary anticyclones over the plateau

that help suppress cloud and precipitation. Plateau vegetation includes sagebrush and juniper pine trees. Lower elevations south of the plateau, including portions of Arizona, southeastern California and southern Nevada, form the desert southwest and are dominated by brush and cactus. Las Vegas, Phoenix and Palm Springs are major desert cities maintaining lush urban environments by thriving on considerable water withdrawal from the regional rivers and underground water sources.

HUMAN IMPACTS ON CLIMATE

Human-induced climate modification occurs every day in large urban areas of North America, affecting over half of North America's population. Urban air pollution, a product of human activities, is the leading meteorological cause of human mortality (Dockery et al., 1993; Schwartz, 1994). The typical pollution-producing weather scenario is associated with lack of air movement under slow-moving or stalled high-pressure cells, leading to elevated concentrations of urban pollutants affecting people with breathing and heart problems. The quasi-stationary Pacific subtropical anticyclone contributes to permanent pollution problems of the Los Angeles Basin in southern California. While urban air pollution still affects air quality and health, concentrations of most pollutants have generally declined in the United States by about 25 per cent since the 1970s due to federal clean air legislation (US Environmental Protection Agency, 1996).

Urban climate modification takes place due to factors besides pollution. These include artificial heat production in city environments, shading and alteration of air flow by large buildings, and water removal by concrete-paved surfaces and drainage systems. Alteration of urban climate occurs across most climatic elements. For example, air temperatures are typically higher over urban areas compared to the surrounding countryside. This 'urban heat island' can be 5–10°C warmer on calm, clear, winter nights when surrounding rural areas become quite cold. Small increases of 10–25 per cent also occur in mean precipitation downwind of some large American cities and are due to enhanced cloud and precipitation generated from urban pollution particles. Humidity in urban areas is, however, generally lower than that of surrounding areas due to rapid drying of concrete surfaces and efficient drainage of rainwater off city streets. Solar radiation receipt on urban surfaces is typically reduced compared to that on non-urban areas due to interception of radiation by buildings and the extensive shading they produce. Although building arrangement and orientation can produce locations where winds are quite high, their overall structural effect is to increase surface friction, thereby reducing mean wind speeds and producing more calms than occur in rural areas.

The largest potential human-induced climate change is that due to increasing concentrations of atmospheric gases such as carbon dioxide and methane. Carbon dioxide concentration has increased by about 30 per cent from pre-industrial early nineteenth-century levels due to burning of fossil fuels (coal, petroleum and natural gas) in industrial applications, for generation of electricity and for transport. It is further enhanced by deforestation occurring extensively in the tropical rainforests and in North America where old growth forests are turned into lumber exports for world consumption. Carbon dioxide and other greenhouse gases absorb radiant energy leaving the Earth's surface, re-emit the radiation back down to the ground and keep air near the ground warmer than it would otherwise be. This physical interaction has been termed the enhanced greenhouse effect and it is expected to slowly raise the Earth's average surface air temperature. Atmospheric numerical models suggest that climate may warm between 1°C and 4°C by year 2100 (Houghton et al., 1996) when carbon dioxide concentration will be roughly double current levels. A certain amount of warming, between 0.3° and 0.6°C, is thought to have taken place globally since the late nineteenth century, although it is still somewhat difficult to ascertain unequivocally that the warming is entirely due to greenhouse gases (Wallace et al., 1995). The assessment of the intergovernmental panel on climate

change (Houghton et al., 1996) is that factors such as ocean circulation changes and uptake of carbon dioxide in the marine and terrestrial biosphere are not well understood and may play a key role in determining the ultimate extent of greenhouse climate variability. The consequences of global warming, however, will potentially be numerous, including sea-level rises due to melting of ice caps, changes in agriculture and water usage and increased frequencies of heat waves and drought.

SUMMARY AND CONCLUSION

The climates of North America are controlled by seasonally and latitudinally-varying solar radiant heat receipt, the temperature and movement of ocean currents, continental orography and the basic heating differences between land and water which, in turn, affect the distribution of air pressure and winds. Climates over much of the eastern portion of the continent are affected by the Gulf Stream and southerly air flow that brings warmth and Gulf of Mexico humidity northwards. The western portion of the continent is dominated by the cold California Current, and the eastern side of the Pacific anticyclone that migrates latitudinally creating strong seasonal variations in precipitation. Between the coastal areas are the generally higher elevations of the High Plains, the Rockies, and the Colorado Plateau. These climates can vary dramatically with elevation, they still exhibit some seasonality in terms of precipitation, but are generally dry except on windward slopes.

The twentieth century has been witness to remarkable human alterations of the atmosphere and climates of the earth. Urbanization has modified the Earth's surface, airflow and hydrological cycle. Air pollution has modified the urban atmosphere, producing a variety of human health problems as well as efforts on national levels to ameliorate these problems. Production of carbon dioxide through the burning of forests and fossil fuels threatens to produce a global warming of the Earth's atmosphere. Problems of pollution and global warming will continue to challenge the world well into the twenty-first century.

KEY READINGS

Barry, R.G. and Chorley, R.J. 1987: *Atmosphere, weather and climate*. London: Methuen.

Henderson-Sellers, A. and Robinson, P.J. 1986: *Contemporary climatology*. Harlow: Longman.

Houghton, J.T., Meira Filho, L.G., Callender, B.A., Harris, N., Kattenberg, A. and Maskell, K. 1996: *Climate change 1995: the science of climate change. The contribution of working group 1*. Cambridge: Cambridge University Press.

Kalkstein, L.S. 1996: *Climate and human health*. WMO Report 843. Geneva: World Meteorological Organization.

4

NATURE, ENVIRONMENT AND THE CULTURES OF NATURE IN NORTH AMERICA

DAVID DEMERITT

INTRODUCTION

I recently came across an advertisement urging me to help 'save the environment' by buying recycled paper. Such appeals have become commonplace of late as businesses try to cash in on a dawning ecological consciousness with an array of products aimed at 'green' consumers. With its slick, Madison Avenue copy, the advertisement skirts over a deep and abiding ambiguity: what kind of 'environment', exactly, will I be saving by consuming recycled products? Surely, the advert is not referring to the competitive, corporate environment in which paper manufacturers now find themselves, though of course any purchases would also help save that as well. Rather, we assume, almost naturally, that it refers to another, altogether more pristine and essential environment, the natural environment or Nature itself. This association is reinforced by the picture of a deer standing in a forest surrounded by huge trees, conceivably the very same trees and wildlife that I will be protecting by buying their eco-friendly paper products.

Yet for all its apparent naturalness there is something thoroughly artificial about this kind of scene. Bombarded as we are with them, it is easy to be cynical about adverts. No doubt the whole picture was staged: the deer tame, or perhaps not even real at all, merely stuffed; the diffuse sunlight in which it was bathed supplied by halogen lighting; and the misty background due to dry ice. The advertisement simplifies, mystifies some might even say, the relationships between consumerism and the environment. All by itself, buying recycled paper, even unbleached paper recycled from post-consumer waste, is not going to save the environment.

But the duplicity runs much deeper than that. It is possible to trace an extraordinary amount of human history in this way of seeing nature. The photograph trades on conventions of landscape painting to frame the deer and the other elements of the picture in the chiaroscuro light of the middle distance, thereby composing a picturesque landscape according the formal rules of aesthetics (Cosgrove, 1984). What appears as nature is largely art. Even the choice of scene is iconic, the deer in the forest symbolizing the entirety of the non-human world through its deliberate play on long-standing American understandings of wilderness as an unspoiled forest landscape verdant with wildlife. In Britain, by contrast, the windswept moors, home to Cathy and Heathcliffe, might be the preferred locale for representing Nature, while a tropical rainforest jungle, another common symbol of Nature itself, is invested with a very different meaning, overflowing as it is with associations with the intertwined histories of race, colonialism, and sexuality.

Clearly then, when we speak of nature and the environment we do not necessarily all mean the same thing. The very word 'nature' as the literary critic Raymond Williams (1983, p.219) has famously observed, has one of the most complex histories in the English language. Whether we use it to denote the biologically instinctive (i.e. human nature), the essential character (nature) of a thing, an original state of being (the state of nature), or the external, non-human world in its entirety (the natural world), its various and interrelated meanings are socially saturated. That is, they all depend upon a linguistic opposition to that which is said to be cultural, artificial, or otherwise human in origin. Since the cultural references by which nature and the natural are defined change over time and space, so too must ideas of nature. This, then, is one way in which nature and the environment might be said to be social constructions: as historically- and geographically-specific concepts that must be always understood by reference to the different ways in which they have been given meaning.

There is also a second, much more material, way in which nature might be thought of as a social construction. To provide for their needs, people have physically transformed their environments. Farmers, for instance, sow their crops, weeding out pests and other organisms deemed undesirable. The distinction between weeds and crops is influenced in part by matters of physiology, such as protein content, but mostly it is a cultural choice, imposed on the land by the people cultivating it. In this way nature and the environment are literally social constructions, things crafted and physically produced by the people inhabiting them. Because different cultural groups will have different conceptions of nature and will reshape the environment accordingly, it is possible to talk about different cultures of nature, each socially constructing it in different ways (Demeritt, 1998).

An appreciation of these cultures of nature and their historical geographies is essential for understanding the physical environment of North America. Not only is the present condition of the land itself a historical legacy of the different ways in which it has been used and apprehended, but so too are the ways in which we ourselves now perceive and experience it. No less than those who came before us, you and I make sense of the world around us in historically-specific and culturally-patterned ways. Born and raised in New York City, I tend to think of the forest, for example, as a wilderness refuge, a place of resort and recreation to be preserved from human despoliation, rather than as a place of work, as I might had I grown up in the timber-dependent communities of the Pacific Northwest, where logging the woods is a way of life. Neither of these experiences of the forest would have made much sense to the various Native American peoples who inhabited the continent before the arrival of European colonists, or even, for that matter, to Americans of a century ago, whose ideas of the forest turned much more on clearing it to make way for settlement and agriculture. No doubt you bring your own, likely quite different, prejudices and expectations to bear. An appreciation of the historical geography of these concepts of nature can help denaturalize them, showing how what appears as a natural and undisturbed environment is in fact the product of a particular culture of nature.

In this chapter, I want to illustrate these more general points about the social construction of nature and the environment by considering three distinct cultural geographies of the forest in North America. First, I will discuss the different ways in which Native peoples and the European colonists who encountered them in northern New England constructed the forest environment during the late seventeenth and early eighteenth centuries. Native people adapted themselves and the forest around them to suit their territorially extensive pattern of hunting and gathering. European colonists refused to recognize this lifestyle and the cultural landscapes it produced. They saw only wilderness and savagery. This dual construction of nature and society legitimated the dispossession of uncivilized Indians, living in a state of nature, to make way for settlement and the progress of civil society that colonists thought would come by clearing the forest. These ideals of progress and civilization through the settlement of wilderness held sway in both the United States and Canada until the turn of the last century when fears about the complete exhaustion of forest resources led to the rise of the conservation movement and the creation of government forest reserves to conserve what remained of the forest for future use. This is the second geography of the forest that I will explore. Many of the now dominant (and controversial) forest management practices owe their origins to this period when scientific forestry was first institutionalized in government-based forest services in North America. Finally, I will return to the present to consider the contemporary debate about the fate of the remaining old growth forests of the Pacific Northwest. This debate plays on and problematizes important concepts and constructions, such as wilderness, private property rights, and conservation, that figured prominently in the geographies of the American forest discussed previously.

These stories about the forest are important in their own right, but they also exemplify some wider trends in the historical geography of resource use in North America. Though different, perhaps in detail, the broad outlines of this story hold true for soil, fish, wildlife, water, and countless other resources across the continent. Native peoples were displaced from the land by colonial settlers who understood the environment and the nature of its resources in very different ways. Market imperatives have encouraged resource depletion, sparking debates about how best to manage and conserve the environment. The forest, then, is just one site for a much wider set of debates over the environment and the cultures of nature in North America.

CULTURES OF THE FOREST IN COLONIAL NEW ENGLAND

A waste and howling wilderness,
Where none inhabited
But hellish fiends, and brutish men
That devils worshipped.
(Michael Wigglesworth, 1662 (quoted in Smith, 1950, p. 4))

Penned by the New England poet and Puritan Michael Wigglesworth, this view of the forest was deeply foreign to the Mi'kmaq and Abenaki Indians who inhabited (what is now) northern New England and Maritime Canada (Fig. 4.1). What Wigglesworth found dreadful and bewildering they called home. These contrasting experiences of the forest reflected not only cultural differences in the ways the two groups thought about the world around them but also some important ecological ones in how they interacted with it.

The Mi'kmaq and Abenakis practised an extensive territoriality based on mobility through the landscape as resources became seasonally available. Since resources varied across the region, so too did patterns of subsistence. Among Native peoples in the Northeast, probably the most significant economic distinction was whether or not they practised agriculture. By planting maize, beans, and squash it was possible to support much higher population densities, but agriculture also required the investment of more labour and tended, therefore, to restrict the spatial range over which agricultural communities could exploit the landscape.

In general, Native peoples living in northern New England and Maritime Canada were much less reliant on domesticated crops than Native peoples to their south, though agriculture did become more important over the course of the seventeenth and eighteenth centuries as wars interfered with traditional seasonal rounds and concentrated people in pallisaded villages where maize cultivation became necessary to feed the larger concentrations of people. Still, the subsistence of the Mi'kmaq and Abenaki peoples of this region depended much more upon hunting and gathering the bounty of the land and sea. Precise strategies changed over time and differed depending upon variations in the local environment, but as of 1600 the general pattern along the northeast coast was something like this. In the spring, they gathered in small, riverside villages near the coast to feast on spawning fish, migrating birds, shellfish, and other aquatic and marine resources. Spring and summer were times of abundance, with well over half of their annual food supply coming from the marine environment. In September, they moved upriver, away from the coast, to harvest eels and migratory birds and to prepare for the fall hunt of moose and caribou. During the winter months, villages broke up into small family bands that ranged widely over the interior uplands in pursuit of beaver, moose, caribou, deer, and bears, although December and January also provided opportunities to fish through the ice for spawning tomcod or to go sealing. Winter was a lean time, when food was scarce until the abundance of spring started the cycle anew.

Their seasonal round made full use of the landscape – uplands, wetlands, the intertidal zone, even offshore islands were all exploited at some season or other. Native peoples were able to do this because they had an intimate understanding of the ecology of a wide variety of other species. Father Biard, the first of many Jesuit missionaries to live among the Abenakis, was both impressed and disgusted by their ability to find nearly everything they needed to survive in a forest that he took to be a howling wilderness.

The intimate knowledge of Native people marked the landscape of New England with a toponymy that remains with us today. In contrast to colonial place names, such as Jonesport or Gouldsborough, which denoted ownership, or those like Plymouth, Gloucester, and Bristol, which consciously recreated places back in England, Mi'kmaq and Abenaki place names tended to be descriptive or to denote some function. Thus

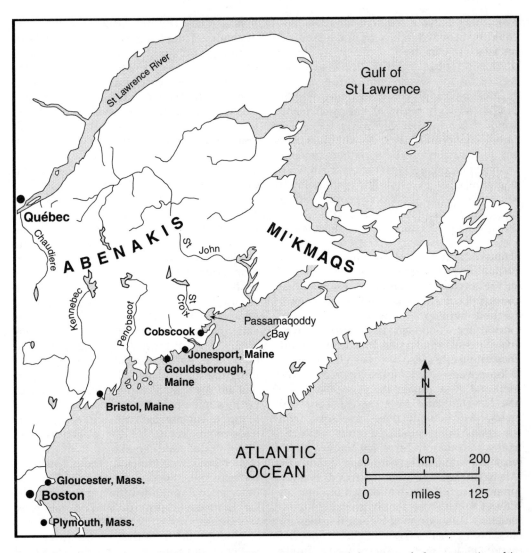

FIG. 4.1 The territory of the Abenaki and Mi'kmaq Indians, with location of places mentioned in the text

the name Cobscook (a river in eastern Maine) denoted broiling rapids, while Passamaqoddy Bay meant pollock plenty place (Eckstrom, 1941). Such names were important because the Native seasonal round depended upon the ability to navigate an extensive territory and identify resources. Instead of moving resources through space from one ecosystem to another, which today allows us to consume supermarket-bought food far from its place of origin, the Abenakis and Mi'kmaq moved themselves around the

landscape. They transformed the many rivers and lakes of the Northeast into highways and the portages between them into the Grand Central Stations of their day. Lightweight and assembled from materials available throughout the region, the birch bark canoe was a crucial piece of aboriginal technology. The mobility it provided enabled Native peoples to make their living from a suite of different resources as they became seasonally abundant within what was a relatively closed system. That is, almost everything that was required for subsistence could be obtained locally from the land and sea (see also Chapter 5).

This is not to suggest that Native peoples feasted, as the seventeenth-century English political philosopher and legal theorist John Locke derisively put it, at 'the spontaneous hand of Nature' ([1688] 1965, p.328). Their forest environment was much more of a social construction than either grasping colonists or present-day romanticism about noble savagism allowed. Fire was an important tool by which the Native peoples encouraged the growth of useful plants and animals. To call them natural resources does not do justice to the way in which they were the intentional products of the way in which the Native peoples cultivated the forest. The use of fire to cultivate particular resources had unintentional effects as well. As a result of frequent burning, the forests were, in many places, open and park-like, with relatively more fire-tolerant species such as white pine than might otherwise have been the case.

European settlers looked upon the white pine and the abundant game populations of the New England forest as the bounty of nature, but actually they were social constructions: things produced out of the particular social relations by which Native peoples engaged with their environment. Among the Abenakis, hunting was controlled by strict property rights, which allocated hunting territories to particular family groups. Hunting practices were informed by origin stories and a complex belief system that connected hunter and hunted through lineage, myth, and ritual. A beaver was not simply a beaver, but also the icon of an Abenaki lineage line whose ancestors demanded certain thanksgiving rites to ensure successful hunts. Such cultural practices served to ensure that game species and other natural resources were not over-exploited (Merchant, 1988).

None of this made any sense to English settlers. By refusing to recognize Abenaki property rights, they 'both trivialized the ecology of Indian life and paved the way for destroying it' (Cronon, 1983, p. 57). As they advanced further and further into Mi'kmaq and Abenaki territory over the eighteenth century, settlers disrupted the delicate ecological balance on which Native life depended. Their fields encroached on Native hunting grounds; their livestock foraged in the forest, competing with deer and the other animals on which Native peoples relied, while white trappers poached relentlessly in Native hunting territories killing off all the game they could find. Typical was the 1764 lament of an Abenaki chief that trespassing settlers were 'destroying the breed of Beaver' and leaving 'impoverished many Indian families' (Eckstrom, 1926, p. 77).

Abenaki complaints about white trespassers carried little weight with colonial settlers or their courts. Accustomed to the closely-tended tillage and pasture of the English countryside, Wigglesworth and other English settlers looked upon the forests of New England as waste: that is, wild, uncultivated, and unenclosed land from which no one could be excluded. The fiends and brutish men who haunted the forests were doubly disqualified from any claims to own them. As devilish infidels who did not recognize the Christian God, Native peoples had no legal standing, while according to Locke's influential theory of property, their failure to enclose and cultivate the forest left it open to seizure by any claimants who domesticated and improved it by mixing their labour with the soil. These two legal pretexts, one founded in religious law, the other in civil, provided the intellectual basis for European claims to land in the New World.

Through this vision of property, colonial settlers gradually reconstructed the New England landscape in the image of old England. As a concept, the doctrine of 'improvement' gave settlers a powerful lens through which to judge not only the land, but also the people on it. Colonists read the landscape in moral terms for signs that it had been cultivated, an act charged with moral, religious, and legal significance. Improvement was thought to be the foundation for claims not just to property but to humanity itself. Since Native peoples did not cultivate the land according to the norms of the English countryside, they had no culture and could be legally dispossessed.

As a practice, 'improvement' involved settlers in a thorough reordering of existing ecosystems in New England. It meant clearing the forest and cultivating the soil so as to subdue nature and render it (or her, in the gendered language of the day) fruitful in the fashion of old England. The amount of the back-breaking work involved forced settlers to focus their labour on much smaller and more clearly-defined parcels of land than had the Abenakis. Their practice of gathering plants and other resources from many different ecosystems as they became seasonally available was much less laborious but required substantially more land. By contrast, farmers operating from fixed locations exchanged commodities through a market. As a result, the abundance of different species on the land came increasingly to depend upon how their values were priced in the marketplace. Through this process of market exchange, apparently distant forces, such as changing fashion styles, could determine the fate of beavers (whose pelts made the felt hats popular among Europe's growing urban middle classes) and other species as much as local processes of ecological adaptation. Within their plots of land, farmers allocated land uses in response, roughly, to the relative costs of land, labour, and capital. The result, compared to Native practice, was intensive, specialized land use that reduced biological diversity and increased particular outputs. Essentially farming earned the higher yields of domesticated plants from the harder work of cultivating them.

It is important, however, not to exaggerate the differences or the degree to which agriculture liberated settlers from the vagaries of nature. Settlers were just as dependent upon the seasons and the land as the Abenakis. They simply managed time and space differently. For settlers in New England, as for the Abenakis, the annual season began with the spring thaw when it became possible to plough the soil in preparation for planting. Draught animals helped with this hard work, and their manure helped replenish soil nutrients depleted by repeated cropping of the same land. Thus, in a fundamental sense, colonial settlers placed themselves at the centre of the ecosystem, deciding what would go

where; the mark of the Abenakis on the land had been much lighter. Summer meant weeding, to protect monocropped fields from competitors, and haying, to provide winter feed for domestic animals, while fall was harvest time when stores were replenished for winter. Winter was the time to clear more land or log the forest for lumber and firewood.

Gradually, over a lifetime of winters spent clearing the forest, the landscape of New England was 'improved'. For settlers the recession of wilderness was progress. It marked the advance of something else that could be called civilization. This American civilization was founded upon relatively free access to land. Coming from Europe, where most people were beholden to landlords for ground rent and other entailments, colonial settlers dreamed of owning landed property enough to provide farms for their children and keep their families free from the yoke of dependent wage labour.

There were a number of deep and abiding contradictions in this desire for independence through the ownership of property. Independence was a highly restricted category, excluding as it did women, black slaves, and children whose labour was owned by their husbands, masters, and fathers. As long as there were Native lands to appropriate, it was possible to purchase this freedom without paying for it. But when this 'free' land ran out, the only way to secure the independence that land afforded was to produce for the market, and thereby risk dependence upon it, by selling the fruits of one's labour or even that labour itself for whatever they would fetch.

As a result, colonial New Englanders were enmeshed in a much more open economic system than the Abenakis and Mi'kmaq. Through the fur trade, Native peoples also found themselves increasingly connected to far off places through the exchange of commodities, but they were much less reliant upon imported goods than colonial New Englanders. Much of their cloth and textiles, most of their metal tools and nearly all of their manufactured goods were imported from Europe, while sugar and molasses came from the Caribbean, wine from the Canaries, tobacco from Virginia, and, increasingly over the course of the eighteenth century, cattle and flour from New York and Pennsylvania. To pay for these imports colonists had to sell something in return. Fish from the oceans, lumber from the forest, and the so-called invisible earnings from the foreign trade of wily merchants and sea captains were among the most important sources of foreign exchange in colonial New England.

The demands of the market led settlers to treat the forest and the landscape more generally as an assemblage of commodities: things that could be exchanged over long distances for other commodities from other ecosystems. Capitalist production for exchange left a heavy imprint on the forest. Where it was not transformed into an agrarian landscape of farms and fields, it yielded pine logs that could be driven downstream for export and exchange. Eventually, this led to the wholesale deforestation of New England. By 1860, census statistics indicate that over half of the total land area of Connecticut had been converted into improved farmland; percentages were only slightly lower in the other New England states. What was left of the forest was under increasing pressure from logging.

THE SCIENCE AND CULTURE OF FOREST CONSERVATION

Conservation does not imply the withdrawal from use, but on the contrary, it implies the fullest utilisation, which, in the case of forests, pre-supposes reproduction (H.N. Whitford and Roland Craig, 1918, p. 185)

This definition of forest conservation may sound strange today, but a century ago logging the forest and conserving it did not seem like the contradictory objectives that they are now so often deemed to be. What has happened in the last century is that the meaning of environmental conservation has changed. Now, environmentalists frequently object to any use of the forest, rather than, as a century ago, opposing only the abuse of the forest. As I will describe in the next section,

this transformation has had profound implications for contemporary debate over the fate of old growth forests in the Pacific Northwest, but to appreciate its full significance it is important to understand something about the science and the culture of forest conservation against which the contemporary environmental movement is reacting.

The forest conservation movement with which Whitford and Craig (1918) identified themselves came to the fore in North America at the turn of the century in response to growing concerns about the exhaustion of forest resources on the continent. As settlers raced across the continent during the nineteenth century they repeated the kind of ecological 'improvements' that had first been visited upon the forests and Native peoples of New England. Most chalked this up to progress. Indeed, in the United States and Canada progress itself was almost synonymous with the progress of agricultural settlement. Gradually, however, the environmental impacts of this wholesale ecological revolution began to tell. In 1864, George Perkins Marsh published his landmark, *Man and nature*, which warned of 'man's ignorant disregard of the laws of nature' and highlighted, in particular, the deleterious effects of forest clearance on the microclimate, soil (increased erosion and siltation) and river systems (increased amplitude of streamflow resulting in more floods and more frequent summer low water periods). The expansion of settlement onto the arid Great Plains magnified concerns over the potential of widespread land clearance to change rainfall patterns, hinder irrigation efforts, and exhaust the lumber supplies needed by settlers on the treeless Prairies (see Chapter 20 of this volume).

These environmental changes were happening at the same time as a series of cultural and economic ones were transforming the face of North America. Immigrants were pouring into new industrial cities like Chicago, Cleveland, and, in Canada, Hamilton, Ontario. The misery and class strife that marked these industrial cities dashed the Jeffersonian vision of the United States as an agrarian republic of independent family farmers and fuelled fears about the corruption of the country as it became like Europe: fatally divided between pauperized masses and a decadently-wealthy ruling class. In 1893, Frederick Jackson Turner famously announced the closing of the American frontier, where, in his hugely influential theory of American history, the availability of 'free' land provided the rough equality of circumstance that made Americans American. Canada was not immune to these anxieties, although its strong Tory tradition gave them a slightly different inflection. The long-settled countryside of southwest Ontario was plagued by many of the same environmental problems that Marsh had identified in his native Vermont, while the slums of Montréal and Toronto were filling with unruly immigrants such that many English Canadians, as well as arch-conservatives among Québec's Franco-Catholic elite, feared for the social order as Canadian cities came to resemble Manchester, England, or worse, the Lower East Side of New York City.

The American forest conservation movement was founded in the context of this widespread anxiety about the price of progress. While Marsh and others warned of the local environmental problems being caused by land clearance, a number of people began to worry about the possibility of a timber famine caused by the total depletion of the forest. Newspapers headlines repeatedly warned, 'The End of Lumber in Sight'. There were good reasons for alarm. Census statistics and the maps based upon them (Fig. 4.2) charted the inexorable decline of the nation's forest. They helped to unsettle a long-standing conviction that the natural resources of the continent were, for all practical purposes, limitless. Government maps and statistics reflected the new ways in which the forests of North America were being simultaneously united and liquidated through an emergent, continental-scale economy, in which changes in the demand for lumber in New York City were felt in the forests of the South, the Pacific Northwest, and the Lake states. But statistics did much more than this. Census statistics were instrumental in constructing a sense of a unified and coherent national forest that was being threatened and needed to be saved. This is one important reason why the conservation movement took off much earlier in the United States than in Canada. Because they were able to imagine the forest in these national-statistical terms as something they shared in common, Americans were much quicker to protect it than their neighbours to the north.

These concerns led to a series of government programmes and reforms in the United States. The federal government played a leading role, both because of its pre-eminent position as custodian of the public lands outside of the original 13 colonies and because the prospect of nation-wide timber famine seemed to demand a national response. In 1880, Congress created the US Forestry Division (which was renamed the Forest Service in 1905) to study the problem of forest depletion and potential solutions to it. Soon after, in 1891, Congress created the first forest reserve, reversing a century of public land policy in the US by specifically setting aside federal lands from pre-emption by or sale to settlers. The national forest system expanded rapidly thereafter, encompassing nearly 150 million acres (c. 61 million ha) by 1910 (Williams, 1989, p. 425). Many states also followed suit, taking tax-defaulted lands to create state-owned forest reserves.

But as Whitford and Craig (1918) firmly insisted, conservation was about much more than simply setting forest lands aside for the future; it also meant using the forest more rationally and efficiently. The first step in this practical programme of what US President Theodore Roosevelt (1905) called 'perpetuation...by use' was to protect the forest from destruction by fire, a chronic problem vividly dramatized by the death toll of the terrible Peshtigo, Wisconsin (1871), Hinkley, Minnesota (1894), and Yacolt, Washington (1903) forest fires. The expansion of organized fire suppression received a big boost from the Weeks Act of 1911, which provided federal matching funds for state forest fire protection programmes. In addition to fire protection, practical forest conservation depended upon the metaphorical transformation of the forest from a mine, to be cut-over once and then abandoned to fire and waste, into a crop and renewable resource

(Fig. 4.3). Conservationists also compared the forest to a savings account yielding an annual dividend of arboreal growth. This metaphorical construction of the forest appealed to lumber companies and other private forest-owning interests. It highlighted the difference between destructive timber-mining, which depleted the supply of natural

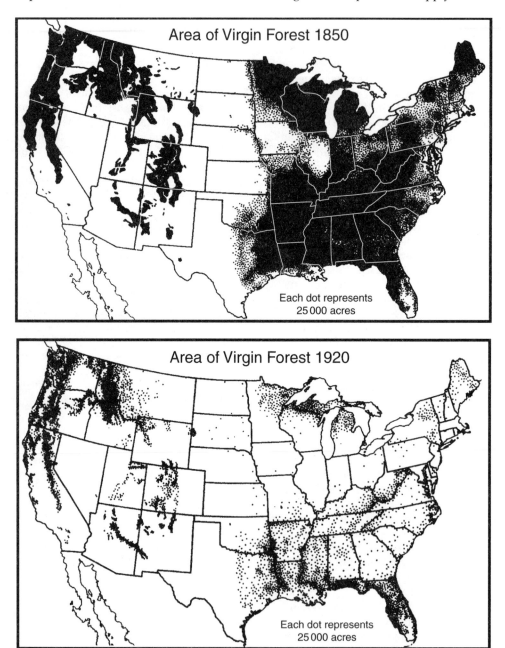

FIG. 4.2 Maps of the shrinking area of 'virgin' forest suggested the urgency of conserving this endangered resource
Source: Greeley (1925)

capital, and conservative forestry in which, as one turn of the century forester put it, 'only the interest [was] taken...[and] the principal of the investment [was] retained' (Cary, 1898, p.161).

Although turn-of-the-century conservationists used these metaphors to promote scientific, sustained-yield forestry to a sceptical forest industry, the relationship between the forest conservation movement and the

Timber as a crop

Starting a lumber crop; transplants in a lumber company's forest nursery

Wild grass was once the only source of hay. After a while hay became valuable enough to be treated as a cultivated crop.

Until recently wild timber was the sole source of forest products in America. Now timber is beginning to become valuable enough to be cultivated – to be cropped.

So enters forestry.

It is useless to bewail the "destruction" of the wild forests. They could not be preserved in large areas (outside of public ownership) any more than the wild grass of the prairies could be preserved.

Forestry is a matter of economics, not esthetics. It is practiced only when practicable.

It now promises to become practicable. Hundreds of forest operators are feeling their way toward the new era on a hundred million acres of forest land.

But one thing is sure – forest products must be patronized if they are to be produced by private enterprise. Forest use and forestry are united twins.

National Lumber Manufacturers Association
702 Transportation Building, Washington, D. C.

Fig. 4.3 Both as metaphor and as practice, the construction of the forest as a crop was central to scientific forest management

forest industry was far from cosy. Conservation crusaders like Gifford Pinchot, the second chief of the US Forest Service, railed against the common practice of cut-and-run logging. With little fixed investment and no long-term interest in the forest or the communities it sustained, lumbermen 'mined' the forest, selfishly maximizing short-term profits by stripping the forest without any regard for future reproduction. Once the wealth of the forest had been liquidated, owners picked up stakes and moved on to new, virgin lands, taking their money with them and leaving an impoverished landscape of stumps, boarded-up mills, and unemployed workers behind them. As this pattern of exploitation repeated itself from coast to coast, the nation seemed to be threatened by timber famine. The question for conservationists was whether this irrational behaviour could be cured by education or whether it was endemic to capitalist exploitation of the forest. Since a tree crop required more than a human lifetime to mature, many conservationists believed that individuals could not be trusted to treat the forest like a crop and manage it conservatively in the long-term public interest. Some called for government regulations on private forest land use, while others insisted that only government ownership would guarantee the conservation of the forest. Only the government, it was argued, had sufficient financial resources, patience, and concern for the long-term public interest to make the necessary investments in scientific forest management and wait a century for a return on them.

This programme of government regulation and ownership was bitterly resisted in the United States. Westerners, in particular, resented Washington bureaucrats, who often seemed more responsive to eastern urban elites and their dreams of unspoiled wilderness than to the practical needs of local people struggling to carve a living out of a harsh western landscape. Westerners howled that the federal government acted like 'an alien landlord' denying individual settlers the right to develop and improve the public lands, as colonial settlers had done in New England centuries before (quoted in Williams, 1989, p. 420). This line of fiercely individual-

istic and libertarian western protest continues to echo today in the so-called Sage Brush rebellions against governmental, and especially federal, environmental regulations. Deeply rooted in the resource-dependent areas of the mountain West, it plays on regional and class antagonisms against a conservation movement associated with the expanding sway of distant experts and urban-based sentimentalism over land use decisions in the West.

Large corporate forestry interests responded to the rising conservation movement more ambivalently. Though eager to cooperate with government fire protection and research programmes, corporations were generally opposed to government regulation. But rather than reject conservation outright, they turned its logic to their advantage. This was possible because corporate managers and conservationists shared so much in common (Hays, 1959). Both placed great stock in the ability of scientific experts to re-engineer social and economic life along more efficient and socially beneficial lines. In the case of forest management, large companies argued that they could be trusted to conserve the forest without any government interference because their massive investments in plant facilities meant that their interest in the long term sustainability of the forest and of the mill towns dependent upon it was identical to that of the general public. Since multinational companies like International Paper were among the first to hire professional foresters and actually to institute the scientific measures conservationists advocated, it was difficult for them to disagree.

The debate over forest conservation in Canada played out somewhat differently than in the United States, despite the fact that many of the ideas, institutions, and even individuals involved came to Canada directly from the US. For one thing, land was a provincial responsibility, so although Canadian conservationists were in contact with one another, the movement was much less nationally unified than in the US, where Washington's authority over public lands in the West provided a focus for nation-wide attention. There were common issues though. The question of raw log exports to the US, which made a fuller use of Canadian forest resources economically feasible but hindered the development of Canadian mills, was hotly debated in provincial capitals from coast to coast. Provincial governments had a say in this question of foreign trade, because lumbering in Canada had, for the most part, been based upon timber licences rather than outright land sales (as in the US), enabling provincial governments to ban exports of logs from Crown lands. With most of their forest land in public ownership, Canadians did not feel the same urgency as conservation crusaders to the south who believed that only public ownership would guarantee the protection of the public interest. In actual practice, however, this did not necessarily turn out to be the case. Many provincial governments, like British Columbia, were so financially dependent upon timber licence revenues that they could not afford either to regulate logging rates or invest much in the way of forest conservation measures (Gillis and Roach, 1986).

A much deeper contradiction ran through the forest conservation movement. Although public ownership and sustained yield management were widely celebrated as the very essence of forest conservation, these do not necessarily ensure the conservation of the other public goods derived from the forest. The crop analogy provided a basis for protecting future forest supplies, but it was harder to justify the protection of non-market public goods derived from forests in terms of sustaining crop yields or natural capital. Consider flood abatement, which was one of the benefits of federal forest reserves widely touted to otherwise sceptical western farmers and Senators. Flood abatement is what economists call an externality. Since lumbermen do not capture the benefits of it, flood abatement is not something that strict profit-maximizers have any incentive to consider. If it happens at all, it is only as an unanticipated by-product of conservative timber harvesting. Likewise, the scenic and recreational value of forests are also unpriced public goods, whose social value the market, and thus the idea of conserving natural capital, does not account for. It was not until the 1976 National Forest Management Act that the US

Forest Service was formally charged with taking non-economic forest values into account in preparing its management plans.

The inherent conflicts between these various forest values and uses are familiar today, but at the turn of the century they were not so obvious. Then, the sky was the limit for scientific forestry and forest conservation; their potential seemed as limitless as it was untested. Far from being at odds, forest production and forest preservation seemed like two sides of the same coin. Turn-of-the-century environmental groups like the Sierra Club and the Society for the Protection of New Hampshire Forests, founded, respectively, by the wealthy scions of San Francisco and Boston society, were organized both to preserve the scenic beauty of mountain forests and to ensure that they were managed scientifically to provide a sustained yield of forest products.

If these now seem like contradictory objectives, it is because there is growing concern about the naked instrumentalism involved with managing the forest like a crop or a liquid financial asset. The metaphors of the forest conservation movement were deeply practical in their implications. They demanded a material reconstruction of the forest, no less sweeping in its extent than that of colonial settlers who sought to improve the land by clearing the forest and planting it to annual crops. In managing the forest as a crop, foresters had to decide what would grow where. As a result, non-merchantable species, long ignored by loggers, were suddenly reconstituted as competition. Foresters experimented with herbicides and other techniques to favour the growth of merchantable species over their competitors.

Their confidence in science has led foresters to treat the forest as an assemblage of individual objects that can be managed more or less in isolation from one another. They sought to maximize the yield without much thought to how larger quantities of a merchantable species would interact with the rest of the forest. In the spruce-fir forests once inhabited by the Abenakis, selective logging for spruce has transformed uneven-aged stands dominated by mature red spruce into much simpler stands with two age classes: a canopy dominated by balsam fir and those mature, red spruce left as 'seed-trees' from previous rounds of harvesting and a dense understory of suppressed firs, which reproduces more prolificly than spruce. As this fir-dominant understory matures, it becomes vulnerable to infestation by spruce budworm, an insect whose preferred food, in fact, is balsam fir. At periodic intervals, growing more devastating with each occurrence, spruce budworm outbreaks have devastated the forest, killing most mature fir and many spruce trees as well. This high mortality, combined with massive clear-cuts to salvage vulnerable stands before they are attacked, has created the ideal habitat for spruce budworm to thrive: large areas of dense, even-aged, fir-dominant forest stressed from fierce competition for canopy space and thus less resistant to attack. Similar problems plague the forests elsewhere in North America, leading foresters increasingly to question the wisdom of a science focused on lumber production without regard to the wider ecological effects of this management strategy (Langston, 1995).

BATTLE FOR THE OLD GROWTH

In Wildness is the preservation of the World (Henry David Thoreau, 1937, p. 672)

Nowhere has public dissatisfaction with the aims, methods, and theories of scientific forestry been greater than in the Pacific Northwest. In the last decade, the forests of this region have become the focal point of a fierce and increasingly international controversy. At issue are the remaining unlogged stands of coastal rainforest, remnants of an old growth forest that once stretched from northern California up through Oregon, Washington, British Columbia, and southern Alaska. In this rain-soaked environment, native tree species can live more than a millennium, reaching gargantuan proportions. The terrific size and age of these spectacular forests has attracted the interest of many. While visitors are inspired by the scenery, lumber companies prize the old growth for the quality and volume of fibre locked up in these forests.

After more than a century of industrial logging, the saws are finally reaching the last uncut stands of coastal rainforest in the Pacific Northwest. In response, environmentalists have organized to save the old growth from apparent destruction. 'Save the ancient forests' has become the rallying cry for an international campaign. Its highly successful appeals unite metropolitan publics in New York, Toronto and overseas in Japan and Western Europe in a shared concern for and with the doings in the remote mill towns and valleys of the Pacific Northwest. The campaign has transformed a local debate over the forests of the region into a global struggle. Standing against them are residents of the region's many timber-dependent communities who fear for their livelihoods, as well as the forest products companies who, naturally enough, oppose any conservation measures that will interfere with the bottom line.

This debate has conventionally been cast as one of jobs versus the environment, but this over-simplifies things considerably. While environmentalists are often blamed for unemployment in the mill towns, in fact it is technological modernization, allowing more work to be done by far fewer workers, that is responsible for most of the job losses in the industry (Grass and Hayter, 1989). As for the shortages and rising price of raw logs, these too have been a long time in coming. This fall-down effect, caused by the exhaustion of old-growth timber before enough second growth timber is ready to replace it, was first predicted more than 50 years ago (Langston, 1995). That said, there can be no mistaking the fact that the debate over the forest pits two very different visions of the forest and its future. Indeed, the contentious question of how to manage and conserve the forest turns largely on representing its true and proper nature. Thus, exploring the competing constructions of the old-growth forest provides an important lens for exploring the stakes in this battle for the forest.

In defence of continued logging, forest products companies echo much of the rhetoric of turn-of-the-century conservation pioneers like Gifford Pinchot. The companies insist that while clear-cut logging may appear to be 'an alarmingly destructive practice' it is in fact 'biologically sound and the simplest, safest, and least expensive way to log and regenerate the forest' (MacMillan Bloedel, n.d., p. 3). Far from marking the end of the forest, clear-cutting, according to the Great Northern Paper Company (n.d.), 'is usually the start of a new forest. In fact, a clear-cut filled with seedlings is a very young forest.' This point is made more subliminally in *Future forests*, a public relations pamphlet recently issued by the British Columbia based MacMillan Bloedel Company (n.d.). The cover shows a seedling sprouting from a stump (Fig. 4.4), while its contents exhaustively detail the great lengths taken by the company to ensure, as then company president Ray Smith (1990) put it, 'we have a forest for tomorrow.'

The assumption throughout is that the forests are there to be used. 'Forests are one of nature's most precious gifts. The working forests that today's society uses to build shelters, provide fuel, enjoy the written word, and package food, are being replaced with a healthy new crop' (MacMillan Bloedel, n.d., p. 12). This is a rhetoric that would have sounded familiar to Pinchot and other turn-of-the-century conservationists, but it is tempered by much greater attention to the diversity of forest values and uses that modern forest management must now sustain. *Future forests*, for example, outlines MacMillan Bloedel's multifaceted efforts to promote recreation on its lands, protect wildlife habitat, maintain biodiversity, even to combat global climate change by storing carbon in young, growing forests.

These promises are guaranteed by appeal to rational science. Once, the mere invocation of science was enough to legitimate the details of whatever foresters did in the woods, but forest companies now employ sophisticated public relations techniques to reassure an uneasy public that forestry is both scientific and sustainable. Again, *Future forests* is typical. It is full of statistics, graphs, figures, and photographs, all designed to convince readers that the company, as 'custodian of the forest', is both willing and able 'to protect and manage the trees, plus all the other values of the forest for the benefit of present and future generations' (MacMillan Bloedel, n.d., p. 11). Through intensive scientific management, careful breeding of 'superior seeds for planting', and technological advances in utilization, we are assured, the productivity of the commercial forest can not only be sustained but actually increased. This vision of a forest improved through science echoes the old dreams of colonial settlers and progressive-era conservationists to dominate nature, even as it creates the space to 'set aside [areas] for wildlife and other special reasons' in addition to the 'considerable areas of old growth preserved in BC's six million hectares of federal and provincial parks and other special reserves' (MacMillan Bloedel, n.d., pp. 8, 10). What could be more reasonable?

The emphasis on expertise serves both to legitimate company practices and to disqualify potential critics as either uninformed, emotional, or extreme. It appeals to a singular and undifferentiated 'public' with an undifferentiated interest in the forest. Unquestioned is the assumption that in future this phantom public 'is going to require even more forest products' (MacMillan Bloedel, n.d., p. 14), and thus that the appropriate criterion for evaluating the company's forest management is the narrowly technical one of whether its scientific practices can deliver the goods. Sidelined are wider moral and political questions about whether more toilet paper is actually such a good thing, whether the forest should ultimately be merely a means to some higher human end, or indeed whether the even-aged industrial forests imagined by MacMillan Bloedel and its foresters are even forests at all.

These are some of the pressing questions raised by the campaign to save the old-growth forests from logging. Often, the nature of this imperilled forest is rendered in the language of ecological science. Environmentalists speak of ecosystem integrity and stability, of niches and habitats, biodiversity and genetic erosion. These forest properties were long ignored by the scientific silviculture of Pinchot and its narrow focus on the immediate use value of nature. By bringing these ecological relations into view, the science of ecology helps create a space for

FIG. 4.4 Forestry companies insist that logging the forest is not inconsistent with conserving it
Source: MacMillan Bloedel (n.d.)

insisting that the value of the forest exceeds the scaler's board-foot measure. It provides a socially legitimate way to distinguish between the rich habitats and biological diversity afforded by 'true' old growth forest and the biologically impoverished monoculture plantations that the lumber companies call forests, with their rows and rows of uniform trees grown from the same seed stock and thinned so as to maximize the production of merchantable fibre.

In concert with the Endangered Species Act (ESA), the 1973 federal law authorizing sweeping measures for the protection of species endangered by extinction, the vision of ecology has provided a powerful tool in the struggle to protect old-growth forest in the United States. When the northern spotted owl was officially declared endangered in 1990, the US Forest Service was forced to draft a plan setting aside millions of hectares of federal forest land, once slated for logging, to protect the old growth habitat of the spotted owl (Dietrich, 1992). Although Canadian environmentalists deploy the same ecological concepts and rhetoric, they are much less powerful without the ESA to give them legal force. As a result, environmentalists in British Columbia have had to rely much more on headline grabbing techniques such as road blockades and public demonstrations in their struggle to convince the public of the importance of saving old-growth forest.

While ecology has provided powerful reasons to save the old growth, environmentalists also play on older, romantic ideas about wilderness and the feelings of awe and reverence it inspired. Whereas for Wigglesworth wilderness represented Satan's home to be redeemed by improvement and cultivation, this attitude was completely reversed by the mid-nineteenth century, such that Thoreau could imagine wilderness as God's own temple, a wondrous and unsullied refuge from a fallen civilization. Environmentalists appeal to these sentiments and the long tradition of nature writing behind them. The old-growth forests are described as 'nature's living cathedrals'. Their spectacular 'columns and arches' put us in touch with 'the heart of existence' and let us 'feel the harmony of a thousand symphonies in which every living thing works together for mutual survival' (Canada's Future Forest Alliance, 1993, p. 36). Such deeply felt moral and spiritual beliefs about the value of wilderness present a potent challenge to the instrumental view of nature as something there simply to be used.

But there are troubles with this wilderness idea. It places the forests of the Pacific Northwest outside and before American history. The rhetoric of primeval and pristine with which old-growth forests are so often described leaves little place for Native peoples to have had a history of their own before colonial contact brought about the steady retreat of wilderness or, for that matter, much of a future either. If wilderness is defined as pure and untouched, then by definition humans can have no place in the forest. The rhetoric of virginity also sexualizes the feminized body of nature. In these ways, the environmentalist view of the forest as wilderness is remarkably similar to the empty and untamed landscapes that early colonists dreamed of mastering. The only difference is that it reverses the polarity of these dualisms, mourning rather than celebrating, as Wigglesworth did, masculine conquest of virgin forest.

Concern with pristine wilderness and its loss tends to lead to a politics of preservation without much concern for people in the here and now. This, of course, is why workers in the timber dependent communities of the Pacific Northwest feel so threatened by this kind of environmentalism. It is

> very much the fantasy of people who have never themselves had to work the land to make a living – urban folk for whom food comes from a supermarket...and for whom the wooden houses in which they live and work apparently have no meaningful connection to the forests in which trees grow and die' (Cronon, 1995, p. 80).

What is worse is that the wilderness ethic contributes to an all-or-nothing attitude about the environment. By longing for the pure and untouched spaces of the old-growth forest where they don't live, people in North America tend to disavow any responsibility for the heavily-urbanized environments in which they do.

Thanks to exactly this kind of either/or thinking, Americans have done an admirable job of drawing lines around certain sacred areas (we did invent the wilderness area) and a terrible job of managing the rest of our land. The reason is not hard to find: the only environmental ethic we have has nothing useful to say about those areas that fall outside the line. Once a landscape is no longer 'virgin' it is typically written off as fallen, lost to nature, irredeemable. We hand it over to the jurisdiction of that other sacrosanct American ethic: laissez-faire economics (Pollan, 1991, p. 188).

CONCLUSION

Nature cannot pre-exist its construction (Donna Haraway, 1992, p. 296)

Through a discussion of the historical geographies of the forest in North America, this chapter has explored the ways in which nature and the environment are socially constructed as both cultural concepts and cultural products. This is important because so many environmental debates presume that nature is simply given and thus that our ideas of it, whether as natural resource or wilderness refuge, must be natural as well. They are not. As both concept and product, nature embodies an extraordinary amount of human historical geography. Cultural conceptions of nature play an important role in identifying what counts as a natural resource. The Abenakis understood the forests very differently from the colonists who displaced them, who in turn would have found it difficult to make any sense of the present old growth forest debate. These cultural concepts lead to different cultural practices that play an important role in shaping and materially constructing the nature of the forest.

But if nature is a cultural construction, it is not something made under conditions entirely of our own choosing. Foresters struggling with insect infestations have learned from long experience that they are not free to make the forest in any way they choose. Other actors matter as well. This realization was crucial to both the ecological critique of old growth logging and to the new found concern of Macmillan Bloedel and other forest companies with forest characteristics and processes beyond the accumulation of merchantable fibre. This recognition of the importance of cultural practices in materially shaping the forest is welcome, but unfortunately it has proven much harder to denaturalize the concepts of 'wilderness' and 'natural resource' organising the Pacific Northwest forest debate. These cultural concepts of nature, no less than the cultural practices altering it physically, have played an important role in constructing the nature of the North American forest. Ideas of wilderness underwrote the dispossession of Native peoples and the establishment of property in the forest first as something to be improved and then later as something to conserve. Now, they are helping to advance the campaign to save the old growth forest. An appreciation of the different cultures of nature and their historical geographies is thus essential for understanding the changing physical environment of North America.

KEY READINGS

Cronon, W. 1983: *Changes in the land: Indians, colonists, and the ecology of New England.* New York: Hill & Wang.

Cronon, W. 1995: The trouble with wilderness; or, getting back to the wrong nature. In Cronon, W. (ed.), *Uncommon ground: toward reinventing nature.* New York: W.W. Norton, 69–90.

Demeritt, D. 1998: Science, social constructivism, and nature. In Castree, N, and Willems-Braun, B. (eds), *Remaking reality: nature at the millennium.* New York: Routledge, 177–97.

Langston, N. 1995: *Forest dreams, forest nightmares: the paradox of old growth in the inland west.* Seattle: University of Washington Press.

Section C

People and culture

NATIVE NORTH AMERICANS

KENNETH S. COATES AND WILLIAM R. MORRISON

Since their first contact with Europeans, Native North Americans[1] have been stereotyped by those who have not known them or have not understood them. They have been the 'noble savages', called into existence to fit the fantasies of a Rousseau; the downtrodden victims of European colonialism; the 'natural men' – the first conservationists, living as one with the environment; the bloodthirsty savages; the intense spiritualists; the shameless drunks, living on social assistance, unable to get or keep a job. These images, partly true and partly untrue, are contradictory, and have coloured their relations with other people in Canada and the United States. Until recent times, they were also under a kind of racial death sentence, for it was widely believed that they were a dying race, doomed to disappear in the face of a superior culture. This belief, at least, has proved false, for their numbers are now growing faster than any other racial group in North America.

PEOPLING NORTH AMERICA

Native societies have always maintained detailed accounts of their origins as a people. Some are consistent with the ideas of modern science, for they speak of a journey from another land, while others reject this idea in favour of a creation story that suggests that they emerged as humans on this continent. In several cultures, it is believed that creation originated with the efforts of animals and birds to create a physical world. The Iroquois, for instance, focus on the Turtle as the being that offered his sturdy back as the foundation for the earth – hence the name Turtle Mountain that many First Nations give to North America. Supernatural beings created both the physical and human world, and established human beings as one part of a vibrant, life-giving environment. The interpretations vary – in one area, the Raven is the key figure, in another it is Coyote, and among the Abenaki of eastern Canada and New England it is Gluskap, the Shaman – but they all unite the human, animal and physical spheres in understanding the world.

The explanation preferred by science, based on some physical evidence and some conjecture, is a story of one of the world's greatest migrations. Sometime between 14 000 and 100 000 years ago – the date is a matter for hot debate – people living in eastern Siberia, pushed northeast by population pressure, and attracted by hunting opportunities, crossed into North America. They came across what is now the Bering Strait; an ice age had locked so much of the world's water into glaciers that the sea level fell, leaving a land bridge several hundred kilometres wide, over which the people came, hunting game, without any sense that they were moving from one continent to another. Two great controversies plague this theory: what route did these proto-Indians choose once they reached North America, and, when did this happen? The traditional answer to the first question is that these people came down the Rocky Mountain corridor from what is now Alaska through the Yukon, between two enormous ice sheets that then covered most of Canada. A more recent interpretation has them travelling south along the Pacific Coast, camping at spots which are now well under water, making proof of this theory difficult. One way or the other, however, archaeology shows that they had reached the southern tip of the hemisphere by 11 000 BP (before present).

But when did they first arrive? This is one of the great debates of modern archaeology. Some insist that the ancestors of North America's Native people arrived 30 000 years ago, or earlier. Yet there is no universally accepted evidence of human presence on the continent before 12 000 BP or so, and the debate remains open.

Another controversy concerns their effect on the continent's fauna. Some scholars suggest that the mammoths, giant bison, and horses which roamed North America were hunted to extinction by newcomers. Others reject this theory, arguing that no species was ever exterminated by Native people; in fact, their spiritual values made such a thing inconceivable. Science is brought into play on both sides, but the argument is as much political and even spiritual as scientific. An allied controversy comes from the belief that all people who lived on the continent before the arrival of Europeans were related to each other, and thus any bones or human remains found by anthropologists are the direct ancestors of living Native people. This has had a chilling effect on the science of anthropology, especially in the United States, where the discovery of human remains has in several instances led to controversy between scientists and tribal representatives.

Whatever the truth of these issues, the fact is that the peopling of North America was a remarkable process, probably accomplished surprisingly quickly – from Alaska to Patagonia in less than a millennium – as the migrants moved across the land, finding new homelands, struggling with neighbours, and learning to cope with the new flora and fauna. We do know when the process stopped, when the last migrants came: these were the last of several groups of the ancestors of today's Inuit, who came to the continent not more than 4000 years BP. Discovering the more temperate lands to be occupied, they occupied the far north, drawn by temperatures warmer than now, which made it possible to survive on Ellesmere Island, as far north as there is land on the continent, before a colder period forced them further south.

Contrary to the standard view of the first peoples as static and unchanging (a view held by some supporters as well as detractors), their societies in fact experienced many transformations over the millennia. New technologies – agriculture, the invention of the bow and arrow, and, especially, technologies particular to cold regions, such as snowshoes, sleds, and the igloo – brought numerous changes, strengthening some groups and weakening others. In at least one region – the Mississippi Valley – a large urban community emerged, flourished, and disappeared, leaving only huge earthen mounds to mark its passing. The nations did not live, as the fantasies of some Europeans and some modern apologists would have it, in perfect peace and harmony with each other. Instead, conflicts developed between neighbouring peoples, alliances were formed, wars fought, peace made, and in some cases the losers were forced to move – sometimes for hundreds of kilometres. The European image of these people is not a picture of their entire history, but a snapshot of them as they were at the time of contact – sometime between the early sixteenth and the early twentieth century, depending on the individual group.

INDIGENOUS SOCIETIES AT THE POINT OF CONTACT

Descriptions of aboriginal culture and society often begin, or at least begin to be detailed, at the point of contact with Europeans: with Columbus in the 1490s, with Central and South America in the early sixteenth century, moving eventually to the Arctic in the early twentieth century. The problem here is that change preceded contact: well before the newcomers reached many groups, the lives of these people had already been changed, in ways large and small, by European technology and European diseases. When Europeans arrived in remote places, ready to record the ways of the indigenous people, they found them using trade goods obtained from Native intermediaries, and often found their numbers severely reduced by diseases such as smallpox. These accounts therefore are of societies already well into a process of change. The shortcomings of the written record caused by this situation have, however, been somewhat overcome by a renewed confidence in indigenous oral tradition, the value of which has long been accepted by anthropologists and ethnographers, and more recently by the courts in Canada and the United States.

What the written and oral records show is that indigenous societies were highly diverse. In the far north, for example, the Inuit lived in small extended family groups, isolated from others for much of the year. In the desert lands of the Southwest groups such as the Hopi and Zuni lived in pueblos – sizeable settlements of several hundred people, with hierarchical societies and complex local economies. The Iroquois of the Great Lakes region lived a semi-settled life, produced a good proportion of their food from farming, and developed an intricate political and diplomatic system that governed inter-tribal relations in the region until the Europeans arrived to upset it. On the western Plains lived societies dependent on the buffalo for survival; not the 'Plains Indians' of Hollywood movies, however, for these people had no horses before Cortez brought them to Mexico; instead, they travelled on foot, using dogs as pack animals. Along the northwest coast, groups such as the Haida and Salish lived in one of the richest non-agricultural areas in the world, flourishing from harvests of salmon and use of the giant cedar trees that were the hallmark of their physical and spiritual worlds. To lump all these people into the category 'Indian,' with the thought that they were all the same, is as absurd as the suggestion that everyone who lives in Africa is culturally and racially identical. One thing all these peoples did have in common, however, was that they were all well adapted to their environments. Indeed, it was the great variety of environment in North America that made their cultures so different.

Environmental and spiritual dimensions

Native North Americans were intensely spiritual people, and in a way that Europeans had not experienced since before the Christian era. They saw themselves as part of the physical world, not separate from or above it. Their world was filled with spirits and with complex interactions between the human, animal, plant, and spiritual world. Spiritual leaders, often called shamans, interpreted these interactions for the people, providing assistance and healing for those in conflict with the spirits. Contrary to the beliefs of some present-day ideologues, these people were not 'perfect' environmentalists; like all human societies, they made both long- and short-term mistakes, and were quite capable of wasting or over-harvesting resources, underestimating their impact on the environment, and causing disruption to their surroundings. But on the whole they worked in remarkable harmony with their environment, largely because they had no cultural imperative to dominate it. Their spiritual

beliefs explained harvesting practices, and at the same time controlled them, thus maintaining an ecological balance.

Only recently have non-Natives truly come to appreciate and understand the depth and complexity of Native spirituality. This discovery has led many outsiders to seek to learn from Native elders. The great popularity of writings of elders such as the Sioux Black Elk (whose deep Christianity is often omitted from descriptions of his ideas) represents a dramatic reversal in attitudes towards indigenous cultures.

Sustenance and livelihood

The first peoples did not only live with their environment; they drew their sustenance from it. Only a few groups developed settled agricultural practices, but all cultures drew their food and much else from their surroundings. Harvesting rested on an intense familiarity with the physical world. Native peoples knew which plants provided food and which ones contained poisons. Through many generations they learned about the migratory patterns of birds, fish, and game animals. They knew where the harvesting of roots and berries was best, and how to use the products of the forest. They knew that if they over-used the resources in one year, they would suffer hardship in the next.

Hardship was not unknown in the indigenous world, although as a number of early commentators noted, it was probably less common than among the peasantry of Europe. A commitment to sharing, combined with various social and cultural means of limiting population size, enabled most groups to maintain a rough equilibrium between the number of people and the available resources. There were many different ways of organising the distribution of resources. In the marginal zones, people tended to live as extended families and to rely on skilled hunters to provide sufficient food for the group. In more settled areas, hierarchical social systems gave leading individuals the power to distribute wealth and resources. Among the West Coast peoples, for instance, the potlatch emerged as a means of distributing resources, with the dispensing of gifts establishing a debt between giver and recipient.

North American cultures did not as a rule work at producing an economic surplus, nor, as a rule, did they support (as did European and Central American cultures) a non-productive priestly or aristocratic caste. There was a good deal of trading, usually between neighbouring groups, but sometimes for goods that came a long way from their source. Coastal people, for example, traded fish and other ocean products (fish oil in particular) for animal products and furs produced in the interior. Luxury products such as shells for ornamentation could travel a very long way, from the Gulf of Mexico to what are now the Canadian Prairies. But for the most part, Native groups subsisted on what was available to them locally, and developed means of sharing their resources within their societies.

Social structures

The first Europeans believed that the social systems of indigenous people in North America were not particularly complicated, and they tended to look for parallels to their own societies; thus the early descriptions talked much of Native 'kings' and 'priests'. It would take many generations for the newcomers to understand and appreciate the real workings of indigenous societies, some of which were very different from European models. Some cultures, for instance, were matrilineal, organizing inheritance through the mother's line, not the father's. Marriage alliances and political relationships were, in some groups, determined through complex clan relationships, and intricate kinship structures strongly influenced social and personal connections. The power to make decisions varied widely, depending on the customs of the group and on the situation. A skilled hunter, for instance, might control the group during a time of harvesting, while a person with perceived spiritual power might have great influence in a time of crisis. In other groups large, complex, and lengthy meetings, involving both men and women, might be required to achieve a consensus before important action was taken.

Inter-tribal relations were also highly variable. Some parts of the continent had a well-developed system of foreign relations and diplomacy, the best example being the Five (later Six) Nations Confederacy of the Iroquois, which rested on a complex series of treaties and decision-making bodies. In other places, notably on the Great Plains and the Far North, the groups were perennially hostile to each other, yet even then often developed trading arrangements that worked even in time of conflict. Warfare was commonplace, as different groups vied for favoured territories or settled personal scores. But, in contrast to the European wars of the time of contact and after, indigenous battles were often more symbolic than bloody. The counting of *coup*, for example, involved striking an enemy but not killing him. Small raids, often involving the capture of women and children and the taking of slaves, were much more common than pitched battles.

By the time the first European arrived, the indigenous people had been living successfully on the continent for millennia, and were confident in their ability to survive in this world and to deal with whatever lay beyond it. But the Europeans did not see them this way; to Columbus and his successors, asserting their technological superiority, the 'Indians' were savage, barbarian, primitive, and heathen. This attitude reflected Europeans' belief in their own superiority, while at the same time providing a convenient rationale for the conquest, subjection, and dispossession of the continent's original inhabitants. It also laid the foundation for the official policy towards Native peoples that persisted until very recently.

THE CONTACT PERIOD

Native North Americans were not initially impressed with the newcomers, though their ships and technology (especially firearms and metal goods) attracted great interest. This was partly because north of Mexico the Europeans often seemed so helpless; from Virginia to the Far North, the early explorers and settlers often were ludicrously ill-prepared for the New World, and survived their initial years largely with the help of the local Native people. It was from the Native people that the colonists learned how to feed, shelter, and often clothe themselves, and, especially, how to move about the land.

In most cases the newcomers were initially welcomed, or at least not attacked, by the indigenous people. Soon, however, it became obvious that the cultures and aspirations of the old and the new inhabitants were bound to clash. What the Europeans wanted from North America – gold, silver, land for agriculture held in private hands – was incompatible with the hunter-gatherer way of life of the Native people. Along the eastern seacoast of what was to become the United States, for example, the Europeans, after an initial period of adjustment, began to move into the interior and clear the land. The result was a series of 'Indian wars', in which the Native people had temporary successes but which resulted in their eventual defeat. Some moved west, some survived on reserves, and some groups were actually extinguished altogether. The original population of the indigenous people can only be estimated, but a reasonable figure for what is now Canada and the US would be three to five million; by the end of the nineteenth century their numbers had been reduced (predominantly by disease but also by warfare) by 90 per cent.

The fur trade

The fur trade, whatever one may think of it as an exploiter of the indigenous people, was at least the one economic activity of the newcomers which worked in partnership with these people rather than seeking to displace them. From the point of view of the newcomers, it served not only as a profitable activity, but was also the main engine for the exploration of the continent. The major fur companies – the London-based Hudson's Bay Company and the Montréal-based North West Company in Canada, and John Jacob Astor's American Fur Company in the United States, raced across the continent, mapping the land, establishing posts, and pioneering trade and transport routes.

The fur trade created its own economic world, resting on indigenous labour. Unlike the settlers, who saw the Native people as impediments to progress, the fur traders depended on them, and tried as much as possible to convince them to maintain their relationship with the land and the environment, though always with the goal of profit. The trade was, in the words of one historian, a 'mutually beneficial symbiosis'. Traders got furs that were worth many times the cost of the trade goods, while the Native people were able to dispose of furs that were surplus to their needs in return for the products of European technology. At least, that was the rosy view of the trade. The darker side involved what happened when over-harvesting occurred, with the use of alcohol in trade, and the loss of independence that resulted when the Natives became dependent on these goods, and when they were bound to the traders by ties of perpetual debt.

Fur traders established relations with the Native people that were closer than those of any other group of newcomers. Because the trading system operated with a handful of traders living in the midst of hundreds of indigenous people, inter-racial sexual relationships flourished, most of them of a short-term nature. Native leaders often sought to have a trader take a daughter or other female relative as a wife, in order to strengthen the bonds of the trading relationship. One result of this was a sizeable mixed-blood population, which on the northern plains developed into a unique society – the Métis, offspring of French traders and aboriginal mothers, who coalesced in the nineteenth century into a self-styled 'nation', and remain a political force in the Canadian West.

The fur trade declined with the gradual erosion of the resource, particularly in eastern and central North America. Over-hunting brought on economic hardship and dislocation for the Native people, and when the fur traders were followed by settlers, as they were in the older districts, the Native people became marginalized and dependent on government assistance. Only in the Far North did the fur trade continue to dominate the economy into the mid-twentieth century, and in some areas, despite the activities of animal rights activists, it remains important to this day.

The biological encounter

The greatest harm to the indigenous peoples of North America came not from settlers or fur traders, but from disease. Before the beginning of the sixteenth century, Europe and North America had had little biological connection; a good number of plants and animals existed on one continent but not on the other. The same, unfortunately, was true of a number of diseases. What Alfred Crosby (1972) called the 'Columbian exchange' was a complex phenomenon. The New World provided useful natural products to the Old – the potato is a well-known example – and the Europeans re-introduced the horse to North America, something that was to revolutionize the culture of the people who lived on the Great Plains.

There were hundreds of other transfers, in a process which went on for centuries, but the most profound and damaging one was the introduction of European diseases for which the indigenous people had no natural immunity. There were many of these, and even the ones such as chicken-pox and measles that were minor problems in Europe were serious scourges in North America. The worst plague, however, was smallpox.

Much of the devastation wrought by smallpox took place out of sight of the newcomers, as infected Native people, fleeing the disease in their tribes, spread it to other communities far from the frontier of settlement. When the Europeans ventured into the interior, they often found the land almost deserted – it was this phenomenon that led to underestimates of the original Native population.

The scale of the devastation has been much debated. Although it is unlikely to have reached the 90 per cent mortality claimed by some, it is well documented that individual groups did suffer the loss of 50–75 per cent of their numbers over short periods. The implications of this devastation went beyond the human misery it caused. The inability of traditional healers and medicines to cope with the illnesses called into question the efficacy of the treatment, the shamans, and by implication the whole indigenous culture, while, in contrast, the Europeans were relatively unaffected. The diseases seemed to hit the young and old with special severity, and the loss of the elders of a tribe struck at the oral traditions of the First Nations. An epidemic that killed 75 per cent of the population also carried away a good deal of its history and cultural knowledge, to a degree that can never be known, causing irreparable damage.

Military alliances and government relations

Popular accounts, when they deal with First Nations, rarely focus on disease and loss of culture. Rather, it concentrates on violence and conflict. The image of Native warriors – Sitting Bull, Geronimo, Pontiac, Cochise, and many others – is one of the most enduring in North American history, particularly that of the United States, and has done much to shape the general understanding of indigenous culture. Perhaps it also explains the notable prevalence of Native imagery in the naming of athletic teams – the Atlanta Braves, Florida State University Seminoles and dozens of others. Native people figured prominently in the military history of North America, especially in the early years of European contact. There were, in the first place, the local struggles in the expanding settlement frontier. More important, though, was the role played by Native groups in the military and diplomatic alliances and the wars of the European powers – the British and French, and to a lesser extent the Spanish and Dutch. Indigenous people played important parts in every war fought in North America until the early nineteenth century, the War of 1812 being the last in which their participation affected the outcome. Once the continent's political boundaries had been fixed, however – something that the War of 1812 settled generally, and which was mostly complete by 1850 – the importance of Native people as military allies disappeared. Groups which had performed vital services during the colonial wars, and who had signed treaties of mutual assistance with colonial governments found themselves relegated to the military sidelines. Once allies, they now became wards of government.

The final period of military activity

occurred in the second half of the nineteenth century, during the settlement of the trans-Mississippi west of the United States, when farmers, miners, and ranchers intruded on Indian lands, and it became clear to the Native people that they could no longer avoid conflict with the newcomers by moving out of their way, for there was nowhere to move to. Group after group of indigenous people found in this period that the assurances of the American government, even when written into a treaty, meant little in the face of manifest destiny and the land-hunger of settlers. If, as was the case in the Black Hills of South Dakota, gold was found on Native land, the violation of their rights was swift and irresistible. In such a case, even when a treaty and the right of law enshrined Native rights, the government, after an ineffectual attempt to honour its obligations, moved to protect the intruders from the attack of the 'savages'. The result was a lengthy series of land confiscations and armed conflicts which usually ended in the surrender of Native land. An egregious example is the 'Trail of Tears', an episode which saw the US government order the Cherokees of Georgia, who had made a particular effort to become 'civilized', to leave their lands and march 1100 km to worse land in Oklahoma.

In Canada the process was gentler, although the final result was not much different. The government of the new Dominion of Canada (1867) recoiled at the American example, as much from its expense as from its inhumanity, and took steps to prevent violence by setting up small reserves for Indians rather than the large ones in the United States, and policing both Natives and settlers with the North West Mounted Police, an elite force more disciplined and efficient than the American Army. The closest Canada came to an 'Indian war' was a short rebellion staged by Métis in 1885, in which only a handful of Indian groups participated, and which was easily suppressed.

Both Canada and the United States adopted a policy of separatism and containment towards Native people, who were put on reservations (in the US) and reserves (in Canada). The avowed goal was to protect them from the evils of society while they followed a path of growth and assimilation which would eventually enable them to become full members of society. The reserves also made them easier to control, kept them out of the public eye and centralized them for purposes of education and acculturation. Many officials in the nineteenth and early twentieth centuries felt that Native North American cultures were doomed to eventual death, and that the reservations were a kind of cultural hospice which would ease what was an inevitable process.

Missionaries, education, and social change

From the earliest days of contact, missionaries had come to North America to Christianize the Native people. Despite incidents of martyrdom, notably of Jesuit priests at the hands of the Iroquois in the late 1640s, the missionary effort was largely successful, at least in terms of a count of converts. Communities of 'praying Indians' appeared in New England and many thousands of converts were secured by Protestant and Roman Catholic missionaries across the continent. Some indigenous people themselves became missionaries to their people. Missionary activity, like everything else associated with Native–white relations, is subject to controversy. Some scholars have claimed that the effect of the missionary effort on indigenous culture was as bad as that of European diseases, while others have pointed to the ameliorating effect of missionaries who tried to protect Native people against the worst effects of cultural change.

The missionaries did play a crucial role in the general attempt to acculturate Native people. They believed that European social mores and Christianity were inseparable; thus they preached not only the gospel, but also urged indigenous people to adopt proper work habits, manner of dress, sexual conduct and public behaviour. The laboratory for social change was the school – most schools for Native children were run by missionaries – and it is the schools that are now receiving the strongest criticism as agents of what is sometimes termed 'cultural genocide'. The schools, particularly the residential schools, were places where Indian children could be separated from their parents and purged of their Native heritage. It must be remembered, however, that many indigenous people sought out Christianity and encouraged their children to attend school so that they could adapt to a changing world. That the schools occasionally housed a paedophile, that the children were sometimes harshly punished, denied the right to speak their own language and in general treated poorly, illustrates the degree to which the goals of missionaries and government were missed.

NATIVE NORTH AMERICANS IN THE CONTEMPORARY PERIOD

Native North Americans carry the legacy of their past into the present. Decades of racism, segregation, and paternalism have left a mark on First Nations across the continent. At the same time, however, these people are experiencing a cultural renaissance, one fuelled by rage at past injustices and optimism about their growing strength and self-assurance.

The barriers to success continue to be high. Indigenous people continue to experience exceptionally high rates of unemployment, poor health (especially diabetes, which is endemic among them), incarceration in prisons (more than 80 per cent of women in prison in the Province of Saskatchewan are Native; only 10 per cent of the province's population is Native). Most reserves have no economic base except government transfer payments, and the few which have achieved spectacular wealth, because they have built gambling casinos or because they profit from oil discovered beneath them, are only exceptions to this fact.

It is difficult to know whether Native social difficulties are caused by poverty, or whether the reverse is true. Whichever it is, the First Nations continue to suffer from alcoholism – not historically endemic among them, but having become so in the past few generations – as well as from drug abuse. Related problems such as foetal alcohol syndrome have created social difficulties that will linger for many decades. A great deal is made by social activists of the problems of Native communities, perhaps too much at times, for the constant bad news serves to reinforce a stereotypical picture of these people as crisis-ridden victims, unable to cope with the modern world. For many of the thousands of Native people who live in cities rather than on reserves the difficulties are compounded: in many cases they suffer from the same problems as their friends and relatives on the reserves, but they must live without the sense of community and the support groups that the reserves offer.

On the cultural front, the reality is mixed. On the one hand, the modern era has seen a re-emergence of Native culture. Novelists, poets, playwrights, actors, and artists have emerged in Canada and the United States – people such as Thomas King, Grahame Greene, Jeannette Armstrong, Ward Churchill, and Scott Momaday have achieved national and international reputations. Cultural activities have been revived – sun dances on the prairies, potlatches on the west coast. Schools run by indigenous groups are reinforcing the cultural resurgence. But in a very important way the outlook is not good, for the use of indigenous languages continues in most cases to decline; the majority of them are endangered, some are spoken only by a handful of people and some have actually disappeared in this century.

Indigenous people and the contemporary world

Whatever one thinks of the place of Native people in the contemporary world, it is impossible to argue (though some still try) that they are ignored by society at large. Despite making up only a small proportion of the population of Canada and the United States – perhaps 1 to 3 per cent, depending on which definition is used – they figure very prominently in popular culture, political debates, and public awareness. The general population may not know what to do about the 'Indian question', but they are certainly aware that it exists. This at least is an improvement from the state of affairs that existed up to 30 years ago, when the matter was simply ignored.

Until the 1960s, aboriginal issues scarcely registered on the political scene in Canada and the United States. Indian grievances and protests were confined to the Department (Canada) or Bureau (US) of Indian Affairs. The social ferment and increased emphasis on civil rights of the 1960s, however, led to a new activism, and brought a strong support from the non-Native community for indigenous causes. The American Indian movement – assertive, radical and dramatic – played a pivotal role in publicizing aboriginal activism. The stand-off at Wounded Knee in 1972, which resulted in the death of two federal agents, was a high water mark of activism, focusing the attention of the world on indigenous issues.

More effective in the long run than incidents such as Wounded Knee, the occupation of Alcatraz Island off San Francisco and other protests of this kind, was the use by Native people of the courts as a means of pressing their claims against governments. The courts recognized past injustices, ordered that treaties be honoured, and ordered governments to attend to unresolved issues. The results have been dramatic in both countries. In the United States, some Native 'nations' operate their own police and local judicial systems, manage their schools and hospitals and control the natural resources on their land. In Canada, existing treaty rights have been 'entrenched' in the constitution and the negotiation of treaties in northern regions has resulted in a series of remarkable gains for Native people, culminating in the creation of Nunavut, a huge political entity made up of the eastern part of the Northwest Territories, where the Inuit make up 80 per cent of the population (see Chapter 22).

THE FUTURE OF NATIVE NORTH AMERICANS

In the past 30 years, Native North Americans have won impressive political and legal victories. In the United States they have secured recognition of tribal sovereignty and auton-

omy. In Canada, they have won a measure of self-government and some groups have signed treaties as a result of contemporary land claims which are vastly more advantageous than those signed in previous centuries. There is evidence of an impressive reassertion of indigenous culture. Aboriginal writers, usually telling stories of the contemporary world rather than of pre-contact times, have found a literary voice and a sizeable audience. Native artists in film, television, dance, painting, and sculpture have found strong markets, though consumers still prefer 'traditional' art forms – Inuit soapstone carvings of the seal hunt, west coast masks – than in purchasing art which is not distinctively 'Native' in motif.

Across what the Americans sometimes call 'Indian country' there are abundant signs of social reconstruction and community development. It can be seen in tribal schools and colleges, Native studies programmes at universities, successful businesses run by indigenous people, large cultural gatherings, language programmes designed to revive dying tongues and health programmes which include traditional medical practices. At the same time, however, there are less positive aspects of Native life. Despite the language programmes, the use of indigenous languages continues to decline. Generation gaps have been opened between older people, who were often raised on the land with only periodic contact with non-indigenous society, and the younger generation that has grown up in the midst of the mainstream consumerist society. Bridging the gap between the two, particularly in a culture that has traditionally placed great value on the wisdom of elders, is often difficult.

For those who like to see a glass as half-empty rather than half-full, there is plenty of social pathology in Indian country to make a case that Native people are in a dismal state. All the key indicators of health and well-being show that Native communities are worse off than non-Native settlements. Suicide rates, especially among the young, are many times the national average, as is alcohol and drug abuse, criminal activity, and disease. Native people have a lower life expectancy than

non-Natives, are more likely to be impris-
oned, have on the whole a chequered work
history, are more likely to be born into and
live in single parent families and have less
schooling than the general population. Much
media attention is given to these problems,
which make good copy, and less to the positive and forward-looking accomplishments,
partly because these take place in classrooms,
ceremonial buildings, spiritual gathering
places and private homes of Native people,
all places that few non-aboriginals have
visited.

If there is a lesson to be learned from the
experience of indigenous people in North
America it is that of survival. Despite great
odds, they remain alive. It would be naive to
deny that they face serious difficulties and
challenges, not the least of which is the
potential disappearance of their languages,
the root of their cultures. But they have
already weathered tremendous challenges
and pressures, and it would be equally naive
to doubt that they are quite capable of with-
standing whatever the twenty-first century
has to offer without losing their identity. History suggests that Natives will survive into
the distant future as distinct peoples and that
their relationship with the newcomer populations will be the primary determinant of
the degree to which this future is one filled
with the fruits of optimism and rebirth, or
with those of struggle and despair.

ENDNOTE

1. The nomenclature of these people is complicated. Traditionally they were called
'Indians' and 'Eskimos' by Europeans, but
in many cases they have rejected these
names. Other terms, such as aboriginal
people, Native people, indigenous people
are still in use. In Canada the preferred
terms are First Nations and Inuit.

KEY READINGS

Boldt, M. 1993: *Surviving as Indians: the chal-
lenge of self-government*. Toronto: University
of Toronto Press.
Bordewich, F.M. 1996: *Killing the white man's
Indian: the reinvention of Native Americans at
the end of the twentieth century*. New York:
Doubleday.
Champagne, D. 1994: *Native America: portrait
of the peoples*. Detroit: Visible Ink.
Kehoe, A. 1992: *North American Indians: a
comprehensive account*. Englewood Cliffs:
Prentice Hall.
McMillan, A.D. 1995: *Native people and cul-
tures of Canada: an anthropological overview*.
Vancouver: Douglas & McIntyre.
Sturdevant, W. (ed.) 1978: *Handbook of North
American Indians*. 20 vols to date. Washing-
ton: Smithsonian Institution.

6

POPULATING THE CONTINENT: THE POST-COLUMBIAN EXPERIENCE

RICHARD H. SCHEIN

INTRODUCTION

The recent peopling of North America has involved the migration of millions of people over a span of several hundred years. The reasons for migration are as varied as the number of people who migrate. Nevertheless, in order to grasp the magnitude of migration decisions and outcomes, we can create some working categories or frameworks of understanding for disaggregating the seeming mass of people who have made that decision; and we can understand those decisions at a variety of scales from the place of individual decision making to general shifts in a world economy that might serve to displace people continents apart but living within a shared global economic system.

At the individual level, we usually examine migration decisions through general concepts such as push and pull factors, contagious diffusion, or chain migration. The case of any one person may, of course, vary dramatically. Similarly, we want to be careful when we talk about 'Chinese' or 'Irish' migration to North America that we do not immediately assume that each is a monolithic group and that all Chinese or Irish are alike. We want to avoid the ecological fallacy at all times. Beyond the individual, however, is a different scale of enquiry, one that examines the milieu or historical-geographical contexts of that same individual decision-making. While these contexts ultimately comprise human decision-making processes, they often seem to us somewhat removed from the immediate realm of human agency. We might think of such contexts in terms of wars or economic recessions or enclosure movements or industrialization. This chapter will focus upon that kind of macro-scale understanding of migration to, and within, North America over the last several hundred years. It will start with the expanding European system of imperialism and colonialism which drew North America into its cultural realm, and contributed to the transformation of a continent, in part through the demographic changes of overseas migration (Table 6.1).

Before we turn to that topic, however, it is important to note that North America was not a pre-Columbian *tabula rasa* – an empty continent waiting to be settled by Europeans, Africans, and Asians. As we saw in Chapter 5, the continent was occupied by many different groups of Native Americans or indigenous peoples. It is perhaps more accurate, then, to think of the post-Columbian European control of the continent as a *re*-peopling, or even as part of a *continuous* peopling of North America.

TABLE 6.1 US immigration (1821–1990) and Canadian immigration (1851–1991)

Period*	United States ('000s)	Canada ('000s)
1821–1830	143	—
1831–1840	599	—
1841–1850	1713	—
1851–1860	2598	352
1861–1870	2315	260
1871–1880	2812	350
1881–1890	5247	630
1891–1900	3688	250
1901–1910	8795	1550
1911–1920	5736	1400
1921–1930	4107	1200
1931–1940	528	149
1941–1950	1035	548
1951–1960	2515	1534
1961–1970	3322	1429
1971–1980	4659	1440
1981–1990	7797	1307

Notes: * for Canada decades end in 1861 *et seq.*
The US and Canadian data cannot be aggregated to produce a 'North America' immigration figure, because the US figures include immigration from Canada and the Canadian figures include immigration from the US.

Sources: US: for 1821–1971, Easterlin (1980), Table 1; for 1971–90, Dinnerstein *et al.* (1996), Table 10.3 (corrected). Canada: for 1851–1971, Statistics Canada: *Population and growth components*; for 1971–91, Statistics Canada: *Recent immigrants by country of last residence*, www.statcan.ca

COLONIAL NORTH AMERICA

There is indication that some Europeans had a long-standing knowledge of North America. Norse explorers reached New-foundland around 1000 AD and Bristol fisher-men were seasonally collecting cod off the Grand Banks of Nova Scotia in the 1480s. But it was not until the 'new' continent of North America was brought fully under a European imperial and colonial gaze in the sixteenth and seventeenth centuries that we saw sig-nificant migrations with permanent conse-quences in what would become the United States and Canada.

European imperialism: competition for a continent

The great colonial and imperial expansions of developing European nation states under the impetus of mercantile capitalism from the late fifteenth century onwards very quickly drew North America into the Atlantic realm. Although each conquering country followed its own particular strategies, the post-Columbian peopling of North America gen-erally entailed a several-stage process of European exploration, 'discovery', claiming – through both mapping and physical pres-ence – and settlement. The totalizing results of such European claims on what would eventually become the United States and Canada have been forever inscribed in our collective memories through those school maps which depict the continent awash with colours, each colour representing a specific European claim. For example, we can iden-tify a 'French North America' which appears to cover two thirds of the continent in its reach from the St Lawrence to the Ohio, Missouri, and Mississippi valleys. Similar large-scale claims by Great Britain and Spain, as well as smaller territorial designs by Sweden, The Netherlands, and Russia com-pletely colour in the map, and account for most of the North American land mass. Depending upon the date of the map, the actual pattern of the colours might change, but the overall picture remained the same for several hundred years.

Claiming, controlling, and occupying are not necessarily the same things, however. Beneath those swathes of colour, so to speak, the reality of early European footholds in the North American continent entailed the grad-ual establishment of settlement nodes and regions in a process that Donald Meinig (1986) has divided into two stages: prelude and fixation. The prelude stage began with momentary contact between European spec-ulators, adventurers and explorers and Native Americans and eventually led to sea-sonal contact, occasional conflict, and the trading of diseases as well as goods between the Europeans and Native Americans. A fixed point of contact with a discernible hinterland marked the transition to more permanent European settlement, which cul-minated in formal territorial claims (lines on a map) as well as an expanding settlement district placing increased pressure on Native populations and ecologies as European domestic staples came to dominate an export-orientated economy. With the intention of permanent settlement came settlers, whether commercial speculators looking for trade goods, government and religious officials charged with the political and moral over-sight of the new colony, or colonists them-selves ranging from slaves and indentured labourers to religious exiles to independent merchants, farmers, and craftspeople search-ing for new opportunities. The result was a set of dynamic settlement systems, with identifiable components (villages, towns, and even cities; transport and communica-tion networks; agricultural hinterlands; trad-ing links), peopled by migrants from distant and disparate lands.

As these settlement systems were pushed inland, away from their coastal origins, there was increasing competition and conflict of several types. Within a system, there might be competition for expanding hinterlands. For example, Boston and New York mer-chants vied for control of the Hudson Valley and upstate New York as early as the seven-teenth century. Philadelphia and Baltimore interests both looked to the rich agricultural potential of the Susquehanna Valley by the end of the eighteenth century. There might be competition between systems, where each system represented the tentacles of a European power reaching out to physically embrace that territory tentatively claimed on the ever-changing maps of North America. For example, the juxtaposition of Spanish, French and British interests in the southeast United States turned Florida into a series of fortified Spanish outposts of Havana in the early sixteenth century, and included battles over places such as St Augustine. Great Britain and France began skirmishes over the bounds of empire off the coast of Maine starting in the early seventeenth century, and continued to jockey for continental position until the British finally ousted France officially in 1760. Included in that competition would be the deportation and wanderings of (French) Catholic Acadians, and the re-population of Nova Scotia – formerly Acadia – by Protestant settlers from New England, England, the Upper Rhine Valley, and the European low countries. Finally, there was competition and conflict over territory between Europeans and Native Americans. Meinig (1986) has identified three colonial types of European relationships with indigenous peoples: expulsion, articulation, and stratification. Expulsion in general characterized the British response to Native presence and resulted in a frontier between the two groups, often formalized through treaty or royal decree, such as the Proclamation Line of 1763 by which the King of England marked the westward extent of allowable settlement as a line extending northeast to southwest down the crest of the Appalachian mountain chain. Articulation was most often practised by the early French traders in North America, where sites of exchange served as nodes articulating two separate territorial parts of an intertwined European–Native economic trading system. A de facto policy of stratification created cultural and ethnic mixing of Europeans (usually the Spanish) and Natives, with a resulting social continuum or hierarchy that could be found in places like the southwestern United States. Each of these policies to some extent is with us even today, as different parts of the United States and Canada, as well as different regions within those countries, have divergent policies regarding the rights and citizenship of Native peoples as well as different legal rulings on the standing of Native land claims. At this stage of the story, however, it is perhaps the general policy of expulsion that holds the most importance for a European-controlled re-peopling of North America. Expulsion can mean simply 'pushing back', as was the case in both countries into the twentieth century, and might involve the voluntary 'retreat' of Native peoples as well as forced exile. Expulsion might also entail confining Native peoples' territorial claims to 'reservations' in the midst of European land settlement. These reservations in the east often were slowly whittled away to non-existence by introducing European systems of individual freehold land tenure leading to the subsequent purchase of reservation lands from individual Native Americans. Coupled with the often-devastating impact of European diseases on Native immune systems, the policy of expulsion worked well to clear room early on for primarily European and African colonists in North America (see also Chapter 5).

It is difficult to identify with quantitative demographic certainty exactly who were the first migrants within a European-controlled North America in what became the United States and Canada. We can point to specific settlements, and begin to identify the general causes and rationales for their inception, or we can mark particular overseas events which sent various people and peoples to a new world, and the features of North American sites that drew them.

British colonial settlement

Although British settlement of North America is often traced to the ill-fated Roanoke colony in the 1580s, successful permanent settlement began in earnest in the first decades of the seventeenth century, and included footholds such as those in Newfoundland and Jamestown (Maine) as well as more substantial plantings in places such as Virginia and Massachusetts. Throughout the seventeenth century, the now-familiar map of colonial British North America emerged, focused on major settlement nuclei, which included from north to south, Newfoundland (Portsmouth), Massachusetts Bay (Boston), Naragansett Bay (Providence), the Connecticut Valley (New Haven), coastal New Jersey (Newark), Pennsylvania (Philadelphia), the upper Chesapeake Bay (Baltimore), Greater Virginia, especially tidewater along the James, York, Rapahannock, and Potomac rivers (Jamestown, Williamsburg), the Carolinas (Charleston), and Georgia (Savannah).

These places varied significantly in their origins, function, and population compositions – both in comparative terms (between regions) and over time (as a colony matured). For example, the Virginia settlements were at their beginning primarily a speculative commercial venture, male dominated by indentured servants, with little intention to recreate a new English society. Within a few years, however, the prospect of successful tobacco cultivation and export transformed the colony and attracted more permanent settlers including women and children. The primarily Anglican colony was further transformed in the early eighteenth century through two facts. First, the Act of Union between England and Scotland opened up Glasgow to Virginia tobacco trade that brought many more Scots to the colony. Second, by 1750 up to 40 per cent of the colony was African American as indentured servitude was replaced by slavery (although not all African Americans were slaves). In New England, by contrast, the impetus to reform the Protestant Church of England served as the immediate rationale for planting a new society, although the success of the venture depended in no small part on the mercantile competence of the subsequent settlers. Pennsylvania began as a settlement of English, Welsh and German Quakers after 1681, but by the 1720s these original settlers were joined by Palatinate Germans and the so-called Scots-Irish from Ulster. The Carolinas included significant numbers of French Huguenots (especially after the revocation of the Edict of Nantes in 1685), and by the 1730s white settlers were actively recruited in order to counter a two to one black slave majority in the colony. Although the results were not hugely successful, the campaign resulted, in part, in a smattering of French and German Swiss, English, Scots, Swedes, Welsh, Irish Quakers, and Huguenot settlers.

French colonial settlement

French settlement in North America was part of an empire-building strategy that extended in a grand arc from Brazil and the West Indies to Hudson's Bay, to the St Lawrence Valley and the Bay of Fundy. Although initial attempts at settlement included ill-fated Huguenot colonies in Port Royal Sound and along the St John's River in Florida in the 1560s, as well as attempts by French Basques in the Gulf of St Lawrence in the 1580s, effective settlement by the French came at the beginning of the seventeenth century.

Two settlement nuclei emerged as central to the French colonial presence, both in present-day Canada. Acadia was focused upon Port Royal and transformed a coastal environment by farming in tidal marshes, while a trading post at Québec (City) developed into a settlement at the mouth of the St Lawrence River. Additional scattered settlements in Newfoundland were periodically established, but these most often were heavily male dominated, and many of them really only served as wintering spots for fishing ventures. By the end of the century (c. 1700), the Acadian settlements were well established around the Bay of Fundy, and were gender balanced in their composition, if generally insular in their relations with the rest of the Atlantic realm. Settlement along the St Lawrence extended upriver from Québec City, to include Trois Rivières and Montréal. The St Lawrence Valley was originally colonized as a chartered commercial enterprise, with only a few agricultural settlers. In the 1660s, however, the French government revoked that charter, and over the next 30 years about 5000 colonists were sent by France to establish a solid French Catholic, agricultural population as the base for continental imperial designs. These settlements were almost exclusively focused upon Le Fleuve in long-lot farms organized in a seigniorial socio-economic system focused upon the urban centres of Montréal and Québec.

This region anchored the French penetration of the continent's northern interior, and served as a launching point for a crescent of outposts, forts, and associated small agricultural settlements that included places such as Fort Niagara, Detroit, Kaskaskia, St Genevieve, Biloxi, Mobile, and, eventually, New Orleans. New Orleans (1718) and Louisbourg (1720) were established as the termini of the French crescent. Several thousand colonists primarily from France and the West Indies, including black slaves, first settled in New Orleans; by 1750 it boasted around 10 000 people, mostly French and African, with commercial connections not only to the North American interior, but also to La Rochelle, Bordeaux, and Hispaniola. Louisbourg stood on Cape Breton at the other end of the crescent, the largest fortress town in North America, intended to guard the St Lawrence and the northeastern fishery against British encroachment. That encroachment was very real. Acadia became Nova Scotia (or New Scotland) through the Treaty of Utrecht in 1713. The British established Halifax in direct opposition to Louisbourg. And finally, the 'Great' or 'Seven Year' or 'French and Indian' War culminated with the end of official French control in Canada after the surrender at Québec in 1759. Despite the geopolitical transition, not much really changed for the majority of the people of New France, however, with the notable exception of the Acadians. There were 13 000 Acadians by 1755, when British tolerance of their presence in Nova Scotia began to waver. Before the end of the year, half the Acadians were deported. Many wandered the Atlantic realm for years. The most famous of these groups numbered about 1500, and eventually ended up in French Louisiana 30 years later, where they formed the basis for the contemporary presence of those people now known as 'Cajuns'.

Spanish colonial settlement

Spanish incursions into territory eventually included in the United States took place over several centuries, and originated from at least two major centres of Spanish presence in the Americas: Cuba (Havana) and the Valley of Mexico (Mexico City). Official Spanish presence in the New World was codified through the Law of the Indies in 1573, which largely dictated the form and appearance of Spanish North American settlements. Presidios and missions were usually the first outposts of a Spanish presence, underscoring the dual hand of the military and the church for securing Spanish footholds in new territories. Once established, these footholds became the bases for civil communities including pueblos, villas, and ciudades which were locally supported by ecomiendas, haciendas, and ranchos. These civil communities blended European and Native American crops and agricultural practice – for example growing wheat alongside native maize, or raising cattle and horses – and were marked by a social hierarchy based on perceived 'racial' mixing. More regionally, and over the course of several centuries, the Spanish impress in what is now the US focused upon several distinct imperial concentrations: Florida, beginning in the sixteenth century and including coastal strongholds as well as settlement along the St John's River and overland to Tallahassee; New Mexico, beginning in the seventeenth century, concentrated along the upper reaches of the Rio Grande River; Arizona and Texas, primarily originating as eighteenth-century settlements; and California, beginning in the late eighteenth century and focused by 1800 along a series of coastal missions, from San Diego to San Francisco, which declined in significance after 1821.

Dutch colonial settlement

The most significant Dutch presence in North America was to be found in the Hudson River valley, (re)named for Henry Hudson whose explorations after 1609 claimed the Hudson and Delaware River valleys for his employers. New Amsterdam was established as one of many coastal ports in the Dutch Atlantic trade, known mostly for its organization by the Dutch West India Company after 1621, and its focus upon sugar and slaves. Colonization in the immediate hinterland of New Amsterdam also took place to tap into a native fur-trading system and to attempt agricultural settlement in support of that enterprise. The settlement was quasi-feudal in its organization into a Patroon system, where land was granted to a man (the Patroon) in exchange for his agreement to furnish tenant settlers within a specified time period. The Dutch hold on the resulting New Amsterdam to Albany settlement axis was tenuous, and when the British claimed 'New York' in 1664, there was little Dutch opposition to the exchange of ownership. The New York area, however, did flourish as one of the most polyglot places in the Atlantic realm, especially in the lower reaches of the Hudson Valley around the city itself. In addition to the original Dutch settlers (in places like Harlem and the Bronx) there were Flemish (Bergen), Huguenots (New Rochelle), Scots (Perth Amboy), New England Baptists and Quakers (Shrewsbury), African slaves, free blacks, Germans, Jews, Swedes, and Norwegians at least.

An African presence

The African slave trade was a critical component of European colonial and imperial expansion into a greater Atlantic realm, and resulted in millions of Africans being forcibly and involuntarily displaced to North and South America. The slave trade was central to the famous Atlantic Trade Triangles that connected Europe, the Americas, and West Africa. Although estimates vary, it is likely that 20 million Africans were taken into slavery, and over 10 million of those made the trip to the Americas alive during the slave trade, which began within a decade of Columbus's voyages and finally ceased by 1870. The majority of African slaves in the Americas landed in Central and South America and the Caribbean, but a significant number were sold to North America, too. Almost as soon as there was European settlement in what became the United States and Canada, there was an African presence. It is likely that the first slaves (in the US) were sold at Jamestown (Virginia) to English colonists by the Dutch. The importance of slave labour in a labour-scarce land became especially apparent by the end of the seventeenth century, when the rate of African migration to the English colonies outstripped the rate of white migration by more than two to one. Although African (and eventually African American) slaves were found throughout the colonies, by far the greatest concentrations were in the South, where the agricultural labour demands of tobacco, indigo and rice promoted slavery. In addition to African slaves, there were significant (although much smaller) numbers of 'free blacks' in the colonies.

It is important to note that European migrants to Canada and the United States are recoverable in the written record by their country (and often sub-national region) of origin. We can draw upon ship's logs, church records, deeds, censuses on both sides of the ocean, and other official documents in order to verify the numbers of Scots or Germans or Dutch in a given location. It is a near-impossible task to do the same for African migrants. While there are specific points of origin along the West African coast that are known as slave-trading ports, these ports were termini in established trading networks that tapped the African continent's interior, and often masked a slave's regional origins. Once 'on board' an African slave became, in the eyes of European traders, owners, record keepers, and citizens, simply 'black' or 'African', with no provision made to delineate in any sense the area or nation of origin. Thus a group of involuntary migrants became a monolithic, racialized category, known simply as African, despite the fact that within Africa there were cultural, social, and political distinctions which helped to mark an individual's identity in much the same way as similar categories worked to identify European migrants. Once in the Americas, the process continued, with first-generation and later-born African Americans simply categorized in census documents as 'slave' or 'free', with no ethnic distinction made as was so often the case with the Irish, the French, the Spanish, and so on.

So far, this brief account of the post-Columbian peopling of Canada and the United States brings us well into the eighteenth century, and to the eve of the American Revolution. After this time it becomes convenient to differentiate more specifically between the two countries. To be sure, the *effectively* settled Euro-American dominated areas of both Canada and the US at the end of the eighteenth century were still largely confined to the more easterly portions of the continent, despite European territorial *claims* which covered the whole continent. And over the course of the next century and a half, the bounds of each nation-state would change significantly, incorporating more and more of the North American continent into their jurisdictions. But where the shifting boundaries of imperial and colonial claims were contested between European powers in North America before the eighteenth century, the American Revolution marks a turning point, after which the boundaries we know today start to become fixed and, more importantly, the definition of those boundaries became increasingly determined by the territorial designs of the US (after 1783) and Canada (after 1867) rather than by European claims.

Before we follow that distinction between Canada and the United States, however, it is useful to summarize the predominant ethnic and racial characteristics of North America north of the Rio Grande on the eve of the American Revolution. Meinig (1986) has called the territory at this time a 'complicated mosaic of places and peoples' and he has provided us with a glimpse of that mosaic, reproduced here (Fig. 6.1). Such a diagram is necessarily a simple, albeit crucial, descriptive enumeration of a varied lot, and is deserving of significantly more interrogation than space here permits. Still, it gives us sense of the regional mosaic, and its organization is summarized by Meinig (p. 214) in the following passage:

> The selections and labelling of ethnic and religious groups have been governed more by their identities in America than in Europe or Africa (Indian remnants are not included). All Englishmen not otherwise classified have been grouped under 'Anglican,' a term with both national and religious connotations. The West Country is the only English region recognized; such persons were mostly Anglican in religion but are not here categorized as such.

Because French regional identities do not seem to have been clearly sustained overseas, only religious categories are given, except for French-speaking Swiss and Walloons who migrated and settled amongst those of another tongue. The many German groups pose a problem. Palatines, Alsatians, and various Calvinists have been labelled as 'Rhenish-Reformed'; Mennonites, Schwenkfelders, Dunkards, and others, as 'Pietist' sects; the Moravians are listed separately because of their clear identity in Georgia.

An independent United States

The United States was born in 1776 through the Declaration of Independence from Great Britain. The Revolutionary War followed that declaration. The surrender of the British General Cornwallis in 1781 concluded the war, leading to the formal establishment of the United States in 1783. The immediate impact of the war and Independence on US population patterns was perhaps most dramatically felt by loyalists and Native Americans. As many as 100 000 colonists loyal to the British Crown left the former colonies, now United States, as a result of the revolution. They were a varied lot, included not only those people traditionally considered 'British' but also other European and African ethnicities, and were scattered throughout the Atlantic realm. Perhaps most famous among the resultant post-revolutionary loyalist settlements were those in Upper Canada explicitly established by Governor Simcoe, and also several locations within the Canadian Maritime provinces. Native Americans were affected in at least two ways by the revolution. First, a number of tribes, such as the Mohawk in upstate New York, actually sided with the British during the war and were subject to the territorial appropriations of war. Second, there was nothing to preclude continued *de facto* policies of Native American expulsion more generally by the new US, and those Native groups who had sided with the colonists, such as some

	Hudson Bay	Newfoundland	Nova Scotia	New Hampshire	Massachusetts	Connecticut	Rhode Island	Canada	New York & E. Jersey	Delaware Estuary	Pennsylvania	Maryland	Virginia	North Carolina	South Carolina	Georgia	Florida	Louisiana	Bermuda & Bahamas	Barbados	St. Domingue	Texas & Lower Rio Grande
ENGLISH																						
Anglican	X		X	X	X	X	X		X	X	X	X	X	X	X				X	X		
Puritan & Sep.				X	X	X			X		X				X				X			
Baptist					X	X	X							X								
Quaker					X	X	X		X	X	X	X	X	X								
Catholic												X	X									
West Country		X																				
SCOTS																						
Lowland									X				X		X	X						
Highland														X	X	X						
Orkney	X																					
IRISH																						
Ulster (Scots-Irish)				X	X				X	X	X	X	X	X								
Quaker															X							
Catholic	X								X		X									X		
WELSH																						
Baptist						X					X			X								
Quaker											X			X								
FRENCH																						
Catholic		X						X										X			X	
Huguenot									X						X	X						
Swiss		X													X	X		X				
Walloon									X													
GERMANS																						
Rhenish-Ref.			X						X		X				X			X				
Swiss			X								X				X	X						
Moravian																X						
Lutheran			X						X	X	X	X			X							
Catholic											X											
Quaker											X											
Pietist											X											
Salzburger																X						
FLEMISH									X													
DUTCH									X	X												
SWEDISH										X												
FINNISH										X												
SPANISH																						
Peninsulares																	X					X
Isleños																	X					X
JEWS																						
Sephardic							X		X		X				X	X						
AFRICANS																						
English					X	X	X		X	X	X	X	X	X	X	X			X	X		
French																		X			X	
Spanish																	X					X

Fig. 6.1 European and African peoples in selected American areas *c.* 1750
Source: after Meining (1986), Table 1

members of the Iroquois Confederacy, also found themselves territorially dispossessed through treaty negotiations with the new federal government. In both cases, the end result was similar: a general loss of land leading to migration based on attempts to confine Native Americans to reservations, either on site or further west; and a general policy of extending individual freehold land tenure practices within reservations in order to expedite potential land transactions.

US transformations to modernity

On the eve of the nineteenth century, the United States was poised for dramatic transformations of its society, polity, and economy – transformations critical to understanding the continually-shifting patterns of settlement and occupation of the new nation. By the early twentieth century, several themes and events would stand out as central to

populating the continent: the annexation of territory as part of the US's east to west frontier expansion, culminating with the Superintendent of the US Census officially declaring the frontier 'closed' in 1890 (and prompting historian Frederick Jackson Turner [1893] to proclaim that the existence of the frontier was the most significant fact of American history); the shift from a mercantile to an industrial capitalist economic system; the growth of cities and the increasing urbanization of the population; and several large-scale migration trends, including east to west movements as the country physically expanded from the Atlantic to the Pacific Coast, rural to urban migration, south to north movements especially of African Americans, and the introduction between 1820 and 1930 of over 30 million overseas immigrants to a nation whose total population in 1790 was not even four million (Table 6.2).

TABLE 6.2 US and Canadian populations, 1820/51–1996

US/Canada Year	United States ('000)	Canada ('000)	US + Canada ('000)
1820/21	9 638	—	—
1830/31	12 866	—	—
1840/41	17 069	—	—
1850/51	23 192	2 436	25 628
1860/61	31 443	3 230	34 673
1870/71	39 818	3 689	43 507
1880/81	50 156	4 323	54 479
1890/91	62 948	4 833	67 781
1900/01	75 995	5 371	81 366
1910/11	91 972	7 207	99 179
1920/21	105 711	8 788	114 499
1930/31	122 775	10 377	133 152
1940/41	131 669	11 507	143 176
1950/51	150 697	14 009	164 706
1960/61	179 323	18 238	197 561
1970/71	203 302	21 568	224 870
1980/81	226 546	24 343	250 889
1990/91	249 398	27 297	276 695
1996/96	265 284	28 847	294 131

Sources: Encyclopaedia Britannica and Canadian Encyclopaedia; for 1990/91 and 1996 the sources were the US Bureau of the Census and the Statistics Canada website www.statcan.ca

The colonial population of what was now the United States had been restricted to relatively close proximity to the Atlantic Coast through any number of factors, including environmental limitations, royal proclamations, economic and social ties and necessity, and the presence of Native Americans and other European colonial claims beyond the crest of the Appalachians. Many, if not all, of these things changed in the century after the American Revolution, and the nineteenth century has been known ever since as the era of 'manifest destiny,' based on a modernizing America's teleological claim to the right of continental domination and settlement (see Chapter 1). The gradual imperial process of stretching the nation across the continent involved both annexing territory and continually migrating people, captured in newspaperman Horace Greeley's now-famous catch phrase for the century, 'go West, young man.' The admonition was largely accurate, for the American frontier in the nineteenth century was largely male dominated, although demographic characteristics varied from place to place depending upon the specific patterns of natural resource exploitation and economic practice (such as fur trading, mining, timber harvesting, or prairie farming) and whether the frontier was an individualistic one (such as the California Gold Rush) or community based (such as Mormon Utah). In all cases, of course, the settlement of a US frontier necessitated the continued practice of Native dispossession, and the 'other side' of manifest destiny involved a century long series of Indian land transfers based on battles, wars, negotiations, treaties, relocation migrations, and Native American confinements.

Territorial annexation

In addition to extinguishing Native American land titles, the process of US territorial annexation involved settling disputes based on previous British colonial land grants as well as conflicting claims on the continent still held by the British, the French, and the Spanish. Many of the British land grants which established the original 13 colonies-now-states simply extended their northern and southern boundaries westwards indefinitely, towards inland territories unknown. Many of these claims also overlapped, so that as the new states looked to expand across the Appalachians there were inevitable conflicts over territory, especially since most states were cash-poor after the revolution, and at the end of the mercantile period of American capitalism land was one of the few avenues to great wealth for both private individuals and corporations, and to fill state coffers. For example, New York, Connecticut, and Virginia claimed overlapping parts of what are now Ohio, Illinois, and Indiana, and those claims had to be first mediated, then settled before individual settlers could move into the territory and gain title to their newly-purchased land. Most of these colony-now-state claims impinged upon the trans-Appalachian US territory east of the Mississippi River, a region eventually ceded to the federal government. The majority of the region (major exceptions were Kentucky and Tennessee) was locally divided by the now-famous chequerboard survey of the Township and Range system (see Chapter 1), and regionally divided, after several proposed plans, into the contemporary pattern of state boundaries.

Similarly, European imperial claims on the trans-Mississippi west had to be settled before the United States could fulfil its manifest destiny. The territorial growth of the US included negotiations and purchases between 1783 and 1853 which established the present territory of the contiguous 48 states (excluding Alaska and Hawaii; the former was purchased from Russia in 1867, the latter was annexed in 1898). The Louisiana Purchase from France in 1803 brought the majority of the Mississippi, Missouri, Platte, Arkansas, and Red River valleys under US ownership and control. In 1821 Florida was purchased from Spain. Between 1845 and 1853 much expansion took place – Texas was annexed (1845), Oregon Country was gained by extinguishing British claims (1846), and the rest of the Southwest, including California, Nevada, Arizona, and Utah, was gained by cession from Mexico (1848) and the Gadsden purchase (1853) (see National Geographic Society, 1988, pp. 94, 96,

100, 104, 106; Paullin and Wright, 1932, pp. 32–3).

Many of these 'frontiers', of course, were already well settled, not only by Native Americans, but also by previous incursions of French, Spanish, Mexican, and British colonial practice. These settlements were incorporated into the expanding United States, even as the majority of the population growth came through east-to-west migration of native Euro- and African Americans and European migrants. The railroads were essential in effecting the settlement of the American West. Distances were enormous, and in the early years the prairie grasslands were deemed the Great American Desert, something to be got across as speedily as possible on the way to the California Gold Rush, the settlements in Oregon and Washington, or perhaps the remnants of Spanish/Mexican settlements in the upper Rio Grande Valley. A series of federally-sponsored route surveys in the 1850s undertaken by War Department engineers attempted to best locate the route of the first transcontinental railroad. A middle route was settled upon and a golden spike joined the Central Pacific and Union Pacific Railroads at Promontory, Utah in 1869. By the end of the century several other routes were completed, including the Great Northern, the Northern Pacific, the Atchison, Topeka and Sante Fe, and the Southern Pacific Railroads. Railroads were subsidized through federal land grants, totalling almost 81 million ha (200 million acres) by 1871. Sales from land ensured an immediate capital subsidy for the railroad and provided a ready market for railroad services by settling the lands with farmers. In order to promote sales, railroad companies became town planners and entered the land office business. They advertised 'back East' and in Europe for settlers, offering incentives such as free or cheap steamboat and railroad passage to the new lands in order to entice potential migrants.

AN URBAN-INDUSTRIAL NATION

Even as the United States was expanding westwards, it also was transforming itself into an urban and industrial nation. The seeds of an American industrial revolution were planted in the New England textile mills at the end of the eighteenth century; by the early nineteenth century specialized manufacturing centres such as Lynn and New Bedford in Massachusetts were emerging, with over 50 per cent of their populations engaged in industrial production. From these beginnings, an integrated national, industrial economy emerged in the years after the Civil War. Between 1865 and 1920 national railroad mileage increased from 35000 to over 400000 miles (56350 km to 644000 km). The American Gross Domestic Product (GDP) increased sixfold in real terms while the share of agriculture in the GDP dropped from 35 per cent to 13 per cent and the percentage of American workers engaged in farming was reduced by half. Capital investment in American manufacturing during this period increased over 1000 per cent in real terms, a pattern exemplified by steel production which rose from under 20000 long tons per year to over 40000000 long tons per year between 1867 and 1920.

In concert with this industrial turn, the United States moved towards becoming a nation of city dwellers. In 1790 only five per cent of the American population was counted as urban, totalling about 200000 city dwellers in the US. On the eve of the Civil War (1860) the nation had just experienced the most dramatic increase in its urban population (measured as a percentage of total population): for the second straight decade urban dwellers increased their numbers by 36 per cent, making the total of US city dwellers a little over six million, accounting for just under 20 per cent of the total population. The 1920 census recorded that for the first time a majority of Americans lived in urban places, and they numbered more than 50 million urban residents. David Ward (1971) has shown us that there were regional patterns to the nineteenth century urbanization of the US. The Northeast urbanized first; a majority of northeasterners lived in cities by the 1880s, while the north-central and western parts of the country did not achieve majority urban populations until 1920. The South, in part because of its lack of an industrial base, never exceeded a 20 per cent urban population figure in the nineteenth century.

As the number of US cities grew, so did their absolute populations. In 1790 no US city had a population of over 50 000. By 1860 16 cities exceeded that size, and three cities exceeded a quarter of a million people. By 1920, three US cities topped a million inhabitants, with another 22 boasting more than a quarter of a million residents. Urbanization patterns were closely linked to transport and communications networks as well as patterns of industrial capital investment as the US urban system developed throughout the nineteenth century. From its primarily (north)eastern seaboard, mercantile origins, the US urban system extended inland along the route of the Erie Canal and into the Ohio–Mississippi Valley in the early nineteenth century; it reached out to incorporate San Francisco on the West Coast, as well as cities along burgeoning eastern railroad lines, such as the Pennsylvania Railroad and the New York Central Railroad by the end of the Civil War; and it had solidified and coincided with a northeastern 'core' region centred on an area roughly defined by Boston, Milwaukee, St Louis, and Baltimore by 1910 (with significant 'outliers' along the West Coast).

MIGRATION AND SETTLEMENT PATTERNS

Within this very general framework of manifest destiny, urbanization, and industrialization, we can begin to see patterns in the continuous peopling of the United States throughout the nineteenth and early twentieth centuries. First, as suggested above, significant numbers of people were moving within the US, both from rural areas to the cities, and from the East to the West. Of the seven million Americans living west of the Mississippi in 1870, 40 per cent had been born east of that river, and clearly the burgeoning of American urban places depended upon farm-to-factory migrations. There also were involuntary migrations within the United States. In the early part of the century the domestic slave trade became especially significant after slave trading with Africa

was made illegal in 1808, and the movement of African American slave populations mirrored, in part, the westward movement of the cotton belt in the US South. On the eve of the Civil War, eight million African Americans were held in slavery, while an additional 500 000 were designated as 'free blacks'. Similarly, many Native Americans including Cherokee, Seminole, Chickasaw, Choctaw, and Creek were a part of 'removals', such as the Trail of Tears, in which Indians deemed in the way of white settlement expansion were relocated to a formally designated 'Indian Territory', which by 1907 had been reintegrated into the US as the state of Oklahoma (see Chapter 5).

Second, the period between 1820 and 1924 saw the arrival in the United States of millions of overseas migrants (Table 6.1), mostly drawn from Europe but including people from around the world. These migrants primarily embodied the various effects of European industrialization, famine, religious intolerance, and war as they sought new homes and new opportunities in the US. Ward (1971) has described the patterns of migrant origins and destinations for the Europe-to-US case. In general, immigrant origins shifted over the course of the nineteenth century from Western and Northern to Southern and Eastern Europe. This has prompted the convention of dividing the period into two waves, before and after 1880, which are often referred to as the 'old' and 'new' migrants. While the absolute numbers of 'old' migrants remained relatively constant over the full time period, the numbers (and thus the percentage of the total) of 'new' migrants increased dramatically. This change in the perceived character of migrants, from the more-familiar English and Irish and German to the less-familiar Italian and Austro-Hungarian and Russian, often prompted nativist sentiment in the US. In fact, that sentiment peaked in the 1920s when two pieces of legislation effectively curbed most immigration for over 40 years. Immigration restrictions enacted in 1921 and 1924 established a ceiling on immigrant totals, and established quotas within that ceiling based on a percentage of immigrant populations extant in the US in 1910 and 1890. By pushing back the date to 1890 for establishing baseline figures in determining quotas, the legislation had the effect of privileging earlier (or 'old') migrants over 'new'.

Immigrant destinations also changed over time, in response to the development of transport connections, US industrial labour needs and farming opportunities, as well as the social networks and wealth of various immigrants themselves. The earliest nineteenth-century European migrants primarily depended upon already established trade connections. Irish migrants, for example, often arrived in the US via Great Britain, where established lumber trading connections to the Canadian Maritimes, Portland, and Boston meant that many Irish eventually were concentrated in the New England

states. Germans, on the other hand, often arrived via the Bremen to Baltimore tobacco trade route, or the Le Havre to New Orleans cotton connection. Once in the US, they generally had more money than their Irish counterparts, and were able to pay for further transport into the Mohawk and Ohio valleys, or up the Mississippi into the upper Midwest. By the 1850s and especially after the Civil War, agricultural opportunities in the Midwest plains as well as newly-developing midwestern cities claimed the greater proportion of immigrant flows, although the New England and mid-Atlantic industrial centres still attracted a share of the total immigrant population. After 1890, as the US urban-economic core region solidified, a greater proportion of immigrants again made

TABLE 6.3 Composition of immigration flows in Canada and United States at various periods

Canadian immigration

Period	Source area %		
	Europe	Asia	Other
1946–1950	84.5	1.2	14.3
1951–1957	89.1	2.3	8.6
1958–1962	84.5	4.3	11.2
1961–1970	68.9	9.2	21.9
1971–1980	36.9	28.3	34.8
1981–1990	27.6	46.1	26.3

United States immigration

Period	Source area %		
	Europe	Asia	Other
1851–1860	94.4	1.6	3.8
1881–1890	90.8	1.3	7.9
1921–1930	60.3	2.4	37.3
1941–1950	60.1	3.1	36.8
1951–1960	52.8	6.0	41.2
1961–1970	34.0	12.7	53.3
1971–1980	17.2	36.3	46.5
1981–1990	9.1	36.1	54.8

Sources: Canada: for 1946–1962, McCann (1982), Table 2.4; for 1961–1990, Statistics Canada Cansim Series, Tables D27, D36, D41. US: for 1851–1970, Easterlin (1980), Table 3; for 1971–1990, Dinnerstein *et al.* (1996), Table 10.3

their way to the industrialized northeastern and north-central states.

In addition to the dominant patterns of European immigration, significant numbers of migrants from other parts of the world also arrived in the United States during this time period. Perhaps most notable was the presence of Asian immigrants arriving on the Pacific coast. Chinese arrivals beginning in the 1850s supplied the West with a labour force, including railroad workers, critical to that region's development and integration into the national economic fabric. Because they were seen as 'competition' for jobs, the Chinese were the first US group to be subject to immigration quotas, in 1882. By the end of the nineteenth century, Japanese immigrants joined the Chinese primarily on the West Coast, and although their arrival continued until the general reform acts in the 1920s, through the Gentleman's Agreement Act in 1907 the Japanese government agreed to work towards curtailing Japanese emigration.

By coupling the timing of immigrant origins and destinations, it is possible to reconstruct some very general regional patterns of immigrant concentration in the nineteenth century United States. For example, the Irish became associated with New England, and especially Boston. Although generally widespread, there were significant concentrations of Germans in the Ohio and Mississippi valleys, in cities such as St Louis, Cincinnati, and Milwaukee. Scandinavians became associated with the agricultural production of the upper Midwest. Italians and 'Austro-Hungarians' were concentrated in mid-Atlantic coal and steel producing districts. Within major cities, ethnic enclaves arose, including 'Chinatowns' in places like San Francisco and Seattle, or 'Poletown' in Detroit.

The effect of the 1920s' immigration acts, the fact of the Great Depression, and the special consideration given to refugees from US allies during the Second World War reduced twentieth-century immigration to a trickle of its former existence, with preference given to those immigrant groups already present in large numbers. Over five million people had immigrated to the US in the decade between 1911 and 1920, almost nine million between

1901 and 1910, prior to the acts. Between 1930 and 1940 approximately half a million made a similar journey. That pattern began to turn around again when the Chinese Exclusion Acts were repealed in 1943, and especially after the 1965 Immigration Reform Act, which maintained the practice of an immigration ceiling but eliminated quotas. Since that time, US immigration has been predominantly urban-bound, and has been dominated by arrivals from Asia, Latin America, and the Caribbean, and in the decade between 1981 and 1990 the numbers of foreign arrivals in the US topped pre-1920 levels for the first time (see Tables 6.1 and 6.3).

The Asian component of the post-Second World War US immigration pattern includes Chinese, South Asians from India and Pakistan, Koreans, Filipinos, and, especially after the US involvement in Vietnam in the 1960s and 70s, significant numbers of Vietnamese, Laotians, and Kampucheans. The category 'Hispanic' is somewhat misleading. It is a general term applied to Americans with Spanish heritage, and its use tends to mask the vast differences of origin, custom, history, class, refugee status, economic status and other distinguishing characteristics of the population so designated. Most significant among the so-called Hispanic migrant population are those immigrants from Mexico, who vastly outnumber recent arrivals to the United States from any other place on earth. Mexican immigrants, as might be expected, are largely clustered in the US Southwest, especially in border regions. Through a similar geography of proximity, half of all Cuban migrants to the US live in Florida. Over half a million Cubans have arrived in the US since Fidel Castro's rise to power in 1959. According to recent data from the US Immigration and Naturalization service (INS), the top 10 countries of birth for US immigrants between 1981 and 1996 are Mexico, the Philippines, Vietnam, China, Dominican Republic, India, Korea, El Salvador, Jamaica, and Cuba. A more recent (1996) accounting would note the presence of migrants from Eastern Europe, especially countries of the former Soviet Union. Missing from these numbers are Puerto Ricans who are US citizens and not subject to immigration restrictions. They have been a significant presence, especially in New York and Chicago, since the 1920s and in increasing numbers after the advent of cheap air fares in the 1950s. In addition to legal immigration, there is concern in some quarters about contemporary illegal immigration. The INS estimates approximately five million illegal immigrants currently reside in the US, the majority of them from Mexico, followed closely by migrants from Central American countries and from Canada.

In addition to the general patterns of immigration noted above, there have been many twentieth-century shifts in the pattern of internal migration, as particular areas of the United States have gained or lost populations. These would include the general movement of many African Americans from south to north and west which began as a rural exodus from southern sharecropping in the nineteenth century and culminated with the twentieth-century search for jobs in many industrial cities nationwide. Additionally, there has been a post-Second World War 'Rustbelt' to 'Sunbelt' movement of jobs (see Chapters 10 and 11) and people, and a significant return migration to places like Appalachia and the Ozarks, regions whose labour force left after the Second World War in search of jobs but whose people have now retired and are looking to 'return home'. Within cities, the most obvious pattern of population redistribution has been the massive suburbanization of the population. Cities in the US resemble doughnuts, or what Peirce Lewis (1983) calls the Galactic City, as people and jobs have left the central city in favour of suburban locations (Chapter 13).

THE DOMINION OF CANADA

Much like the United States, nineteenth and early-twentieth century Canada underwent dramatic social, political, and economic transformations including a change of status from colony to Dominion. Without reducing the Canadian experience to a derivative of the US it is nevertheless possible to draw parallels in the significance of territorial consolidation and expansion from east to west,

internal migration and international immigration, urbanization, and industrialization as central to the Canadian experience and the re-peopling of the country (see Table 6.2).

Territorial consolidation and expansion

The existence of the independent United States after 1783 led to a series of treaties which established the boundary between the new nation and British North America, including what became 'Canada'. That boundary would be subject to negotiation and treaty for the next 70 years. Despite the designation of a vast territory generally north of the 45th parallel east of the Mississippi after 1783, and north of the 49th parallel west of that river after 1842–46, European effective settlement of Canada in 1800 was confined to its eastern domain: Maritime settlements primarily in coastal Nova Scotia but also along the coasts of Newfoundland, Prince Edward Island, and Cape Breton; New Brunswick settlements focused upon the St John River; and settlements up the St Lawrence, especially between Québec City and Lake Ontario, with pockets 'up river' along the shores of Lake Ontario and Lake Erie. That eastern portion of British North America known as Québec was formally divided in 1791 into Upper and Lower Canada. The rest of the continent north of the US was still the domain of indigenous peoples, although it was marked with European trading posts, especially those of the Hudson's Bay and North West companies. In fact, as late as 1865 the European settled portion of 'Canada' was confined to good agricultural lands within 100 miles of the US border, and east of Lake Huron. In 1867, the British North America Act created the Dominion of Canada, a federal state consisting of Ontario (Upper Canada), Québec (Lower Canada), Nova Scotia, and New Brunswick. In the next three years, the Dominion acquired Rupert's Land and the Northwest Territories from the Hudson's Bay Company and combined them to form the Northwest Territories, effectively claiming all that territory north of the US and west of the original four provinces (excepting British

Columbia and Alaska) for Canada. Between 1870 and 1873, Manitoba was created as the fifth province, a treaty with the US allowed British Columbia to become the sixth province, and Prince Edward Island became the seventh province. By the end of the century, the Northwest Territories was divided into districts, with the exception of the Yukon (which was separated in 1898), and only present-day Newfoundland remained outside the Dominion. It was not until the twentieth century that the rest of the country was politically divided into the now-familiar map of Canada: Saskatchewan and Alberta became provinces in 1905; Manitoba, Ontario, and Québec were extended northwards in 1912; the contemporary Northwest Territories was reduced to comprising the districts of Mackenzie, Keewatin, and Franklin by 1920; Newfoundland finally joined the Dominion in 1949 after a bitter debate and a close vote (see, for example, Kerr et al., 1990, Plate 2; (see also Fig. 1.1, this volume).

Despite the territorial acquisition of half a continent, as late as 1891 the effectively settled Euro-American portion of the Dominion was restricted to a Canadian ecumene extending from Windsor/London to Halifax, with pockets of settlement westwards in southern Manitoba, in Saskatchewan and between Edmonton and the US border, and in British Columbia extending inland from Victoria and Vancouver. These settlements were held together by a Halifax–Vancouver transcontinental Canadian Pacific railway, and various feeders, completed by 1887.

Approximately 100 000 Native Americans at this time were scattered across the country, generally having either surrendered their lands, been confined to reservations, or existing as mobile populations in the northern extremes of the country. The nineteenth century was one of largely British Isles migrations to the Canadian colony-then-Dominion. Between 1815 and 1865 over one million immigrants from the British Isles arrived in Canada. Irish immigrants dominated in the early years of that period, while their numbers declined in favour of English and Scots arrivals after 1860. The population of the country doubled between 1851 and 1891, from almost two and a half to almost five

million, but overall the population change was largely due to natural increase rather than immigration. The generally late (by US standards) westwards movement of the Canadian frontier as well as the attractiveness of the nearby industrializing United States led to more people leaving than arriving in Canada between 1850 and 1890. It has been estimated that as many as 1.3 million people left for the US during that time; and perhaps as many as 2.8 million left Canada for the US between 1840 and 1940.

Urban–Industrial Canada

Canada also was transforming itself into an urban–industrial nation by the end of the nineteenth century. By 1900, 35 per cent of its population was urban, compared to only 13 per cent at mid century. An industrial heartland had emerged between Windsor and Québec and the provinces of Québec and Ontario accounted for over 80 per cent of the nation's value-added by manufacturing in 1900. Montréal was truly the Canadian metropolis, although Toronto was quickly gaining ground. There were over twenty significant urban industrial places in Canada by 1900, with Montréal and Toronto topping the list. Measures of manufacturing value for Canadian cities indicate that both Montréal and Toronto each surpassed their nearest urban competitors by four to one and three to one ratios respectively. In the twentieth century this pattern was consolidated and expanded, as Canada became one of the fastest-growing economies in the world. Canada's industrial economy was transformed to a truly national one, still concentrated in central Canada but also visible in cities across the continent. By the post-Second World War period, the pattern persisted, although Toronto had surpassed Montréal as Canada's economic capital (see Chapter 16).

Migration and settlement patterns

Meanwhile, the re-peopling of the West caught up with the designs of national territorial expansion. The agricultural incorporation of the Canadian Prairies of Alberta,

Saskatchewan, and Manitoba represent a continuation of the great nineteenth-century frontier expansions already underway in the United States. Land divided in 160 acre (64.75 ha) parcels as part of a Township and Range survey system (see Chapter 1) was made available to settlers, who arrived especially between 1900 and 1930, although the marginality of many of the lands meant that out-migration was already occurring in some places by the 1920s. Although slightly more than half of the Prairie population was of British ethnic origin, more than 40 per cent were settlers from other European countries. Group migration was common in this part of Canada, and pockets of settlers with German, Scandinavian, French, Ukrainian and other European ethnic origins could be found across the Canadian West. In addition to the wheat produced on the Prairies, the western frontier also was transformed by mining activities and timber production.

More generally, European migration to Canada picked up at the end of the nineteenth century, especially between 1909 and 1913. These immigrants moved to rural and urban areas, although urban destinations predominated in the East, while rural ones were favoured in the West. In addition to European immigrants, and the internal westward movement of eastern Canadians, migrants arrived on the Pacific coast from Asia. They came in much smaller numbers, however, especially after the Chinese Immigration Act in 1885 which placed a head tax on Chinese immigrants; a tax that was as high as $500 per person by 1904. Like their US counterparts, Asians provided a cheap labour force for the Canadian West that eventually was subject to racist practice as Chinese and Japanese Canadians were seen as a threat to Euro-Canadian national dominance. The numbers of immigrants to Canada are smaller than their US counterparts during these same general time periods. Rarely did annual immigration to Canada exceed 200 000 people, except for the peak years of 1907 and 1910–13. In 1913 over 400 000 immigrants arrived in Canada (see Table 6.1). Total Asian immigration before the First World War rarely topped 10 000 people per year; fewer than 5000 per year was more

the rule. The Canadian population equation was also modified by the almost half a million Canadians who left the country for the US between 1890 and 1914.

Since the Second World War, the Canadian Far West, especially British Columbia, has seen the greatest positive population change overall, while many of the prairie province regions have experienced a net out-migration. In terms of international migration, British Columbia and Ontario have seen the most significant changes. Levels of immigration recently have risen and are predominantly urban-bound – in the early 1990s about 220 000 immigrants were arriving annually, an increase of 80 000 per year over the post-Second World War annual average (although the pattern has been one of 'peaks and troughs', including a peak in 1957 when 282 000 people arrived). Over the post-Second World War period, source origins also have changed. In the 1950s, European migrants still predominated (over 80 per cent); by 1994 only 17 per cent were from Europe (see Table 6.3). Asian immigration now predominates. In 1996, seven of the top ten countries-of-origin for Canadian immigrants were Asian, and they accounted for almost half of all immigrants that year. Migrants from Hong Kong far surpassed any others in absolute numbers.

SUMMARY

The continuous peopling of North America from the first presence of 'indigenous peoples' to the most recent arrivals on American shores has involved migrants from almost every region of the globe, making the United States and Canada perhaps the most wholly 'immigrant societies' in modern history. Individual reasons for migration and settlement vary almost infinitely, as do the mechanisms and processes by which people have made their way to North American shores as solitary travellers, or within family and community groups, or as eclectic collections of forced migrants such as indentured labourers and slaves. The general contexts of those migrations, however, can be captured as broad-brush political, social, and economic trends. The origins and destinations of migration flows and the resultant settlement patterns can be broadly conceived over a several-hundred year time-span beginning (in this chapter) with the jockeying for continental position by European imperial powers.

Similarly, the legacy of that continuous peopling of North America remains with the two societies today, as the two countries of Canada and the United States articulate their unique immigrant historical geographies with the contemporary cultural demands of a 'post-colonial' world as well as with an increasingly integrated world political-economy. Based in part upon the North American immigrant past, a number of social, political, and economic issues face the North American people in the future. The problematics of immigration policy, guest workers, racialized populations, Native land rights, assimilation and segregation, cultural identity, political power and access to resources, pluralism, and democracy, all can be better understood in light of the continuous and diverse peopling of North America. Perhaps the experience of the past can lead toward a more enlightened road to the future in both the US and Canada.

KEY READINGS

Gentilcore, R.L., Measner, D., Walder, R.H., Matthews, G.J. and Moldofsky, B. 1993: *Historical atlas of Canada, Vol. II: the land transformed, 1800–1891.* Toronto: University of Toronto Press.

Harris, R.C. and Matthews, G.J. 1987: *Historical atlas of Canada, Vol. I: from the beginning to 1800.* Toronto: University of Toronto Press.

Kerr, D., Holdsworth, D.W., Laskin, S.L. and Matthews, G.J. 1990: *Historical atlas of Canada, Vol. III: addressing the twentieth century, 1891–1961.* Toronto: University of Toronto Press.

Meinig, D.W. 1986: *The shaping of America: a geographical perspective on 500 years of history, Vol. I: Atlantic America, 1492–1800.* New Haven: Yale University Press.

Meinig, D.W. 1993: *The shaping of America: a geographical perspective on 500 years of history, Vol. II: continental America, 1800–1867.* New Haven: Yale University Press.

National Geographic Society 1988: *Historical atlas of the United States.* Washington: National Geographic Society.

7

DIVERSITY AND CONVERGENCE: CULTURE AND LANDSCAPE IN NORTH AMERICA AT THE START OF THE NEW CENTURY

HENK AAY

INTRODUCTION

Except for some vague differences between the United States and Canada, outsiders' geographical images of North America contain little regional and cultural variety; rather, such images are highly simplified and stereotyped with a pervasive sameness. More than other countries, the US as global political and economic leader has actively transmitted its identity along with its landscape dimensions to the rest of the world (this is not true for Canada). American movies, television programmes, magazines, popular music, fast food, dress, software, along with a host of other goods and services, have been aggressively exported to and enthusiastically welcomed by the rest of the world. In turn, non-North Americans view the US and Canada by the light of such cultural exports writ large. Of greater relevance to the perceptions of the cultural landscape is the sending abroad and international diffusion of elements of the contemporary American built environment such as shopping malls, skyscrapers, motels, single-family dwellings, and freeways. Their spread and acceptance contribute to an Americanization of cultures around the world, to an international placelessness, and to the impression that such landscape elements represent the gist of the continent. As a result, to the outsider North America is often viewed as a largely homogeneous region of superhighways connecting widely separated sky-scraping cities; everywhere similar auto-orientated, single-family dwelling suburbs circled around shopping malls; deteriorating, crime-ridden, minority-dominated inner cities; enormous, technologically advanced and specialized factory farms; human activities firmly in control of the natural world and, above all, a region dominated by the landscape equipment of *economic* interests.

REGIONAL DIVERSITY AND CONVERGENCE

This perception does correspond to a discernible trend of greater cultural conver-

gence over the last twenty-five years. Actually, such a trend is much older and dates from the mid-nineteenth century when nation-wide industries, marketing, transport, and media were established. Together, mass communications, mass production, big business, and central governments have been powerful agents in fashioning more of a mass culture today with standardized tastes, behaviour, products, and artefacts. These, of course, readily affect the cultural geography of North America by reducing place-to-place diversity. No matter where one finds them, suburbs, industrial and office parks, shopping centres, downtowns, strip malls, freeway corridors and roads, airports, and the new rural built environment are largely identical in design and architecture and bear little or no relationship to their cultural or physical settings. Whether you are in Charlottetown (Prince Edward Island), Drummondville (Québec), Alexandria (Virginia), or Salt Lake City (Utah), the planned shopping mall is designed to look and feel much the same, often with many identical branded goods and familiar retail stores. The same is true for other aspects of culture besides those related to the built environment. Speech patterns, dress, music, sport, foods, and religion, to name a few culture traits, are each forfeiting their regional/local expressions and have been replaced by nation- and continent-wide common habits. For example, the distinctive accents of the American South have become progressively more modulated, and country music, once also centred in the South, has spread to the entire continent.

As North America moves into the twenty-first century, will this trend towards a placeless cultural region and landscape broaden and accelerate? Will regional cultural distinctiveness become more and more a minor and trivial theme overwhelmed by an aspatial mass culture? Is North America more assailable by such influences because its internal cultural differences lack the depth and significance of older countries in Europe and Asia, and because both Canada and the United States are cultural mosaics more easily submerged by new nation-wide landscape trends? Or, are these pressures continuously counterbalanced and intersected both by historic regional personalities (e.g. Québec) and by the formation of new regional constellations (e.g. tropical Florida)? There is more general evidence to suggest that the headlong momentum toward cultural convergence is under restraint from new basic (postmodern) values including an enthusiasm for local landscapes, historic preservation and ethnic communities and traditions (Zelinsky, 1992, pp. 169–77). Into the next century, regional diversity and cultural convergence will often incongruously exist side by side at different levels of North American society. I maintain that such a reading does more justice to the changing cultural geography of North America than one which proposes an inevitable homogenization of the North American scene. The human landscape is one text, one dimension of such cultural diversity and convergence. It must be read to reveal what it says about the tastes, values, technology, and aspirations of the people who fashion(ed) it. The landscape especially includes the everyday equipment of human life, things like houses, gas stations, lawns, golf courses, cemeteries, billboards and barns. These are part and parcel of the workaday world of ordinary people and can reveal something about what kind of people Americans and Canadians were and are (Lewis, 1979).

CULTURE REGIONS OF NORTH AMERICA

The cultural heterogeneity of North America is packaged in different types of overlapping regions at a variety of scales (Fig. 7.1). Wilbur Zelinsky (1973, 1992), a founder of new approaches to American cultural geography, helpfully distinguishes *traditional* from *voluntary* culture areas. A *traditional* culture region is formed over a long period of time in an environmental setting by people who fashion and come to share a common worldview and way of life. Their culture shows its face in all areas of human existence, from livelihood to art to religion and includes a distinctive visible landscape. At the macroscopic level, examples would include the French, Irish,

Spanish, and Polish nations. At the microscopic level, North American examples would take in both rural Acadian and Amish areas as well as urban Haitian and Cuban ones. Any such *traditional* area, of course, is never static: assimilation, interactions with other areas and internal events and inventions are always bringing change. A *voluntary* region, on the other hand, is formed by people with similar interests moving to a common destination region (see below).

The *traditional* North American regional culture areas such as French Canada, New England, the South, the Midwest, and the Canadian Prairies, directly derive from European and African cultures through voluntary and involuntary migration. With their arrival and diffusion in North America, hybrid culture regions were formed out of the interaction of these motherland cultures with each other, with new natural environments, with new livelihoods, and with larger external economic and political influences. Depending on distinctness, location, competition and place in the political economy, traditional culture areas may continue to be vibrant, influential, growing and maturing into the present and future (e.g. Hispanic, Mormon, French Canadian culture areas) or they may lose vigour, become more inconsequential, relict or overlapped by new cultural constellations (e.g. New England, Midland, and Native American culture areas). Nevertheless, these *traditional* culture areas still remain quite relevant for understanding North America's regional cultural variety today because, even in the presence of Native American cultures, they operated on a relatively clean slate and played such a formative role in etching the initial patterns. For example, settlement systems, field patterns, road networks, and vernacular architecture are largely unerasable and enduring and, therefore, vital features of any *traditional* culture area today and into the future.

Traditional culture regions

At the highest level of culture considered here – the nation – the significant differences between Canada and the United States come into view. While there are the obvious

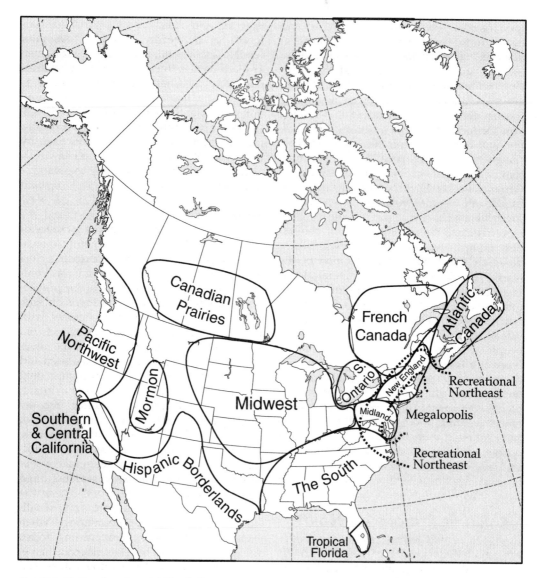

Fig. 7.1 Cultural regions of North America

differences such as Canada's English and French founding nations, ethnic mosaic, and small, American border-hugging ecumene, and the US's melting-pot and contributions from African and Hispanic traditions, at a global level of cultural differentiation the two countries are much the same. Both are derived from the larger European culture area, especially from northwestern Europe and most especially from Great Britain. Both are among the most economically advanced countries, sharing a faith in technology, control over nature, mobility and material progress. English-speaking Canada retains more of an underlying British cultural flavour than the US. The US, on the other hand, has witnessed more autonomous, home-grown cultural development than Canada which, apart from French Canada, has found it difficult to define Canadian culture except in reaction to the US. The penchant for violence, the institutionalization of militarism, a garish and unfettered market economy, an energetic nationalism, an unrestrained individualism and a missionary zeal for the American way of life – all these distinguish American from Canadian culture. To varying degrees these culture traits also play themselves out on the landscape. Crossing the international border from Michigan into Ontario, I am always struck, amongst other things, by a more tempered market economy (few billboards) and by a greater accent on the social over the individual dimension (more land-use planning and more attention to aesthetics).

At the next lower level, *regional* cultural differences within Canada and the United States are highlighted. This level accounts for a great deal of North America's cultural heterogeneity and therefore needs a more elaborate and careful exposition. What follows is a region-by-region description and interpretation of North America's cultural diversity.

Native culture areas
Before European explorers and settlers arrived in North America, the continent was entirely occupied by distinct culture areas of native hunting and gathering and agricultural societies, each adapted to the opportunities and limitations of particular natural settings. These cultures represented wholly original ways of life which contrasted sharply with European forms of social organization, livelihood, relationships to land and nature, and religion (see Chapter 5). European contact decimated Native American populations (probably by more than two thirds) and devastated their cultures through the introduction of Old World diseases, warfare (both with Europeans and with other tribes), dispossession of tribal lands and exile to new areas, as well as through abandonment of their cultures by assimilation to European ways. Today, most aboriginal people live on hundreds of widely scattered 'reservations' ('reserves' in Canada) which, with some notable exceptions such as northern Canada (including the new territory of Nunavut) and the pueblos of New Mexico and Arizona, are more the backwater areas of the nation than culture areas with distinctive ways of life. In the cultural geographical

diversity of today, the often unrecognized legacy of indigenous cultures throughout North America consists of tens of thousands of place names, the origins of many present-day urban sites and transport corridors as well as a number of domesticated food and non-food plants such as corn and tobacco. Currently, the Indian land ethic is a significant contribution to North America; many environmentalists regard this ethic as an alternative to the prevailing exploitative mind set.

French culture areas
There are several reasons for the presence of a distinctive French Canadian culture area. Most importantly, a large, contiguous area within the St Lawrence Lowland was completely and initially occupied by French settlers who developed a distinct rural culture, isolated from the commercial highways of the continent (Harris and Warkentin, 1974). When the agricultural land base in the lower St Lawrence Lowland became exhausted and French Canadians migrated to surrounding Anglo culture areas, such as New England, they soon became assimilated. Smaller French (Acadian) culture nodes took root early in today's Nova Scotia, New Brunswick, Prince Edward Island and Louisiana. Additionally, the continuing presence and power of French Canadian culture was sealed when Québec became a separate province in the Canadian Confederation, when Canada officially became a bilingual country, and when French Canadian society became one of two founding nations of the state. These achievements guaranteed the political future and power of French Canada.

Language, collective proximity and a common environment have been the key to maintaining a distinctive French Canadian matrix of long-lot land division, Roman Catholicism, settlements, institutions, habits, and attitudes. The natural environment of a cold winter climate, the infertile soils of the St Lawrence Valley and the physical hardships of the surrounding inhospitable Canadian Shield and Appalachian Mountains also left deep marks on French Canadian life. During the last quarter century, the threats to Québec's cultural independence within the

Canadian confederation have led to a growing French Canadian (or more specifically Québec) nationalism and a determination to establish a separate country. A culture area established more than 300 years ago may yet develop into a sovereign state.

Atlantic Canada

Consisting of the provinces of Newfoundland (excluding Labrador), New Brunswick, Nova Scotia and Prince Edward Island, Canada's Atlantic culture area is the product of a number of interacting factors: close affiliation with the sea; thin, rocky and acidic soils which restricted agriculture to a few favoured regions, such as Prince Edward Island and the Annapolis Valley in Nova Scotia; a chronically-lagging economy based on primary resources which together with geographically isolated communities has served to preserve folk traditions more than anywhere else in Canada; an original geographical mosaic of Acadian (French), English, Irish and Scottish communities which has endured into the present without the further cultural widening from other European immigrants and nations around the world characteristic of the rest of Canada; and, continuing out-migration to elsewhere in Canada (especially Ontario) and to New England which has served to fix an image of the region in the eyes of the rest of Canada.

Acadian culture, especially in northern New Brunswick, western Prince Edward Island and Cape Breton Island, has become a significant part of the identity of Atlantic Canada. Long inconspicuous, a new nationalism has brought Acadian culture out into the open. United by language (French is an official language), unique history (expulsion and return), Roman Catholicism, indigenous literature and music, and a strong co-operative movement in the local economy, Acadians distinguish themselves from 'Anglo' as well as from Québécois culture. The vernacular Acadian cultural landscape is more challenging to spot. Differentiated from the rest of Atlantic Canada, the Acadian landscape includes large, brick-and-stone Roman Catholic churches, more marginal agricultural land, a distinctive flag based on the French tricolour and other tricolour mark-

ings decorating many homes and businesses, and the adjective evangeline added to the names of many businesses and institutions (*Evangeline* is Longfellow's epic poem about the separation, due to the expulsion, of a young Acadian woman from her lover).

More generally, Atlantic Canada's culture includes a strong rural tradition along with a human-scale countryside: small fishing harbours, coves and outports encircled by equipment sheds, dwellings, vegetable gardens, fishing gear and fish plants; small churches serving many different Protestant denominations; distinctive French-speaking Acadian communities; local ethnic communities structured by kinship; active Irish, Scottish Highland and Lowland dialects in a country where other anglophones speak standard English; rich musical idioms, humour and theatre drawn from its folk traditions and successfully exported to the rest of Canada (Ennals and Holdsworth, 1988). To the outsider Atlantic Canada stands for a region with simpler, slower and traditional ways of life that express orthodox values. This, in part, explains its popularity as a tourist destination. Such a perception, however, masks the region's real and continuing economic and social hardships (see Chapter 19). If Québec becomes a separate state early in the twenty-first century and further isolates Atlantic Canada from the rest of the country, these hardships may well intensify and weaken the cultural well-being of this region.

New England

If French Canadian culture largely remained bound to its original zone of occupance, New England culture became thoroughly expansionist and, as a result, developed into one very influential source region for North American values and habits. The hearth of Massachusetts, Rhode Island, Connecticut, Vermont, New Hampshire and Maine became the staging area for a stream of English settlers into a northern latitudinal belt via the Mohawk River into New York State, the Great Lakes states, and eventually into the Pacific Northwest. Moreover, the aftermath of the American Revolution prompted many New Englanders to move to

British controlled Upper Canada (Ontario) and Nova Scotia, areas that were already overwhelmingly British in their cultural orientation.

The six-state core area was nationally pre-eminent in the century following the American Revolution across most of the range of human endeavour: manufacturing, engineering, maritime trade, finance, education, literature, theology, science and civic values. Today, the southern part of New England retains this national pre-eminence as part of Megalopolis. Within this area a particular (Puritan) understanding of community was laid out on the land, one which still defines the cultural landscape today. A New England 'town' combined an agricultural area with a service centre into a single political unit with the size of the rural area large enough to sustain a church. In the town centre the church and other civic buildings were arrayed around a common. The white buildings around a green common have become a powerful visual icon for New England and for the goodness of small-town America in general. Today, the historic New England village is an indispensable part of the cultural capital of the region's tourist industry.

The initial English settlers who cleared the land for agriculture in this core area encountered stony, marginal soils and a severe winter climate. These limitations eventually caused most to abandon agriculture and to find employment in manufacturing and in the maritime trade of the region or to join the stream of migrants pushing New England culture west across the northern belt of the nation. A difficult physical setting did not limit but, rather, encouraged innovation. The agricultural landscape was replaced with secondary forest and today New England has a wooded look.

The New England cultural template is not as visible or powerful today as it was a century ago as Megalopolis and non-British immigrants have added dramatic new realities to the region. Yet New England culture remains an enduring and original contribution to the present-day cultural heterogeneity of North America.

Midland culture area

This intermediate culture region is much less well represented in the public mind than New England and the South yet it contributed as much to the cultural geography of the Midwest and of the entire United States as the other two. Located in Pennsylvania, New Jersey and Maryland, the Midland area became a door through which for the first time flowed immigrants from the European mainland (especially German and Swiss) as well as from Great Britain, something more difficult in the more closed religious, social and economic worlds of New England and the South as well as French Canada. Even though New England became more ethnically pluriform with the later arrival of the Irish, Italian and Portuguese, the Midland area remains more diverse and it set the cultural pattern for the future ethnic and religious mosaic of the Midwest and Northwest as well as Canada's Prairie provinces. In addition to greater religious tolerance and ethnic variety, the Midland region also boasted more economic freedom; in this way, too, it is a forerunner of the US's larger cultural identity – liberty, tolerance and the unfettered pursuit of economic gain. One very influential form of this economic freedom was the grid-iron town, first adopted in the Midland area. European urban forms and New England towns remained along the eastern seaboard but the grid-iron town spread throughout all of North America. Its morphology was everywhere the same and therefore ideal for the speculator, surveyor and real estate agent laying out and expanding new towns and selling lots in newly-settled areas. The cultural traits of the Midland region were carried west, like those of New England, although along different physical routes, into a more southerly zone blending there with other traits to form the Midwest and contributing to the culture of other regions.

The South

The cultural characteristics of this much larger traditional region are more home-grown than the others which cultivated and projected certain British or French derived habits. Here, British settlers, African slaves, an unfamiliar humid subtropical environment and a plantation economic system combined to produce an original cultural cast. Southern culture eventually encompassed a region from Kentucky to northern Florida and from Virginia to Texas along with many sub-regions, such as French Louisiana, the Bluegrass, the Ozarks and Appalachia (Chapter 21). A strong sectionalism and isolation within the nation developed early in the South due to the institution of slavery, defeat in the Civil War, and a slow and punishing economic, social and political reconstruction as well as the lack of immigration from other countries following initial settlement. This sectionalism and isolation helped reinforce many common traits that emerged from the human and natural history of this region: two societies segregated by race, chronic poverty, local rural allegiances, an inefficient plantation economy, a small town society, distinctive regional politics, strong and individualistic adherence to evangelical and fundamentalistic Christianity (especially Baptist), gracious hospitality, distinctive black and white dialects, regional foods such as grits, turnip greens and ham hocks, and music such as gospel, blues, jazz, and country . The South of today (the new South) is quickly losing its national isolation and is becoming more like the rest of the nation. Legal racial segregation has been dismantled, agriculture has become quite specialized and mechanized and is no longer the dominant economic sector; industrial employment has grown rapidly and has become far more diversified; urbanization is approaching national levels. Some parts of the South have taken on the traits and landscapes of the national culture (for example, Atlanta (see Chapter 14) and the southern Piedmont); other parts (rural Mississippi) are more reminiscent of the old South. At the same time, some of the South's cultural products, such as stock-car racing and zydeco music, have become widely disseminated. This familiar two-way process is levelling the cultural differences between the South and the rest of the nation and will continue to do so into the next century (see, again, Chapter 21).

The Midwest

By the late nineteenth century, the newly emerging rural societies located between the technologically developed and sophisticated eastern seaboard and the wild and uncultured mountainous West were first regarded as a distinctive region (Shortridge, 1989). While the boundaries and precise cultural meaning of this region have shifted over time, the pastoral ideal has remained central to the Midwest's personality. Nationally, the Midwest was and is viewed as an ideal middle landscape between the vices of the urban and industrial East and the lawless wilderness of the West. Its perceived, stereotyped and manufactured culture traits include modest, honest and industrious yeoman farmers fashioning a bountiful garden, a superior moral character produced by working the soil and by self-reliance, egalitarian communities with deep roots in places working for the common good, and people comfortably in control without overwhelming nature. These perceived qualities made the Midwest into the soul of the nation and a repository of its basic values.

The actual cultural landscape of the Midwest was formed by New England, midland and southern cultures each pushing into their own west beyond the Appalachians at the end of the eighteenth century, first into woodlands, and then, by the 1830s, into grasslands. A broad swathe of the middle of the continent was incorporated into the Midwest: from western New York to the Dakotas (including parts of Ontario and Manitoba) and from Wisconsin to northern Missouri. Each of the seaboard-founding culture areas carried their own familiar crops westwards: wheat via New England, grain and livestock (the corn belt) via Pennsylvania, and cotton via the South. They also innovated new agricultural techniques (such as dryland farming) on the more arid margins of the region's grasslands. Other cultural habits from the founding culture areas were more broadly mixed and widely disseminated throughout the entire Midwest. The Midland's broad ethnic and Protestant religious diversity and its grid-iron towns are found widely in the region. New England's free standing, wooden home, on its own lot, with front and rear yard was settled on the entire region as the basic residence unit in the towns. But the largest influence on the look of the land in the Midwest today was the region-wide implementation of the national Township and Range (see Chapter 1) Survey System. This land survey system engraved a regular and geometric pattern on the entire human landscape of the Midwest as well as on other regions surveyed and settled after 1785; it created a widely-recognized 'national' landscape. The size and outline of farms, the layout of fields, the rural road system, and the location and layout of towns, all fit within this grid, reflecting the outline of the chequerboard survey system.

The image of an American middle landscape recedes somewhat when the heavily-industrialized eastern portion of the Midwest comes into view and, indeed, for that reason, some would exclude states such as Ohio and Michigan from the region. Nevertheless, the Midwest remains a kind of 'average' America, the heartland of the nation (see also Chapter 20).

Southern Ontario

Central to the cultural geography of southern Ontario, part of the heartland of Canada, is the British legacy planted by migrants from the United States loyal to their original home country and by settlers directly from the British Isles; institutions, customs and artefacts were British replicas. Upper Canada (Ontario) was to preserve a more orthodox British political and cultural presence in North America in distinction from the cultural experimentation in the US. This initial cultural cast is an enduring feature of the landscape especially in rural areas and small towns. A rectangular survey plan was laid out by the British across southern Ontario in advance of agricultural settlement; it defined the entire transport system, the roads that farms faced, the geographical pattern of forest clearance and remnant wood lots, the disposition of fields, the location and layout of towns, and the sites of country churches and schools. Like French Canada, this Britain overseas was hemmed in and could not easily expand: the Canadian Shield to the north, the Great Lakes and the US border to the west and south, and French Canadian culture to the east. A British-derived cultural landscape developed with place names, cemeteries, one-and-a-half-storey houses, gable-roofed barns and Victorian main streets. Into especially the urban part of this general matrix settled twentieth-century immigrants first from continental Europe and today from all over the world, transforming a once solid, conservative British city like Toronto (Chapter 16) into one of the most ethnically diverse urban areas on earth. Southern Ontario will experience the contributions and problems of these diverse and juxtaposed national traditions as it moves into the twenty-first century. The British rootstock together with many other contributing national rootlets are combining to change Ontario's cultural identity.

The Canadian Prairies

While settlement in the United States spread westwards from the Mississippi without a break, the Canadian Shield interposed a 2000-km barrier to the extension of the Canadian ecumene. Amongst other things, this resulted in delayed initial settlement, in a lack of environmental transition to a drier, grassland and plains setting, and in a much greater sense of separate identity from Canada's heartland than in the US's Midwest. As in the rest of North America's Great Plains, here the human interaction with the basic, larger than life elements of land and sky was the central element in the cultural experience. This experience with a new and unfamiliar land spawned a distinctive and enduring Canadian Prairie literature and art. The initial Anglo-Canadian and Protestant cultural elements of this region were very much under the control of the federal government in Ottawa; this resulted in a more restrained and less individualistic western frontier than in the US. The arrival early in the twentieth century of hundreds of thousands of non-British immigrants, especially Germans, Swedes, Mennonites, Jews and Ukrainians, started to change the cultural character of the region, yet it remains more Anglo-North American than the Midwest. The iconography of the Canadian Prairies includes a number of complementary images:

flat, empty spaces; spacious skies; a human landscape dwarfed by the scale of land and sky; a chequerboard of wheat fields; grain elevators and prairie towns isolated against the vast expanse of the land (Rees, 1988) (see Chapter 20).

American Wests

Outside of and elsewhere within North America, the West still conjures up cowboy country, a Hollywood- and Madison Avenue-created fabled semi-arid and mountainous land populated by heroes who conquer nature and represent freedom, masculinity and strength. The reality is quite different for the culture of the early western cattle industry. There is no single American West but an assortment of different Wests. To a significant degree this is because the physical geography from Alaska to the Mexican border separated nodes of population one from another by uninhabited mountain and desert.

One very early American West with enduring cultural impact is the Hispanic Borderlands stretching from California to Texas. Like all founding cultures, its influence on the landscape was far greater than its initial population, and this has intensified and been modernized with the more recent rapid immigration of Mexicans and others from Central America. Spanish presidios, missions and grid-iron towns with a central plaza were nuclei around which later cities like Tucson grew. Spanish colonial architecture gave a distinctive look to the built environment, one that has been extended into the present by Spanish-looking residential and commercial architecture. Spanish–Mexican dress is the principal component of the American cowboy look. Spanish ranch boundaries persist as present-day property, road and political boundary lines. And, finally, Spanish place names cover many natural and human features.

For the millions of Mexican Americans who have more recently settled in the United States, this early Spanish legacy ties them more closely to the area. At the same time, present-day Mexican artefacts, habits and institutions are flooding the Hispanic Borderlands, and also ethnic islands like Chicago and New York, and, indeed the US in general:

restaurants serving enchiladas, quesadillas, salsa and Mexican beers, embroidered Mexican dresses and broad brimmed hats for men, 'norteno' music, Spanish-language radio and television stations, Spanish-language education, Mexican–American Catholicism and Protestantism, and Mexican–American politics. With Mexico steadily drawn more closely into the North American economy and with legal and illegal immigration from Mexico and the rest of the hemisphere showing no signs of moderating, the sway of Spanish–Mexican culture over the Hispanic Borderlands and the rest of the US (not Canada) will continue to widen dramatically into the twenty-first century.

Another American West is the Mormon culture area of Utah, eastern Nevada and Southern Idaho with outliers elsewhere in the West and, indeed, throughout North America. Here, in this semi-arid intermontane region, a narrowly-defined religious community isolated itself from the American mainstream and formed a deep attachment to its sacred and appointed land. In distinction from the grazing economy of most western lands, the Mormons overcame an inhospitable environment and raised crops on irrigated land. The cultural characteristics of this core region go beyond religion and land use; there exists a general Mormon cultural landscape. Nucleated agricultural communities come with wide grid-iron streets orientated to the cardinal directions and with distinctive barns, agricultural equipment and activities in the towns themselves. Mormon cemeteries, irrigation systems, chapels, temples, and place names round out the landscape.

Other Wests include some of the largest reservations for indigenous cultures, as well as relict (some now tourist) and active isolated mining communities with their boom town, male-orientated, individualistic ethos and ramshackle landscapes. And a new upscale interior West (a voluntary region, see below) is also emerging, one of celebrities with large estates and private airstrips, exclusive ski resorts and hobby ranches (Riebsame and Robb, 1997). Of course, the original West included the Pacific Rim. However, up until the Second World War, southern and central

California as well as the Pacific Northwest did not stand out from the national norms and develop into distinct regional cultures. These culture areas are better described as voluntary than as traditional regions. It is to this emerging, often blurred cultural overlay that we now turn.

Voluntary culture regions

One universal North American trait is the freedom of personal mobility. Heavy, continuous and unrestricted migration to economic, retirement, amenity and recreational destinations has re-sorted populations into new emerging culture areas overlaying, eroding and even effacing traditional ones. These are classified as voluntary culture regions because they result from many, often like-minded individuals spontaneously deciding to relocate to a common destination region. Some voluntary culture regions are quite well established (southern California), others just emerging (Pacific Northwest). Some are quite stable and continuous (tropical Florida), others, more seasonal (recreational Northeast). Many voluntary culture regions, such as military enclaves, college towns, different types of recreation and pleasure-seeking locales, and retirement zones, are found at a more local scale and, therefore, beyond the compass of this chapter. Nevertheless, a chain of similar small voluntary culture regions exhibits its own way of life, including individuals with common outlooks and behaviours, a built environment reflecting the nature of the dominant activity, and associated customs, institutions and rituals.

Southern and central California

At the national scale, the most influential voluntary region is southern and central California. In the twentieth century millions of people from all over the United States settled here, attracted by a warm, sunny and healthy climate, by economic opportunity, and by the image of a desirable lifestyle and of a different and ideal America. Especially after the Second World War, southern California became a cultural hearth for the nation and the world: the Hollywood movie industry as well as the television and music industries,

fashion, fads, architecture, and automobile culture. Today, migration from the rest of the country has been replaced by immigration from all over the globe (especially from Mexico and Asia) making urban southern California a postmodern microcosm of the world and the most culturally diverse region in the nation (Chapter 15). As each new culture group is assimilated, it also contributes its ways of life to the California scene.

Tropical Florida

Tropical Florida is an even newer voluntary region with an as yet unfinished personality but certainly, like southern California, far different from any traditional culture area in North America. The cultural influence of the American South is today quite marginal here, and the essential character of the region is now based on two migration streams: one (earlier) of tourists and retirees from North America's colder areas, and the other (later) of Hispanic immigrants, refugees and exiles especially from the Caribbean and South America. The cultural landscape of north- and Anglo-facing tropical Florida is water orientated, tourist and exotic with a Mediterranean tropical look. The cultural landscape of south- and Hispanic-facing tropical Florida is made up of urban concentrations and built environments of those from a range of countries of the larger Caribbean region, such as Cubans, Puerto Ricans, Mexicans and Haitians. Different Caribbean and Latin folkways and artefacts contribute to the rapid cultural transformation of southern Florida; these signal the region's southward orientation, more specifically, its gateway to and economic pre-eminence in the Caribbean and in Latin America.

Megalopolis

In an area from southern New Hampshire to northern Virginia, the traditional culture areas and traits of New England, Midland and the South have been submerged by a rapidly-coalescing urban region commonly called Megalopolis. Like other continental-scale city regions (such as Canada's Main Street from Windsor to Québec City) as well as other metropolitan areas, Megalopolis is an urban culture with all the usual qualities such as high-intensity interaction, a stimulating built environment, great functional complexity, loss of community, a very wide range of personal choices, anonymity, segmented personal relationships, and a fast pace of life. But Megalopolis is more than a typical North American urban region. It is extremely large – more than 50 million people. It is the powerful nerve centre of the United States in banking, public policy, and education. It was and remains an important entranceway and staging area for immigrants and retains a greater ethnic diversity and more distinct ethnic urban landscapes than most other regions. And while its contemporary built environment is like that of all modern North American metropolitan settings, historical environments are more prominent and receive more public and private attention. Since Megalopolis is the cradle of the nation, preserving the cultural landscape of this heritage has a high priority (see Chapter 17).

Recreational Northeast

Wrapped around and functionally linked to Megalopolis lies an extensive recreation region extending north of Maine into Atlantic Canada, west to Lake Ontario and southwest along the Appalachian system which we can refer to as the 'recreational Northeast'. This voluntary region is based on the seasonal movement of vacationers (and students!) from Megalopolis to fishing and hunting venues, colleges, ocean and lakeside beaches, fall colour tours, ski resorts and other winter sports, bed and breakfasts, small-town antique markets, cottages, river canoeing, music camps, hiking trails, and lake and ocean sailing locations, among others. The recreational Northeast possesses two essential qualities: a relatively unspoiled natural environment filled with scenic and recreational resources (e.g. the Adirondack Mountains of New York State), and a rustic and historic human landscape (e.g. New England villages and fishing ports). Both these qualities are in sharp contrast to the environment and life of Megalopolis. As resource extraction and farming have declined in this entire region, recreation and tourism have steadily grown in importance because of the populous and affluent region next door. A culture of leisure and

pleasure defines the region and is transmitted by vacation behaviour, tourist landscapes, a prevailing natural landscape with a rustic human touch, and by visitor communities pursuing common recreational activities with standard paraphernalia and environmental requirements. It must be pointed out, of course, that at a larger scale many other such recreational culture regions exist throughout North America based on sun, coast, mountains, sport, and gambling. Northern Michigan, northern Ontario, the Rockies, and Nevada's urban centres all come to mind.

Pacific Northwest

A last area just beginning to emerge as a distinct voluntary culture region is the Pacific Northwest. Running from central California through Oregon and Washington to Canada's British Columbia, this area did not set itself apart from the national averages until relatively recently. In Canada, a strong British and, in the United States, an earlier New England tradition have been culturally formative. But as a voluntary region the Pacific Northwest has recently started to attract migrants from the rest of the continent (in the US especially from southern California), not for economic but for reasons of perceived life-style and the beauty of the natural environment. Some (Garreau, 1981) have dubbed the area 'Ecotopia'. A relatively unspoiled and lightly settled, scenic alpine region with great ecological diversity, including extensive rainforests (see Chapter 4), serves as one foundation of this perception. Another is that this region more than any other place has found a balance between economic growth and environmental well-being; a managed growth has replaced a no-holds-barred development. Still another foundation of the Northwest as Ecotopia is the pioneering leadership of this region in the modern environmental movement beginning in the 1960s. It appears then that one distinguishing and growing cultural trait of the Pacific Northwest is its environmental ethic (Chapter 18). On other cultural fronts, with perhaps the exception of its tradition of social tolerance, the Pacific Northwest conforms to larger national cultural trends.

CONCLUSIONS

At the continental scale, the cultural geography of North America is a layered covering consisting of *traditional* and *voluntary* regions. *Traditional* culture regions date back to the arrival and spread of Native, European and African cultures. In most cases, the initial landscape impact of these cultures remains quite functional and relevant even though other aspects have shown tremendous change. Culture change in these traditions has been channelled and filtered by their original general dispositions: twentieth century Québécois culture is commensurable with eighteenth-century French Canadian culture even though external influences from a global cosmopolitan order and other national/regional traditions have left deep marks. *Voluntary* culture regions are reshaping the human geography of North America. Entirely new constellations of culture as well as new traits are formed by selective migration from diverse locations of like-minded people to a common regional destination and by the blending and interaction of diverse foreign cultures at such common meeting grounds. The twenty-first century will witness a deepening and maturing of existing voluntary regions as well as the formation of new ones in reaction to continent-wide environmental, economic, social, technological and political change. In response to global warming, perhaps a 'Coolbelt' will form across the northern United States and Canada much like the 'Sunbelt' came about in response to the amenity of a warmer climate in this century. All the while, the traditional culture areas will continue to be active in generating, modulating, guiding and particularizing culture change. At the same time, the drift towards cultural homogeneity will become more evident as a placeless landscape. The built environment along freeway corridors and urban beltways may reveal everywhere identical suburban downtowns, factory outlet malls, office parks and industrial greenhouses; the back roads everywhere lead the way to the diverse and subtle cultural landscapes of North America.

KEY READINGS

Birdsall, S.S. and Florin, J.W. 1992: *Regional landscapes of the United States and Canada.* New York: Wiley.

Conzen, M.P. (ed.) 1990: *The making of the American Landscape.* Boston, MA: Unwin Hyman.

Garreau, J. 1981: *The nine nations of North America.* Boston, MA: Houghton Mifflin Co.

McCann, L.D. (ed.) 1987: *Heartland and hinterland: a geography of Canada.* Scarborough, Ontario: Prentice Hall.

Rooney, J.F. Jr., Zelinsky, W., and Louder, D. R. (eds) 1982: *This remarkable continent : an atlas of United States and Canadian society and culture.* College Station: Texas A & M University Press.

Zelinsky, W., 1992: *The cultural geography of the United States: a revised edition.* Englewood Cliffs, NJ: Prentice Hall. This book has a 26-page annotated bibliography organized by region and subject. Rather than including many of these sources in the list of references, I refer the reader to this excellent bibliography for sources prior to 1992.

8

THE DIVIDED CONTINENT: POLITICAL, POPULATION, ETHNIC AND RACIAL DIVISION

DONALD G. CARTWRIGHT

Canadians have a far lighter sense of national identity than people of most other nations, a condition that Canadian author/journalist Richard Gwynn (1995) has described as a lightness of identity, one that could slip away without being noticed. Gwynn quotes from Lansing Lamont (1994) who stated, 'Maybe Canada is not meant to survive. Maybe it isn't destined to live out its span as a nation' (Gwynn, 1995, p. 8). As a neighbour to the most powerful country in the world, there is a variety of external attractions from that source that make it difficult for Canadians to resist a loss of identity. There are also a number of intranational processes that contribute to this erosion just as there are in the United States, but the major difference is that internal processes of division in America are not as threatening to the integrity of the nation nor to the sense of national identity as they are in Canada. Before we investigate these it is fitting to discuss some of the issues of external relationships that exist between the two countries.

INTERNATIONAL RELATIONS

Canadians have beeen described by one author as 'just like Americans, only less so' (Martin, 1993, p. 44). But Canadians have also spent the past 200 years resolutely not being Americans; in fact they are among the few peoples in the world who, given a chance to become Americans, have chosen not to. The physical features of the continent draw the two societies together with the roughly north to south alignment of the Western Cordillera, the Great Plains and the Appalachian Mountains, while the same features serve to keep Canadians apart (Chapter 2). These do not have the same divisive impact in the United States. The incised river systems of the Appalachians, in general, cut across the alignment of the mountains from east to west or west to east. The headwaters of many of these rivers, or their tributaries, opened gateways through this cordillera with Cumberland Gap perhaps the most famous. Several river routes were enhanced by early, basic roads that bypassed obstructions, such as the Wilderness Road in 1775.

The dendritic drainage pattern of the Mississippi River system provided tributaries east and west that were also used as early transport routes, particularly the Ohio River to the east and the Missouri River that extends well into the Western Cordillera. The latter mountains are not nearly as rugged and formidable in the US as they are in Canada. An impetus that helped to moderate obstructions associated with the physical features appeared in 1787 when the US federal government removed all limitations to interstate trade. Limitations on interprovincial trade in Canada are still present.

Complementing the integrating effect of these physical linkages between Canada and the United States is the distribution of the Canadian population, for almost 75 per cent live within 150 km of a relatively permeable international boundary. For the two populations there are similarities in language, race and religion – consequently cross-border trips for business and pleasure into the US are voluminous. Operating in the opposite direction, American corporations, trade unions, service clubs, professional organizations, sports leagues and cultural groups function easily in Canada for, with few variations, commercial laws, business practices and rules, technologies used, corporate structures and financial systems are the same. With such accessibility to the Canadian economy and society available, it is necessary to defend the cultural integrity of the country through its political systems. There is, however, a long history of harmonious relationships between Canada and the US, and this makes political interference in that relationship a difficult and, at times, a delicate proposition. Boundary maintenance, therefore, becomes a matter of controlling rather than preventing American access to Canadian society (Gibbins, 1994, p. 260).

Cultural influences from Europe, chiefly Great Britain, once had dominance in Canada. With the decline of British sway, at around the turn of the century, developments in the American private sector for the mass circulation of magazines, motion pictures, records, radio and, eventually, television began to influence Canadian society. These features of American culture flowed easily into the northern neighbour unimpeded by distance or, for at least 70 per cent of the Canadian population, by variation in language. Arthur Siegel (1983) described the air waves as 'agents of denationalization by serving as roadways for foreign, largely American, cultural values' (p. 1). To combat these 'roadways', the Canadian government, which has jurisdiction over air waves, introduced the Canadian Broadcasting Corporation (CBC) in the mid 1930s to help entrench cultural sovereignty and enhance national unity. Over several decades, quotas for Canadian content on both radio and television have been legislated and are now enforced through a federal agency, the Canadian Radio-Television and Telecommunications Commission (CRTC). This has not been an attempt to exclude foreign content, but through legislative restriction it is intended to provide protection for the development of Canadian performers and performances. Defensive measures such as these were reinforced through a variety of financial support mechanisms for perfomer and other cultural artisans. Since a large proportion of the Canadian population lives close to the American border, people can make use of technological developments in cable and satellite delivery to circumvent these restrictions on foreign content. Gibbins (1994) summarized the significance of these developments by stating, 'In the future, as in the past, technological innovation is likely to progressively erode cultural barriers between Canada and the United States' (p. 278). Gibbins also discusses the implications of such intervention in the marketplace by the Canadian federal government, for these actions protect collective values to ensure the survival of a distinctive national culture but, in so doing, they also restrict individual freedom. We shall encounter other examples of collective values taking precedent over individual freedoms later in this chapter.

Just as the American cultural ethos permeates the Canadian ecumene, other transactions in the realm of economic integration, since the end of the Second World War, have demonstrated that the border between the two nations has become more 'familial than international' (Gibbins, 1994, p. 283). In January, 1989, after extensive negotiations, a comprehensive Free Trade Agreement (FTA) was ratified by both governments. It is important to note that, even before this agreement was signed, the United States absorbed three-quarters of Canada's exports. Trade already passed easily across the border with 90 per cent of all exports and imports subject to custom duties of less than 5 per cent. Gwynn (1995, p. 38) states that pre-FTA cross-border trade was freer than between any two member states of the then European Common Market. The Canadian–American trading relationship is greater than that between any other two countries, now or at any time in the past (Wonnacott, 1991). In the early 1990s, the US exported more to the province of Ontario than it did to Japan, its second most important trading partner. It is because of this voluminous exchange that there are claims that trade mattered very little to the FTA. It did mean, however, that Canada was transformed from a protected economy into a free-market one, with all remaining tariffs to be abolished by 1999.

Since Canadian business would be required to compete directly with one of the world's most efficient economies, opinion was divided over the merits of the FTA. At the outset, the premiers of Ontario, Manitoba and Prince Edward Island opposed the pact, those of Newfoundland and Nova Scotia were uncertain while the remaining five endorsed it. Eventually all but Ontario supported the agreement, some more firmly than others. Québec's endorsement was enthusiastic because of the new direction that the agreement opened for that province. Leaders of the business community realized that under the FTA they would no longer depend on English Canada for markets or technology (Newman, 1995). Furthermore, wider access to markets would enhance the province's economy, a condition deemed essential to its cultural survival.

Ontario and Québec have the greatest concentration of manufacturing firms in Canada. Consequently these provinces must bear the brunt of increased competition from imports under the agreement. Those in favour of the accord pointed out that losses to producers, including the labour force, will be offset by benefits to the consumer. Eventually the two provinces are expected to gain more than other regions in Canada because they are favourably located to capture substantial sales of manufacturing goods to the US – 'As Canadian firms specialize to service the large US market, they will be able to capture economies of scale and comparative advantage benefits from trade' (Wonnacott, 1991, p. 130).

Ten years after the ratification of the Free Trade Agreement, the share of Canada's total exports that went to the United States had risen from three-quarters to 81 per cent (Canada absorbs about 21 per cent of American exports). Two-way trade with this partner grew faster than Canadian trade with Europe or Asia for the same time period and involves a trade flow of about $1.4 billion a day (Fig. 8.1). There is also evidence that the specialization and efficiency, which Wonnacott described above, is being achieved although some economists predict that we will be well into the next millennium before the full benefits of increased productivity will appear in other liberalized sectors. Many

opponents to the FTA in 1988 still maintain that it has been bad economically for Canada, chiefly because of job losses that are associated with industrial restructuring, but perhaps the most vehement criticism against the agreements has come from those who feared its impact on Canada's cultural domains.

Under the FTA, protective policies and legislation for cultural domains in Canada – publishing, film and video, music and sound recording, broadcasting and cable – were formally exempted from the final agreement; but the Americans retain the right to retaliate against cultural-sovereignty legislation since they insist that culture is a business like any other (Gibbins, 1994; Thompson, 1992). This conundrum illustrates a fundamental difference in the approach to cultural industries in the two countries. In Canada, these industries are mainly national in scale and in large part protected from free market buffeting, while in the United States they are driven by the market and are global in their impact. Culture takes on very different meanings in Canada and the United States. To the Americans, it is their most successful commercial

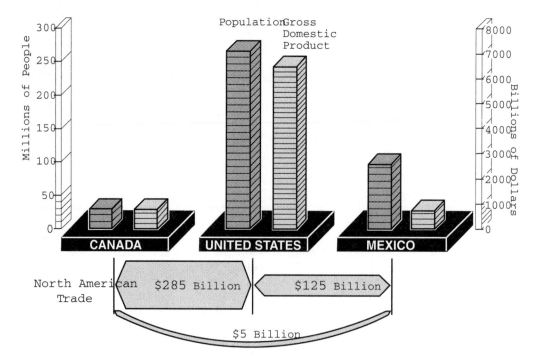

FIG. 8.1 Population, GDP and trade, Canada, US and Mexico, 1996

export; to Canadians it is their most fragile of domestic commodities. Though maturing, Canada remains a putty culture – penetrable and unshaped' (Newman, 1995, p. 96).

Critics of the agreement believe that the American business community will eventually challenge Canadian protective policies and legislation on cultural matters, and the risk of economic retaliation elsewhere will force the government to concede. Furthermore, federal reductions in support programmes for all areas of Canadian culture, throughout the 1990s, have further weakened and exposed the country to what detractors refer to as 'cultural imperialism' from the United States.[1]

There are some irritants between Canada and the United States that are more intractable and have not been resolved in the spirit of accommodation that was generally evident in the negotiations over the Free Trade Agreement. Border disputes have developed offshore as the two countries attempt to clarify their jurisdictions over ocean resources, particularly the fisheries. On the west coast, for example, there is conflict over salmon fisheries that is based on issues of conservation and access to the fish off Alaska's coast. The Canadian government has spent millions on salmon propagation and habitat enhancement, and favours a reduction in fish quotas. The Alaskan fishermen dispute this stating that lower catches will eventually reduce the number of fishing boats and stocks will rejuvenate naturally – the free-market arguments. Although there are legal precedents and a long-accepted international convention that salmon should be caught by the fishermen from the country in which the fish spawn, some Alaskan fishermen maintain that these fish, particularly the lucrative sockeye salmon, are 'pasturing' in their state's ocean waters eating Alaskan resources, therefore they have a right to catch them (Cernetig, 1997, p. A1).

There are also disagreements in several areas over the precise location of the offshore international border, and these have increased in significance since technological developments now allow us to exploit living and non-living resources in ocean space at greater depths and over a wider area. The governments of the two nations cannot agree on the location of the maritime boundary in the Arctic's Beaufort Sea, and this could affect access to offshore oil resources. A similar dispute over the location of the maritime boundary in the body of water between British Columbia's Queen Charlotte Islands and the Alaska Panhandle (the Dixon Entrance) is having an impact on the fishing fleets in that area. Canadian boats have been arrested in the Dixon Entrance for fishing in 'American waters'. These are 'Canadian waters' according to the (Canadian) federal government, an entrenched position that reflects in part that country's annoyance at losing, in 1903, territorial claims in the Alaska Panhandle to the United States.

On the continent, the occurrence of acid rain in Canada is an example of one irritant that has been partially resolved. Just as the physical features of the continent facilitate a north–south contact and interaction between the two nations, atmospheric circulation patterns can carry pollution from the United States well into Canada. Sulphur dioxide and nitrogen oxides, released into the atmosphere by industry and individuals, are carried over the border by southerly winds in the form of acid rain. This type of precipitation increases the acidity of the soil and waterways in Canada, and has a direct impact on vegetation and, consequently, on food chains and the entire ecosystem. After many years of negotiation, President George Bush introduced a policy on acid rain in 1989 that came close to accommodating Canadian environmental concerns and standards. The Acid Rain Treaty between Canada and the US evolved from this policy initiative and, while it does not resolve the problem, it did remove a major irritant in Canadian–American relations. Although issues of conflict remain between the two countries, they do have a record of resolution and compromise that should serve to reduce many others in the future.

Canada has been involved with the United States in the joint defence of North America since the early 1940s. Both countries are members of NATO, and in 1958 they formed the North American Air Defence Command (NORAD), renamed the North American Aerospace Command in 1981. Northern

Canada became the site for distant early warning (DEW) radar sites, for American missile guidance testing and, for a short period, the installation of nuclear warheads under American control. The latter were withdrawn in the early 1970s. The process of joint military preparedness was also an interesting form of sovereignty defence, for through its participation Canada prevented a unilateral move by the US to protect a northern flank that could not be left undefended. As a junior partner, however, Canada also had to recognize a concurrent erosion of sovereignty to American defence policy – the 'sovereignty paradox' (Tucker, 1980). The threat of nuclear attack over Canada's northern territory has diminished with the collapse of the Soviet Union, and Canada has been able to shift its defence strategy to NATO. The need for American military presence on Canadian soil has also diminished with developments in satellite surveillance and airborne warning-and-control systems (Gibbins, 1994).

These are only some of the developments that have drawn Canada into a closer cultural, economic and political association with the United States. It is at times difficult for the government of Canada to pursue policies and programmes of action against American interest – for there are internal divisions in Canada that can counteract an integrated endorsement of such initiatives.

INTRANATIONAL DIVISIONS IN CANADA

Regionalism in Canada: the periphery and the core

The periphery
The regional structure of Canada has been described as a dependency relationship between the hinterland, consisting of the western and the Atlantic provinces, and the heartland of southern Ontario and southern Québec (McCann, 1982). There are of course internal variations within each of these regional blocs. Atlantic Canada, for example, incorporates the provinces of New Brunswick, Nova Scotia, Prince Edward

Island and Newfoundland–Labrador, but there are notable differences among them. In general terms, however, the four provinces have endured an out-migration of people, have a strong reliance on a primary-resource economy, and consequently suffer the vagaries of external markets. They share a heavy dependence on the sea that is not characteristic of other areas in Canada. This region also relies on transfer payments from the national government in Ottawa as a significant portion of its GDP.

Long before Newfoundland joined the Canadian Confederation in 1949, the three Maritime provinces – Nova Scotia, New Brunswick and Prince Edward Island – were early beneficiaries of protective tariffs and favourable freight rates that linked them economically to central Canada and nurtured their emerging manufacturing industries. Other sectors of their economy – agriculture, forestry and fishing – were more vulnerable since they relied on access to foreign markets, particularly to the United States. There was not one common economic interest, however, that bound these provinces together. If the fishery was dominant in one, it was second to forestry or agriculture in the others.

Since the end of the Second World War, this eastern region has become increasingly marginal to the national economy. Consequently a feeling of peripheral neglect from central Canada has emerged as the common element among the provinces. With the decline of the manufacturing sector, replaced by similar activities in central Canada that were closer to major markets, other common characteristics emerged – endemic high unemployment rates and increasing dependency on transfer payments from the federal government. Following the union with Canada in 1949, Newfoundland eventually shared this regional consciousness of being neglected and marginal (Chapter 19).

A severe blow to Newfoundland's economy occurred in the early 1990s with the collapse of the northern cod fishery. Prior to this, the federal ministry responsible for fisheries and oceans had been encouraging the industry to expand in spite of warnings of resource depletion from marine scientists. The same ministry then had to close the lucrative northern cod fishery in 1992 and had to invoke a two-year moratorium on the industry. The closure was eventually extended to other off-shore areas and to other fish species; the moratorium was also prolonged. This event has intensified the feelings of dependency throughout the Atlantic region.

The three Prairie provinces of Manitoba, Saskatchewan and Alberta have a different regional identity than in the past when they were heavily dependent on a wheat economy. It is another section of Canada that experienced a shift in its economic structure after the Second World War. Other resources – oil, gas, potash, coal and uranium – were developed to displace the agricultural sector as the feature-in-common for the region. The prairie economy became more similar to that of British Columbia, and this generated a sense of a large 'western region'. The area continues to rely heavily on natural resources, has a dependency on foreign markets and a shared sense of alienation from central Canada. There is a feeling throughout the western region that the federal government in Ottawa is more concerned with the internal issue of national unity rather than with their need for protection from the vagaries of external markets. This perception of indifference to the economic difficulties of the region has led to the phenomenon known as 'western alienation'. This occasionally erupts as outrage against federal policies such as the National Energy Programme, introduced in the 1980s.

This programme was intended to keep Canadian energy prices below world levels and to promote self sufficiency. It also gave the federal government access to oil revenues, in part, by shifting drilling activities on to federal lands. In the west, this was considered to be a federal raid on provincial resource revenues (Gibbins, 1994). Canadian ownership of the oil industry was to be increased from roughly 10 per cent to 50 per cent by 1990. To this end companies were encouraged to purchase petroleum assets just as world oil prices were about to decline.

The NEP had only one important economic effect. In a few months it stimulated

a massive transfer of Canadian wealth to foreign investors as public and private–companies were encouraged to buy up petroleum assets at what proved to be the highest prices in history. The foreigners must be laughing still at the dimwittedness of Canadian nationalists, buying wildly at the top of the market (Bliss, 1994, p. 269).

The NEP was eventually abandoned in the mid 1980s, but not before it had contributed to the federal deficit and intensified western alienation, particularly in Alberta. The outcome of government policy on this issue was similar to what occurred in Atlantic Canada when the federal government encouraged investment in, and expansion of, the off-shore fishing industry in the 1970s and early 1980s.

Another policy that has not been wholly embraced in western Canada is the federal initiatives on biculturalism and bilingualism. The Prairies have been multicultural in their population composition since the late 1890s. German, Scandinavian, Ukrainian, Dutch, Russian and American immigrants produced a multitude of languages, religions and cultures, with French Canadians only a small percentage of the total population. Furthermore, an intermixed settlement pattern of Europeans on the Prairies was, in large part, an outcome of federal policy in the nineteenth century. The anglophone society in the West and in Ontario had expressed concern that Slav immigrants were settling in large clusters and, if allowed to continue, would be immune to the assimilative pressures into a nation that was intended to be predominantly British in character (Barr and Lehr, 1982). By 1931, only 56.5 per cent of the prairie settlers were of British or French origin compared to 80.1 per cent for the whole of Canada and 82.7 per cent for Ontario. Today, the western perception of the French Canadian population is as a regional minority rather than as a national charter group (Gibbins, 1994).

The Core: Ontario/Québec

The target for much of the discontent in the 'West' and in the Atlantic region is central Canada, which includes the provinces of

Ontario and Québec. It is the perception of most people in the peripheral regions that federal policies and programmes tend to favour the development of the two central provinces. Ontario absorbs this criticism because it is the economic heartland of the nation accounting for more than 40 per cent of the total Canadian economy, 38 per cent of the national population and has a disproportionate share of the nation's corporate taxes. It also contains the country's largest city, Toronto, and the capital city, Ottawa. These features help to entrench the attitude, at least among Ontarians, that their province is the centre of Canadian life. To the peripheral populations, the concept of national interest in Canada expresses a relationship not among regions but between the centre and the periphery. Gibbins (1994) has summarized this relationship:

> In this context, the centre ceases to be a region like the other regions and instead takes on the colouration of the whole. It is the metropolis to the regional hinterlands. . . . To illustrate the point . . . it is conceivable to imagine Québec, Newfoundland or the West separating from Canada, but it is inconceivable to imagine Ontario doing so. Ontario is Canada to a degree that no other region can claim (p. 186).

If the concentration of people, wealth and industry in Ontario and Québec has contributed to economic division within Canada, it is the attempts that have been made to accommodate the needs of Québec that have generated another form of division within the country – this time a cultural division. While the form of regionalism discussed above produces discontent in the peripheral regions beyond Ontario/Québec, this does not manifest in widespread movements to separate from Canada. It is the cultural divisions in the country that bring this threat forth and one to which we shall now turn.

Divisions over national unity

It must be stated at the outset that many of the complaints against the federal government's apparent preoccupation with issues of national unity and the demand from Québec

for protection against the erosive processes that diminish its cultural intergrity are founded in part on ignorance. This ignorance relates to the precarious geographical situation of Canada's French-language population and of the forces that constantly threaten the vitality of French Canadian culture.

The majority of the population in Québec live in the southern portion of the province in a pattern that duplicates the national ecumene (Fig. 8.2). The shift of population away from the marginal areas of the province has been a dominant feature since the end of the Second World War, with the greater Montréal area the recipient of most of these people because of the range of employment opportunities. This urban area is also where the anglophone population (people of English mother tongue) is heavily concentrated, and it is this group who traditionally dominated the corporate world of Montréal (Chapter 16). Prior to the mid 1970s, anyone wishing to advance through the corporate hierarchy had to learn English and be prepared to function almost entirely in that language.

As the people of Québec increased their participation in urban employment, the rate of population increase began to decline as couples delayed starting a family and women entered the labour force in greater numbers. Québec's birth rate became the lowest in Canada by 1971, and the gross reproduction rate at 1.7 was below the population replacement level of 2.1 children per family. The incidence of exogamy also increased, an outcome of the bicultural interactions that developed in the urban centres of southern Québec. English often becomes the language of the home in such a union, particularly when the female is anglophone. With falling birth rates and a trend toward English dominance in the corporate world, the province placed increasing reliance on those entering the province as migrants to bolster the francophone culture. Unfortunately for this scheme, the arrivals were mainly 'allophones', those whose mother tongue was neither English nor French, and the majority of these people gravitated to English as their second language, not French. In 1968, the premier of the province estimated that of the 620 000 immigrants who came to Québec

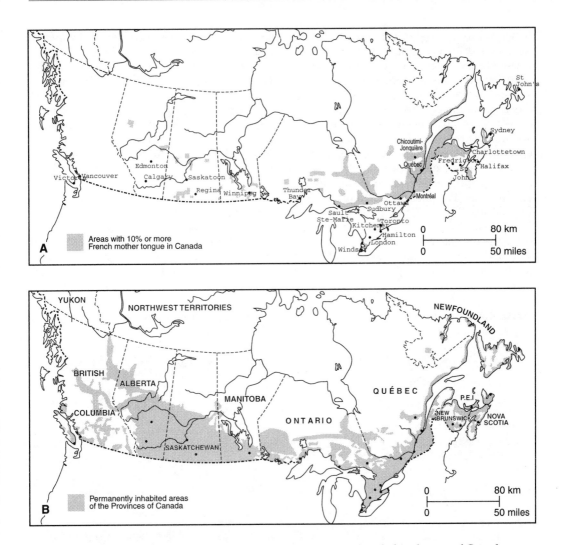

Fig. 8.2 (a) Francophone areas of Canada, 1990s (b) permanently inhabited areas of Canada

between 1945 and 1966, only 8 per cent came from a French mother-tongue source while 80 per cent of the arrivals were absorbed into the anglophone population (Levesque, 1968).

To counteract the advancement of English in Québec, provincial governments enacted legislation in the 1970s that made French the official language of the province. Immigrants and francophones were now required by law to send their children to French-language schools. Businesses had to accommodate employees who wished to work entirely in French, and municipalities, non-English school boards and hospitals were instructed

to use French as the internal language of communication. All public signs and advertisements were to be in French only. This is another example in Canada of collective values taking precedence over individual freedoms.

There is a consensus among researchers in Canada that the use and status of French have increased in Québec since the legislated entrenchment of the language. While it cannot be demonstrated empirically that these improvements are the direct outcome of language policies and legislation, opinion-poll surveys have revealed a strong correlation between provincial intervention and the rising status of French (Laporte, 1984; Bourhis, 1984). The language has prospered in several domains – the arts, the media, education and in commerce – and is now considered to be more secure than it was 25 years ago.

A feature of Québec society that has accompanied this enhancement has been the erosion of the non-francophone population over the same time period (Table 8.1). The number of people of English mother tongue residing in the province declined from 789 000 in 1971 (13.1 per cent) to approximately 626 000 (9.2 per cent) in 1991, a decrease of 20.6 per cent (de Vries, 1994). Another writer has calculated a higher real net loss since the anglophone population of 1971 would have reached 900 000 by 1991 if the exodus had not occurred (Angell, 1996). This has produced a loss of one in three for the anglophone population. The allophone population increased from 6 to almost 8 per cent of the total population between 1971 and 1991, but Angell maintains that 60.5 per cent of these people actually left the province during this time period so that the percentage to total in 1991 should have been higher (Table 8.1).

These demolinguistic processes are contributing to a gradual shift in Québec's population composition in favour of the francophones, and it is predicted that this trend will continue well into the next millennium (Termote and Gauvreau, 1988). This

TABLE 8.1 Population composition by mother tongue in the Province of Québec, in Montréal CMA, and the Ile de Montréal, 1971–1991

Region	Mother tongue	1971 '000 %	1981 '000 %	1986* '000 %	1991* '000 %
Québec	French	4 867 80.7	5 254 82.5	5 435 82.9	5 586 82.0
	English	789 13.1	693 10.9	679 10.3	626 9.2
	Other	371 6.2	421 6.6	444 6.8	598 8.8
Montréal CMA	French	1 820 66.3	1 937 68.5	2 036 69.7	2 139 68.4
	English	595 21.7	520 18.4	493 16.9	485 15.5
	Other	328 12.0	371 13.1	391 13.4	503 16.1
Île de Montréal	French	1 198 63.3	1 051 59.7	1 053 60.0	1 005 56.4
	English	464 22.6	393 22.3	373 20.4	362 20.4
	Other	296 14.1	316 18.0	327 19.6	408 23.2

Notes: CMA = Census Metropolitan Area
* For 1986 and 1991 multiple responses were distributed according to the techniques used by Statistics Canada
Source: for 1971 and 1981, Statistics Canada population data; for 1986 and 1991, Statistics Canada special tabulation

has contributed to a shift in the distribution of Canada's linguistic populations whereby French-speaking people are becoming concentrated in Québec and English-speaking people occupy the rest of the country. This geographical division makes it difficult for the federal government to develop broadly acceptable policies that are designed to enhance national unity.

While the province of Québec must reconcile this loss of people, it must also accommodate a tenacious presence of the English language. Francophones have achieved great success in business and commerce in the past 20 years, particularly in Montréal. Although they are able to function in these domains largely in French, competence in English is still a requirement for those who wish to participate in international trade and finance; a sector that increased, for the province, by 208 per cent between 1981 and 1995 (Ip, 1996). To illustrate this we can draw from two sources.

About seven years after the promulgation of the Official Language Charter (1977) in Québec (the legislation that is responsible for the entrenchment of the French language), one academic commented on the role of English in the province's business community: 'One intriguing finding concerns the status of English as a language of economic advancement in Québec. Mass perceptions suggest that English has maintained much of its status relative to French as the language of business' (Laporte, 1984, p. 75). Ten years later two other researchers reviewed the work that had been done to determine the linkages between language policies and language use in Québec and stated:

In general, then, these various studies discussed appear to show that while some progress seems to have been made in terms of the perceived status of French, when compared with studies carried out prior to language legislation in Québec, English continues to hold significant status (Hamers and Hummel, 1994, p. 147).

Other studies confirm that the English language is still a force of attraction in several domains within Québec. In the face of this, the provincial government in Québec City will continue to interpret its role as guardian of the language and the culture, and will develop programmes and policies to achieve this. One example can be taken from Québec's 'sign law'. This legislation was passed in 1988 and required businesses in the province to use only the French language on exterior signs that identified their location and type of enterprise. It was promulgated, in part, to create a streetscape that would reflect the French-dominant culture of the province. For those who endorsed this legislation, the maintenance of a *visage linguistique* was an essential ingredient to a viable culture – 'If French is to be secure within the boundaries of Québec, it is important to maintain an urban landscape of signs and symbols that reinforces the predominance of French, and clearly indicates to immigrants the linguistic character of the Province' (Whitaker, 1989, p. 3). This became a highly divisive issue within Québec and beyond. The 'sign law' was modified in 1993, but the replacement legislation (Bill 83), while not prohibiting bilingual signs, prescribed specific designs so that the French portion would retain prominence. In the five years that it took to modify the original restrictive law a backlash spread through parts of Canada, and this contributed to the defeat of a proposed constitutional amendment that would have met Québec's concerns about its status in the Canadian federation.

Canada's Constitution had been patriated from Great Britain in 1982, but without the approval of the government of Québec. Five years later, the Prime Minister and the 10 premiers of the provinces signed an agreement that was designed to obtain Québec's full participation in the new Constitution while concurrently protecting its culture. It was to give recognition to Québec as a distinct society, and all provinces were to receive more powers at the expense of the federal government. Designated the Meech Lake Accord, the document, to become law, required the ratification of Parliament and all 10 provincial legislatures. This was to be accomplished by June 1990, but Québec's sign law appeared in December 1988.

Consequent anger against the law emerged in the rest of Canada and was soon directed at the accord. 'It is no accident that objections to the Meech Lake Accord bloomed in some quarters precisely when Québec francophones reaffirmed their majority rights against anglophones on the question of the language on commercial signs' (Thorsell, 1989, p. A14). Resentment was based in part on the federal request to Canadians to embrace the concept of two official languages for the nation while Québec was becoming officially unilingual French and banning English signs. The Premier of Manitoba, formerly in favour of the agreement, withdrew his support in response to Québec's sign law. The Meech Lake Accord was not ratified by all 10 provinces, whereupon the federal Minister of Transport, Lucien Bouchard, resigned his seat in Parliament, returned to Québec and joined the separatist movement. In the late 1990s he became head of the secessionist Parti Québécois and premier of Québec.

The introduction of the sign law was handled awkwardly by the premier of Québec in 1988, for there was no attempt to explain clearly to the people of Canada why the provincial government feared for the erosion of their language and their culture. Nor did the federal government develop a campaign of information and communication throughout the country. The concerns that permeate the province over the inroads that the English language is continuing to make are not well understood in the rest of the country. A former prime minister of Canada has expressed the risks that a country must endure when it is burdened with citizens who are ignorant about their own country:

We operate against a backdrop of profound ignorance about our country. That mattered less when key decisions were made in the old way: in secret by men in suits. In this new era, public opinion rules, and sadly it is often ill-informed. ... A large and diverse nation that is ignorant ot itself is bound to be divided' (Clark, 1996, p. A13).

Since the demise of the Meech Lake Accord Canadians have become obsessed with the issue of national unity.

Is the United States free from such erosive intranational divisions? They are present but these have been less threatening to national integrity than in Canada. Nevertheless, they do contribute to internal tensions and at times appear intractable. We shall investigate two of these.

INTRANATIONAL DIVISIONS IN THE UNITED STATES

Racial divisions in America

If the divisions in Canada have been strongly cultural those in the United States have been more racially focused between the white and black populations. One outcome of this division is urban residential segregation.

No group in the history of the United States has ever experienced the sustained high level of racial segregation that has been imposed on the blacks in large American cities for the past fifty years. ... Not only is the depth of black segregation unprecedented and utterly unique compared with that of other groups, but it shows little sign of change with the passage of time or improvements in socioeconomic status (Massey and Denton, 1993, p. 2).

When the American Civil War (1861–65) ended, 80 per cent of black Americans were in the rural South, but as job opportunities developed in the cities they began to migrate to urban centres in the South and in the North. Whatever skills they had developed on the land, most blacks lacked the education and training to take advantage of urban jobs beyond that of manual labour. A low economic status, therefore, confined them to poor housing. Consequently, in the rapidly-growing cities in the South, such as Atlanta, Richmond and Montgomery, segregated residential patterns were emerging by the mid 1890s (Litwack, 1998). This process did not manifest into urban black ghettos, however, until after 1900 – in the South or in the North.

At the turn of the century, industrialization in northern cities shifted production from the home and small shops to large factories that required hundreds of workers. Tenements and row houses were built near the factories to accommodate the employees, most of whom were rural immigrants from Southern and Eastern Europe. With the outbreak of war in Europe in 1914 this source of labour was cut off, and when America entered the war in 1917 the shortage of workers became acute. To fill this void employers turned to African Americans who, in response, left the South in large numbers. Although some jobs that were formerly restricted to whites were, because of the labour shortage, now open to a few blacks, 80 per cent of black men were in the dirtiest and lowest paid jobs (Thernstrom and Thernstrom, 1997, p. 57). Even northern-born blacks were clustered at the low end of employment, all victims of the racial barriers that closed off better paid and more desirable jobs. It is often assumed that employers were the ones who orchestrated these limitations. In fact it was the trade unions who developed them in an attempt to restrict competition for jobs to whites (p. 58).

When the European labourers arrived in American cities they gravitated to specific ethnic neighbourhoods where they could live among kin and kind. Such enclaves offered the arrivals a temporary zone of security while they got a job, learned English and eventually developed the skills and confidence to disperse into more ethnically mixed neighbourhoods. Such choice for residential change was generally not open to blacks. The residential colour line in the housing market became as intense as in the labour market, a situation that was fostered by a white fear of black neighbours.

> As the size of the urban black population rose steadily after 1900, white racial views hardened and the relatively fluid and open period of race relations in the north drew to a close. . . . Whites became increasingly intolerant of black neighbours and fear of racial turnover and black 'invasion' spread (Massey and Denton, 1993, p. 30).

A variety of processes was developed by the white society to ensure urban segregation. Racial violence spread through cities in the northern states from 1900 to 1920 with blacks who lived in integrated or predominantly white areas the targets. As the violence slowly waned in the 1920s, other devices emerged that were equally effective. Neighbourhood 'improvement associations' were formed ostensibly to promote local security and property values; their real purpose was to hold the colour line. Volunteer members lobbied city councils for zoning restrictions and threatened to boycott real estate agents who sold homes to blacks and white retailers who had black customers. They collected funds to buy out black settlers or to purchase long vacant homes. The 'associations' also devised restrictive real estate covenants that were written into deeds to prohibit the owners and their heirs from selling or renting their property to blacks. A violator could be taken to court for damages. These covenants usually had a span of 20 years and were applied widely throughout the United States until 1948 when they were declared unenforceable by the US Supreme Court.

The percentage of blacks within northern cities increased rapidly in the post-war years – indeed many doubled their black proportions between 1950 and 1970. For example, in Chicago the black population increased from 14 to 33 per cent of the total, and in Cleveland and Detroit the gain was 16 to 38 per cent and 16 to 44 per cent respectively. Similar increases occurred in other northern cities with Washington, one of the largest, reaching 71 per cent by 1970. During this time the federal government passed the Fair Housing Act (1968) through which real estate agents were no longer able to openly refuse to sell or rent a property to blacks. Instead of disappearing, however, such practices became more subtle, but just as effective. Nor did the Civil Rights Act of 1964, which outlawed racial discrimination in employment, contribute to a decline in urban residential segregation. To demonstrate this we need to examine a measurement technique that is applied frequently by social scientists to plural societies.

A standard measure of segregation is the index of dissimilarity, which indicates how evenly two groups are distributed over urban space. The sources on the index range from 0 to 100, with 0 meaning complete residential integration (each census tract, city block, etc., has the same ethnic/racial com-position as the city as a whole) and 100 complete residential segregation.[2] The persis-tence of urban residential segregation is revealed in Table 8.2. Segregation peaked around 1950, and then began to decline slowly until 1970 where it remained high at an average of nearly 80 for northern cities and 89 for those in the South. It has been esti-mated that 70 per cent of blacks would have had to move to produce a non-segregated distribution for most northern and southern cities; for many that figure would have been closer to 90 per cent (Massey and Denton, 1993, p. 46).

A dominant feature of American cities during the 1970s and 1980s was the large population shift to the suburbs, but this trend occurred much earlier for whites than for blacks. An average of 71 per cent of northern whites lived in suburbs by 1980, but only 23 per cent of blacks had made this shift

(Massey and Denton, 1993). Some blacks were able to improve their socio-economic status under the Civil Rights Act, conse-quently many middle-class and skilled work-ing-class members left the ghettos. Black entry into the suburbs in the early 1980s did not bring integration however. It would take more than a decade before this process would begin to appear.

Some outcomes of urban racial segregation
The more affluent blacks who joined the move to the suburbs left behind an isolated and disadvantaged community that did not have the institutions and social resources to compete with the modern white society. Any fluctuation in the economy is felt more intensely in these central, segregated areas, thereby creating an underclass and a geo-graphical concentration of poverty. Housing decay and abandonment, crime and social disorder are magnified. In the poor urban

TABLE 8.2 Trends in urban residential segregation between blacks and non-Hispanic whites in the United States 1950–90

| | % Black | Indices | | | |
	1990	1950	1970	1990	Absolute change 1970–90
Atlanta	67	90	82	68	−14
Baltimore	59	91	82	71	−11
Chicago	39	92	92	86	−6
Cleveland	47	91	91	85	−6
Dallas	30	88	87	63	−24
Detroit	76	89	88	88	0
Houston	28	90	78	67	−11
Los Angeles	14	85	91	73	−18
Miami	27	96	85	72	−13
New Orleans	62	83	73	69	−4
New York	29	87	81	82	+1
Philadelphia	40	89	80	77	−3
San Francisco	11	79	80	67	−13
St Louis	48	93	85	77	−8
Washington	66	81	81	66	−15

Note: indices are of dissimilarity for the 15 metropolitan areas with the largest black population in 1990.
Sources: Massey and Denton (1993); *Statistical abstracts of the United States*, 17th edn, 1997

ghettos drug use, unemployment, welfare dependency, teenage childbearing and unwed parenthood become commonplace. A black male who is unable to find work in or near the ghetto may abandon his family, leaving his wife to support herself and the children.

The children of the poor black ghetto face a bleak future, with the probabilities of entry into society beyond rather dim. Their schools lack the resources that are available in more affluent neighbourhoods, and their teachers must also function as social workers to students who suffer all the disadvantages of a poor, female-headed household. Their distance from the majority is also revealed in the language that has emerged in these areas. Black Vernacular English (BVE) has become progressively more removed from Standard American English (SAE) and its speakers are at a disadvantage in US schools and in the labour market. This is a street speech, but it is not considered by linguists to be a degenerate or illogical version of SAE. It has a consistent grammar, pronunciation and lexicon of its own with no implication as an inferior language. It has evolved in an environment of intense urban residential segregation that blacks have experienced since the early part of the twentieth century (Labov and Harris, 1986). It is also a vernacular that has become more uniform throughout the larger northern cities in contrast to white speech patterns that are more regionally distinct. In recent years BVE has become known as 'Ebonics', a terminology that has been used by some school boards in California.

Children who grew up in the ghetto speaking BVE are at a disadvantage when they enter schools where texts, instructional materials and standardized tests are written in SAE. Lack of success in mastering standard English can lead to frustration and may contribute to the high drop-out rates in the ghetto. Success also has its own problems, for children will often ridicule a classmate for 'talking white' (Lee, 1994, p. A15).

Recent trends

There are some who contend that urban residential segregation for blacks has been declining since 1980. Between 1970 and 1990 the index of dissimilarity has remained relatively unchanged in four large metropolitan centres – New York, Chicago, Cleveland and Detroit – but in other cities the segregation index dropped on average from 84 to 74 over 20 years (Table 8.2). This decline is more pronounced in smaller- and medium-sized metropolitan centres (Thernstrom and Thernstrom, 1997, p. 214).

> However you measure it, residential segregation has been declining in the United States for the past quarter century. Not as rapidly as some observers would like, obviously, but there can be no doubt about the direction of change. Nevertheless, the pace is slow, and it cannot be denied that residential patterns in the United States today remain closely connected to race. The color of your skin still affects which neighborhood you live in, though much less than before (Thernstrom and Thernstrom, 1997, p. 219).

The persistence of residential patterns appears to be associated with a preference to live close to others who have similar attributes. In a survey conducted in Detroit in 1976 almost all blacks preferred a racial mixture of 50 per cent black or higher, but 95 per cent were also willing to live with a black percentage between 15 and 70. In contrast, the tolerance of whites for racial mixing was quite limited. Of the white respondents, 73 per cent were unwilling to move into a neighbourhood that had more than a third black residents. If they lived in an area that attained this level of 'integration', 57 per cent stated that they would feel uncomfortable living there, and 41 per cent would attempt to leave (Farley et al., 1978). The most recent national survey of residential preferences that included both blacks and whites revealed a persistent endorsement for integration. The definition of an integrated neighbourhood, however, differed for the two groups. White respondents were in favour of such intermixture as long as the area was 80 per cent white; for most blacks integration was a 50:50 balance. Only 20 to 25 per cent of whites would accept living in such a neighbourhood (Thernstrom and Thernstrom, 1997, p. 229). These preferences indicate that residential integration will only proceed gradually in the United States, and it is unlikely to reach 0 on the dissimilarity index. However, this form of separation may not be considered a negative element in the American society if the inequalities in levels of educational attainment, the unbalanced structure of the black family, and the rise in black crime rates can be resolved.

Emerging ethnic divisions in the United States

The sort of divisive regional issues that were discussed earlier for Canada are relatively muted in the United States. A more divisive issue that has emerged from a regional context, however, is associated with the growth of multiculturalism in America and its attendant problems of heightened ethnic consciousness and political conflict. In the nineteenth and early twentieth centuries, Europe was overwhelmingly the source region for most immigrants to the US, but since 1965 immigrants have been mostly Latin American and Asian in origin (85 per cent). The regions that have felt the impact of non-European immigration most strongly have been the southeastern and southwestern tier of states (Chapters 14 and 15). Some have interpreted this salient of largely Hispanic people as a threat to the territorial integrity of the US, and are reacting in a manner similar to the Québeckers' response to the 'invasion' of English.

When the birth-rate differentials and the estimated numbers of undocumented migrants are factored into the legal immigrant figures, some demographers have predicted that the country will cease to have a European-origin majority population within the next 80 years (Schmidt, 1993). Federal statisticians have also forecast that by about 2070, the population of Hispanic origin will outnumber the black population in the United States. Presently, 85 per cent of these people are located in only 10 states in the continental US (Fig. 8.3). With such a concentration it is possible for recent arrivals to retain their mother tongue longer as a variety of institutional activities, which function in Spanish, will have sufficient patronage to survive. In the past 20 years, Spanish-language radio

stations, TV networks and magazines and newspapers have proliferated in these states and beyond. In Los Angeles, for example, Coors is advertised as *cervesa* as often as it is as beer (Garreau, 1981, p. 212). In his publication *The nine nations of North America* (1981), the writer/journalist Joel Garreau linked four US states with large Hispanic populations to northern Mexico and described it as a 'nation' that he dubbed MexAmerica. This did nothing to allay the fears of the nationalists, who fear a loss of territory.

> The Southwest [US] is now what all Anglo North America will soon be – a place where the largest minority will be Spanish speaking. It's a place being inexorably redefined – in terms of language, custom, economics, television, music, food, politics, advertising, employment, architecture, fashion and even the pace of life – by the ever-growing numbers of Hispanics in its midst. It is becoming MexAmerica (Garreau, 1981, p. 211).

In the years since this publication appeared, the Hispanic population has grown in other states beyond MexAmerica as job opportunities in large urban centres such as Chicago and New York (and neighbouring New Jersey), and the service industries of Florida have attracted people of Hispanic origin. This dispersion has added to the multicultural character of the most populous regions of the United States, and has generated a divisive issue in the nation – the language policy conflict.

English has never received legal recognition as the official language of the United States either in the Constitution or in laws passed by the Congress. In the formative years of the nation, antimonarchical forces rejected any policy of language manipulation. Moreover, the use of other languages was in keeping with the anti-elitist attitudes and the democratic foundation of freedom in the new country. As the nation increased its interaction with the international community, it was widely assumed that it would be through English. The members of the Continental Congress believed that this would entrench the significance of the language for those Americans whose mother tongue was not English (Higham, 1992). By the nine-

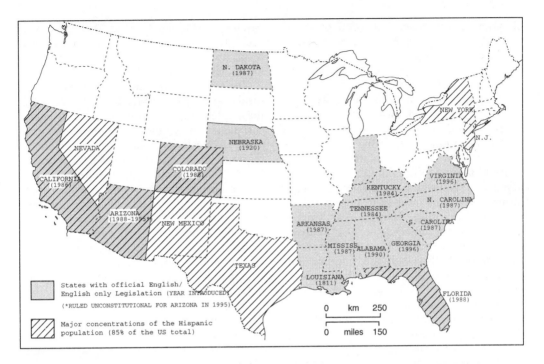

Fig. 8.3 States with official English/English only legislation, 1997, and major concentrations of Hispanic population, 1994

teenth century, Americans relied on the education system, through a common school programme, to foster national unity (Heath, 1992). Through the early twentieth century, German Americans were the largest non-English, non-indigenous group in the US, estimated to be nine million in 1910 (Schmidt, 1993). The potential for ethnic entrenchment of this minority, however, was lost as anti-German feelings that were associated with America's involvement in the First World War contributed to the erosion of German culture and the German language in the US.

In contrast to these earlier patterns of interaction, the increase in Hispanic immigrants in the past 20 years – documented and undocumented – have, in cities in the Southwest, generated ghettos in which Spanish is the predominant language. In Los Angeles, for example, the percentage of the population who speak a language in the home other than English increased from 35 per cent in 1980 to almost 50 per cent in 1990. Throughout the state of California, almost a third of the residents did not speak Eng-

lish in the home; in Santa Fe, New Mexico the number was 40 per cent in 1990. This has contributed to a concern for the status of English among many Americans.

The emergence of cultural pluralism in the United States, during the 1960s and 1970s, accompanied the rise of the Black Power campaign and a new ethnic movement in which rights equality were recognized for all racial and ethnic groups. The rights of individuals, therefore, were derived from membership in these groups (Imhoff,1991). While white European ethnic groups – Polish American, Irish American, Italian American – thereby claimed a recognition that was similar to that won by black Americans during the Civil Rights movement, there were no implications for the English language. However, when political advocacy groups who represented Hispanic people claimed the cultural pluralism rationale for their group rights and group equality there were implications for the state of English in the US. Reactions were soon to follow these claims. Those who fear the resultant moves toward bilingual education and bilingual voting ballots believe that they will lead to ethnolinguistic conflict and hence that they will have the potential to be politically divisive for American society as a whole.

There are several features of this evolving plural society in America that continue to nourish the division over language policy. When most immigrants to the United States came from Europe they remained as settlers and had few opportunites to return to their homeland. The social and economic forces that fostered language transfer to English, therefore, had few countervailing trends. Current immigration not only enhances the Hispanic communities in America, but many can and do return periodically to visit their place of origin, especially in Mexico. Contact with the cultural hearth is not severed as it was for the European immigrants. Thus the usual sequence of assimilation among immigrants and their descendants has been broken.

Senator S.I Hayakawa, a Republican from California and a former resident of Canada, launched the 'Official English' movement in April 1981, when he introduced into the Senate a proposed amendment to the Constitution to designate English as the official language of the country. Although the Congress showed little interest in this issue, Senator Hayakawa and a few associates followed his proposal with the introduction of a national lobby group known as 'US English'. This organization directs its attention to state and local levels of government, as well as to Congress, where it attacks the policies of bilingual education and bilingual ballots, and calls for the designation of English as the sole 'official' language at all levels. By 1997, 20 states had passed laws or state constitutional amendments making English the official or only language of goverment[3] (Fig.8.3). Reaffirming the pre-eminence of English meant reaffirming a unifying force in America and thereby avoiding 'the nightmarish vision of separatist Québec as the inevitable future of the United States if we fail to protect English now' (Leibowicz, 1992, p.104). Equating the situation in the southern United States to that in Canada is a favourite ploy of members of US English. One member, in a letter to US Secretary of State Henry Kissinger in 1985, warned that continued immigration into the US could produce a 'Chicano Québec' in the area bordering Mexico (Martinez, 1988, p.159).

The southern United States has many cultural characteristics of the Hispanic society to the south so that entrants do not encounter a completely new and threatening environment. According to some social scientists, there are Mexicans who perceive the southern US as 'lost territory' and therefore consider the region to be an extension of Mexican space (Stoddard et al., 1983). The situation is further complicated because of the thousands of undocumented workers who enter the United States illegally every year, particularly from Mexico and Central America. There is fear among many Americans that the nation is losing control over its southern border.

US English is not an organization to be taken lightly for, unlike fringe groups, this lobby has a broad membership of over 400 000 and a budget of $6 million annually. Their opponents will bring court cases against them in order to have English-only laws declared discriminatory and therefore something to be overturned. They will counter the mantra of US English by declaring that the unity of the United States does not depend upon the declaration of English as an official language. As long as the current pattern of immigration to America continues, this divisive issue will intensify.

POSTLUDE

Americans are encountering a polyethnic society that is moving away from its traditional composition of immigrants and the customary processes of assimilation. Changes in source regions and immigrant linkages to their countries of origin suggest that a different kind of assimilation is occurring in the United States. Some members of the charter group feel threatened by this and cling to the melting-pot theme as a proven antidote. They endorse a moulding process of assimilation that will maintain a strong, traditional homogeneous American identity – consequently the need to adjust and adapt is clearly on the side of the immigrant. As a nativist approach, it calls for a reduction in immigration and official status for English to sustain a truly 'American stock'. In this context, some social scientists argue that the contemporary conflict over language policy is in fact an ethnic conflict and is not about language at all (Crawford, 1992; Schmidt, 1993).

There are others who welcome the shift and propose a new metaphor for the emerging society as a mosaic or salad bowl – an America of cultural pluralism. The concept of a nation to them is less significant than the development of a cosmopolitan spirit. A weakening of national identity, therefore, is a reasonable trade for the linkages that will develop from group associations, intra-nationally and internationally. Between these extremes there are less intense approaches to the new plural society, but common among this middle ground is the recognition that America will become what the immigrants make of it (Riche, 1991; Aleinikoff, 1998; Suro, 1998).

Their perception of the emerging America is not synonymous with the form of multiculturalism that is endorsed in Canada, but it is founded upon empathy, non-discrimination and active engagement among the constituent cultural groups; it embraces a principle of mutuality. To achieve this, group boundaries must be permeable, which means a society will emerge in which affiliations are flexible. For example, between 1960 and 1990 the number of people who identified themselves as American Indian increased by 255 per cent (Alienikoff, 1998, p. 84). Adherents realize that with entrenched racial divisions in the United States, the society must be transformed markedly before boundaries become truly permeable and mutually-respecting groups develop interaction. There may be disagreement over what it means to be 'American' in this nascent society, but not over allegiance to and development of the nation. It will be a nation of English speakers with a common nationality and understanding among groups, but there will be no need for designation of 'official' language status.

The division in the United States over the form of assimilation and the sort of society that will emerge from the new polyethnism is not as evident in Canada. Although the latter country is experiencing a general shift away from the traditional deux-nations-plus-aboriginals composition and the usual European source regions for immigrants, the new countries of origin are not exactly the same as in the US nor are the patterns of external linkages as intense. As recently as 1971, Canada was a country whose residents were 97 per cent of European origin; most of the rest were First Nations and Inuit. The visible minority population – people of neither European nor Native background – has been rising steadily for a generation. In 1996, it reached 11.2 per cent of the population, almost double the proportion 10 years earlier, with most of the immigrants coming from Asia, Africa and the West Indies. Since 70 per cent of the visible minority population in 1996 were born outside Canada, it would seem that the country must soon face the same issue of assimilation as the US. Virtually all members of the visible minority, however, live in cities with Vancouver (20 per cent) and Toronto (42 per cent) the favourites, and consequently it is misleading to speak of the nation as multiracial. It is these two cities that have become multiracial with, for instance, a diversity in Toronto that is greater than any European city (Chapter 16), while the rest of the country is pretty much the same as it was 50 years ago. When one discusses alterations to patterns of assimilation in Canada, it is in Toronto and Vancouver that they are emerging. It is too early to speculate on the direction that the changes will take in these urban places and what their impact, if any, will be on the rest of the country.

ENDNOTES

1. The Free Trade Agreement was superseded by a North American Free Trade Agreement (NAFTA) when Mexico joined the United States and Canada in 1993 to create a market of 360 million people in the largest trade zone in the world (Dunn, 1995, p. 297). The arrangements within the former agreement were essentially extended to include Mexico. There is, however, an 'accession clause' in NAFTA

that did not exist in the FTA. This allows other countries, or groups of countries, to be considered for membership in the future. The major implication for Canada in this new accord is whether NAFTA, in the future, will affect locational decisions for American investors.

2. If a city has a 20 per cent black population, an even residential pattern requires that each census tract be 20 per cent black and 80 per cent white. Consequently, if one tract is 30 per cent black, 10 per cent of the excess black population would need to move into areas where they are below 20 per cent to nudge the city toward an even pattern. Every census tract is then representative of the racial composition of the city. If, however, we consider the ideal urban city to be totally integrated, regardless of the population composition of the city, the index should be 0, hence an index of 0.30 for a tract means that segregation exists and 30 per cent of the minority in that tract must move throughout the city to eliminate residential segregation completely.

3. Some laws prohibit state and local governments from providing bilingual services to residents not yet proficient in English. Others prohibit state legislatures from passing laws which 'ignore the role of English.' Others state that English is the official language.

KEY READINGS

Garreau, J. 1981: *The nine nations of North America*. Boston: Houghton Mifflin Co.

Gibbins, R. 1994: *Conflict and unity: an introduction to Canadian political life*. Scarborough, Ontario: Nelson Canada Ltd.

Gwynn, R. 1995: *Nationalism without walls: the unbearable lightness of being Canadian*. Toronto: McClelland & Stewart, Inc.

Randall, S.J., Konrad, H. and Silverman, S. (eds) 1992: *North America without borders? Integrating Canada, the United States and Mexico*. Calgary, Alberta: University of Calgary Press.

Suro, R. 1998: *Strangers among us: how Latino immigration is transforming America*. New York: Alfred A. Knopf.

Thernstrom, J.H. and Thernstrom, A. 1997: *America in black and white: one nation indivisible*. New York: Simon & Schuster.

Section D

The North American economy

AGRICULTURE

OWEN J. FURUSETH

The North American countries of Canada and the United States have carved a niche on the world stage as urbanized and industrial nations. Consequently, their significance as global agricultural powers and the importance that agriculture plays in their national economies are often overlooked. None the less, North America has emerged as the world's premier agricultural region in the post-Second World War era, for over the past 50 years its agricultural production has expanded at an unprecedented rate. During this period average agricultural output in Canada and the US has grown by 2 per cent annually.

As we enter the twenty-first century, no other region on earth can match the volume of agricultural output, nor the variety of farm products of Canadian and American farms and ranches. The North American farmers consistently produce food and fibre surpluses. In addition to being world food traders, the United States and Canada have been heavy donors to international food and aid programmes. More critically, prospects for increased agricultural production in the region are quite favourable. The Canadian and US food markets are mature. With slow population growth rates and a high per capita food supply, self-sufficiency ratios for cereals range from 125 to 150 per cent. At the same time, the North American output of cereal crops is growing at a 1.1 per cent annual rate. The combination of a stable domestic demand and increased output means that global exports of cereals are expected to continue to increase. To a great extent, the most significant limitation on North American food production is external – the international demand – rather than internal.

In an increasingly urban-orientated economy, agriculture represents a declining share of the national economies of both Canada and the United States. None the less, the role that it plays in helping to stabilize economic relations should not be understated. In recent years as North America has suffered through deficits with global trading partners, strong agricultural exports have helped to ease export shortfalls. From January to September 1996, for example, agricultural exports from Canada's farms totalled Can$14.9 billion, a Can$4.09-billion surplus over agricultural imports.

Clearly one of the most critical factors affecting the productivity of North American agriculture has been, and continues to be, the infusion of technological innovations into agricultural processes. Many of the technological and scientific improvements that characterize modern agriculture were developed in North American universities and laboratories and tested on the region's farms. As a consequence, the role of conventional production inputs (labour and land) have diminished in their importance, replaced by capital-intensive technology. Whether one embraces or decries advances in agricultural technology, there is no denying that North American agriculturists have pushed the envelope of scientific farming and that Canadians and Americans have been largely willing customers for the output.

In a related fashion, North American agriculture has also led the way in adopting the industrial structured production model. The contemporary global agro-food system, comprised of agri-technology industries, national and transnational agro-businesses and heavy government involvement in agriculture markets is centred in the United States and Canada. Many of the early innovations that evolved into the agri-business model of food production originated in North America. The industrial revolution of North American agriculture is, however, not complete. It has followed an incremental pattern, totally dominating some agricultural sectors and some regions, while not affecting others. Most recently, growing consumer mistrust over the quality of industrially-produced food products has slowed expansion and given pause to a trajectory that assumes total industrialization.

The principal purpose of this chapter is to provide an introduction to complex factors shaping the geography and structure of North American agriculture. The present agricultural scene is a mosaic shaped by history, physical environment, social and economic relations, government, and technology. While it is unrealistic to generalize completely between nations and subnational regions, several broad trends and strategic implications for US and Canadian agriculture are possible. Thus, this text takes up the story of North American agriculture from Europeanization of the continent through to the present with the intent of exploring the development of a critical element of North American society.

CULTURAL AND INSTITUTIONAL ANTECEDENTS

The Old World colonists imported to North America a variety of land tenure and land division systems. While the profit motive undergirded all settlement schemes, each approach was also intended to sustain the colonists who peopled the new lands. Consequently, agricultural development was critical, both to feed the fledgling North Americans and also to generate an income stream. Individually, the Spanish, French, English, and Dutch settlement systems and institutional arrangements varied markedly both in form and organization. Most importantly, the structural differences resulted in an uneven pattern of settlement and success in building economically viable colonies. In general, North America's prototypic farming settlement structure, which is characterized by isolated farmsteads removed from market towns and villages, is largely attributable to these settlement models and land disposal regulations.

The United States

In the United States, history and culture are replete with admiration and respect for the role of the agriculture enterprise in building the American nation-state. Indeed, the farmer and agricultural development have been central tenets in popular culture and public policy since the formation of the country. Over 200 years later, Thomas Jefferson's praise for the independent yeoman farmer, characterized by his hard work and individualism, a steward of natural resources, living on a small, family-operated farm remains a powerful American symbol. The family farm is an American cultural icon that garners

public appeal and commands public policy deference.

Unquestionably, the Northwest Ordinance of 1785 was the most critical template shaping early American expansion. The ordinance operated on two levels: it channelled agricultural development on the frontier, while simultaneously building an organized infrastructure for future settlement and economic growth. The Northwest Ordinance became a land subdivision framework that guided settlement of virtually all the US west of the original 13 colonies. The state of Ohio (the Northwest in the late eighteenth century) was the testing ground for the Township and Range system (see Chapter 1) (Fig. 9.1). The simplicity and uniformity of the Township and Range system eased record keeping, facilitated the alienation and settlement of virgin land, and greatly aided the agricultural development of North America. Along the way, the US Treasury was enlarged by payments relating to land purchases and subsequently taxation matters. Thanks to the efficiency brought by this system, the exploration, settlement, and population of the American West took only five generations rather than Thomas Jefferson's estimate of one hundred (Opie, 1994).

Canada

The development of Canada into an economically-powerful nation-state is commonly linked to the exploitation of natural resource staples. Before Confederation, the focus was on furs and fishery. Later timber resources and minerals emerged as dominant commodities; with the most recent efforts targeting various energy resources. However, as with the southern neighbour, agriculture and agricultural resources were absolutely critical to providing a base for early Canadian settlement and economic stability.

The Precambrian Canadian Shield squeezed agricultural activity and permanent European settlement into a narrow band of land south and east of the impenetrable 'true North' (Fig. 9.2). The lowlands of the St Lawrence River valley and the valleys and lowlands of coastal Maritime Canada provided agricultural opportunities: a matrix

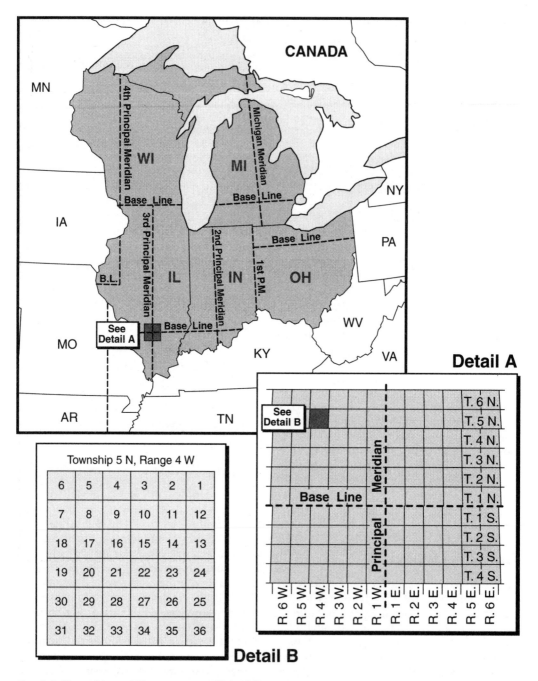

FIG. 9.1 Township and Range system, United States

of suitable soils, climate and drainage conducive to agricultural development. The resulting geographical framework laid by agriculturally-orientated settlement and the farming economy were the basis to establishing the modern Canadian nation-state *vis-à-vis* urban structure, inter-regional economic relations and national imagery.

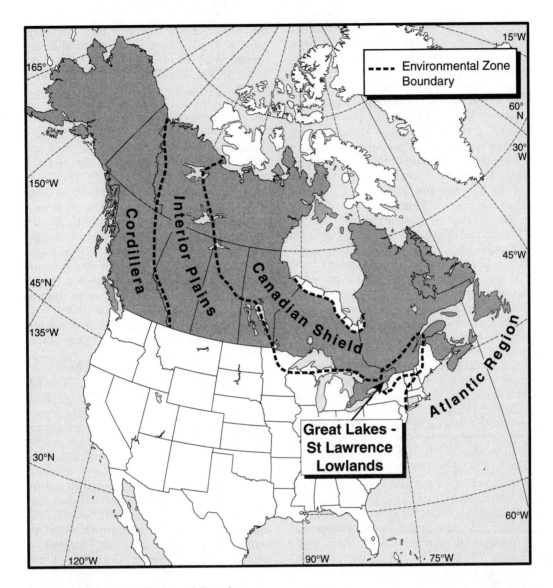

Fig. 9.2 Environmental zones of Canada

The earliest French settlers to Acadia around the Bay of Fundy and lower St Lawrence region were primarily farmers. Farms were self-supporting, every farm had a vegetable garden, most a small orchard, and livestock. Cattle provided most of the Acadian agricultural surplus. Some agricultural specialization developed, primarily small grains and apples, but no major agricultural staples were exported and comprehensive marketing organizations never formed. By 1865, only Prince Edward Island produced a general surplus of agricultural goods.

Over time, a mixed economy evolved. Farmers worked in the forests during the winter months from November to April or May. Timber gangs cut and hauled lumber to frozen rivers, then floated it downstream with the thaw. A similar pattern existed for fishing. With fisheries located close to shore and requiring little capital outlay other than a small boat and a hook and line, farmers could move into the fishing economy easily in order to supplement income and diet. Economically the part-time farmer-fishermen were substantially better off than their full-time brethren.

By the time of Confederation in 1867, the emergence of the Canadian heartland in Upper Canada (southern Ontario) and Lower Canada (southern Québec) was already underway. With a population of over 1.5 million people, urbanization and industrial development was gaining predominance; but farming remained the largest employer in central Canada.

The European settlement of western Canada, particularly the Prairie provinces of Manitoba, Saskatchewan, and Alberta, followed Confederation. The settlement initiative was predicated on three factors: the deposing of Native land rights; the construction of a rail link to eastern Canada; and the establishment of a land survey system. The latter two elements were especially critical to setting up and fostering the agricultural economy. Eventually, the federal government chose the US's Township and Range model for surveying and establishing land records in western Canada. The selection was guided by this system's capability for 'the rapid and accurate division of the prairie region into farm holdings.' The linking of Winnipeg to eastern Canada in 1883, greatly enhanced the homesteading of the West and agricultural development in the Prairies (see Chapter 20). Subsequently, transport costs were reduced, resulting in immigrant homesteaders and farm implements moving west from central Canada while grains and agricultural commodities moved east. Over time, protective trade barriers (the National Policy) as well as other federal actions were used to discourage north–south economic linkings between western Canada and the US that could adversely affect Canadian political unity and economic cohesion.

THE SIGNIFICANCE OF AGRICULTURE

For most of their nationhood, Canada and the United States have relied on the agriculture sector as a linchpin of their economic growth. Anglo-American agricultural development retained predominance until the end of the Second World War. At the beginning of the war, the number of American farms was at its peak of seven million. Today there are fewer than two million farms.

The importance of agriculture to the economic lives of both nations for more than two centuries is a reflection of the synergy between environment and society. The North American nations are blessed with a large and bountiful environment. The food producing resource base is generous. Of the US land area (3.8 million km^2), 40 per cent is classified as farmland, with a production ratio of two hectares. That is, there are two hectares of farmland, including pastures and rangeland, for each American. The proportion of Canada that is classified as agricultural land (680 000 km^2) is substantially less at 6.8 per cent. However, given the lower population, the production ratio is slightly better at 2.2 ha of agricultural land for each Canadian.

Beyond natural endowments, the social, economic and political constructs in both nations created a favourable environment that nurtured and sustained the agricultural enterprise. Farming and rural life are deeply rooted imagery in American and Canadian culture. The opening of the frontier, manifest destiny, and human domination over nature are themes ingrained in the social history and psyche of North America. All were intertwined and aided by agricultural development. In a similar fashion, large-scale immigration and mercantilism would not have been possible without agricultural expansion. Throughout North American history, the governments of Canada and the United States have been active partners promoting agriculture. Indeed, numerous policies and programmes to foster the growth of agriculture, make farmers more efficient, and protect agricultural interests from negative market outcomes have been promulgated.

The growth and long economic primacy of farming in North America reflects a meshing of diverse human and environmental components in a multi-layered complex. While agricultural interests have benefited from external political concerns and economic agents, they

TABLE 9.1 Profile of Canadian agriculture, 1996

	Hectares of farms	% of area in farms	% change in farmland 1991–96	Ave farm size (ha)	% change in census farms 1991–96	% ag land area irrigated	Ave gross farm receipts Can$	Ave capital value Can$
Canada	67 175 000	7.3	.03	244	–1.8	0.01	117 115	567 778
Newfoundland	43 100	0.01	–7.9	59	0.8	*	103 662	251 169
Prince Edward Is	261 750	46.2	2.3	119	–6.8	0.04	158 684	643 262
Nova Scotia	388 000	7.3	–1.1	96	1.0	*	93 739	324 045
New Brunswick	372 000	5.1	0.02	116	–1.4	*	100 602	345 876
Québec	3 392 160	2.5	0.01	95	–6.2	0.09	115 812	606 825
Ontario	5 538 510	6.2	2.8	82	–2.2	0.01	138 932	420 541
Manitoba	7 641 400	13.9	0.01	314	–5.3	0.02	122 012	518 938
Saskatchewan	26 260 875	46.0	–1.1	461	–6.3	0.03	98 700	523 587
Alberta	20 785 180	32.3	1.0	352	3.0	0.02	134 107	680 577
British Columbia	2 491 600	2.7	5.4	115	12.6	0.05	84 842	634 629

Source: Statistics Canada (1997a)

have also helped to shape relationships that impacted the larger societal structures.

Canada

A review of farm characteristics reveals a diversity of scale and operational structure in contemporary Canadian agriculture. The 1996 Census of Agriculture data presented in Table 9.1 provide a snapshot of provincial agricultural status. Regional differences between the Atlantic provinces, central Canada and the Prairies are evident in a number of different areas. These differences arise from environmental variation as well as socio-economic differences.

Nationally, the amount of land in farming uses is relatively stable; however, the number of farms is declining as smaller farming operations are consolidated into larger units. Between 1991 and 1996, the number of farms with gross receipts of Can$100 000 grew by almost 11 per cent. Reflecting climatic and physiographic bounds, the greatest quantity of farmland in Canada, nearly 80 per cent of the total base, is located in the Prairie provinces. Proportionally, agricultural land is most important in the Prairies as well as in Prince Edward Island.

Similar patterns are seen in the size of individual farm units. In general, the smallest farms are located in central Canada, with slightly larger farmsteads in Atlantic Canada, and the largest farms and ranches in the three Prairie provinces. Prairie farmers are also Canada's largest irrigators. Nearly 75 per cent of all the irrigated agricultural land in Canada is in the Prairies, with over 60 per cent in Alberta.

The fiscal status of Canadian farming is less regionally grounded. The largest money earning farms are located in Prince Edward Island (Can$158 684), Ontario (Can$138 932), and Alberta (Can$134 107). The high gross receipts in PEI and Ontario reflect the intensive cropping of high-value commodities, such as fruits, vegetables and speciality crops, on relatively small farms. Urban-orientated market gardens are particularly notable in the Ontario context. Alberta's earnings are similarly linked to output characteristics, with the heavy use of irrigation water offering farmers greater cropping options. The marginally lower earnings on large farms in Saskatchewan and Manitoba reflect the extensive nature of dry-land prairie farming.

The relatively aspatial pattern of capital valuation shows the impact of farm-related expenses including farm equipment and machinery and land prices. In some instances (e.g. Alberta) heavy reliance on equipment and machinery boasts provincial values, while in other cases escalating land values (e.g. PEI and British Columbia) is more effectual in accounting for higher average values.

Structural characteristics of US agriculture

One of the clearest signs of the current revolution in North American agriculture is structural change. Agricultural structure refers to the economic, organizational and fiscal characteristics of farming units. Longitudinal research looking at structural change provides valuable insight into the evolving spatial pattern of agricultural land development.

Research by Wimberley (1987) analysing 20 county-level Census of Agriculture variables for all US agricultural counties between 1969 and 1982 found that counties could be grouped into three structural categories. He named these groups: corporate-commercial farming counties, large-farm-area farming counties, and small family farming counties. These labels were descriptive. The corporate-commercial counties were dominated by farms that were corporately organized, with high gross farm sales and high numbers of hired workers. In turn, they had large contract labour and custom farm work expenses. Large farm area counties were characterized by a high percentage of county land in farms; they had a large number of part-time and tenant farming operations and large investments in machinery and equipment. The third group, small family farming counties, had large numbers of farms per county, large numbers of farms grossing less than $2500 annually, and a high percentage of part- and full-time farmers.

More recently this work was updated using 1982–92 Census of Agriculture data (Thomas et al., 1996). A comparison of simple

TABLE 9.2 Structural change in US agriculture, 1982–92 (county-level)

Agricultural County Type	Scale
corporate–commercial (inc) large-farm area (inc) small family farming (inc)	average number of farms (dec) percentage of land in farms (dec) number of farms earning less than $2 500 (dec) farm size (inc) farm real estate values (inc)
Ownership/Operation	**Operator Characteristics**
number of corporate farms (inc) number of family farms (dec) number of part-time operators (dec) number of tenant operators (dec) number of partnership operations (dec)	off-farm employee operators (dec) operators' age (dec)
Labour Resources	**Equipment Costs**
workers employed (dec) expenses for contract labour (inc) expenses for custom work (inc)	expenses, farm machine expenses (no change)

Notes: (inc) increasing
　　　(dec) decreasing
Source: US Department of Commerce (1984, 1994)

statistics over the period reveals transitional changes in the structure of US agriculture (Table 9.2). The shift away from smaller, lower input farming operations to larger, specialized farming operations is shown in the changes occurring across multiple dimensions of farm structure.

In this latest analysis Wimberley's farm typology was partially modified. The corporate-commercial label remained, but farming-firm counties and small-farm counties were substituted. The farming-firm replaced the large-farm-area label, while small farm replaced small family farming. Statistical analysis was used to cluster farming counties into five groups. A map containing the groupings shows strong areal differentiation in farm structure (Fig. 9.3). Corporate farms are the dominant production units in California and other areas of the Pacific Coast as well as southern and central Florida, southern Texas, and northern Maine. These areas are important vegetable and fruit growing areas, and the correlation between

the corporate farming and these commodities is validated by these data.

Farming-firm agriculture counties are characterized by a high percentage of farmland in the county with younger farm operators having large investments in machinery and equipment. This type of agricultural operation is heavily concentrated in the Midwest and Great Plains states, the traditional grain, beef and hog producing region of the United States. Other areas where this model is strong include the lower Mississippi River valley and eastern North Carolina coastal plain.

The small farm model is less pronounced, but there are distinctive swathes of concentration including east Texas to central Missouri, northern Alabama to western Ohio, and east Tennessee to eastern North Carolina. Small farm counties have large numbers of low-income generating farms with resident operators. The remaining two groups were made up of counties with significant mixes of farming-firm operations

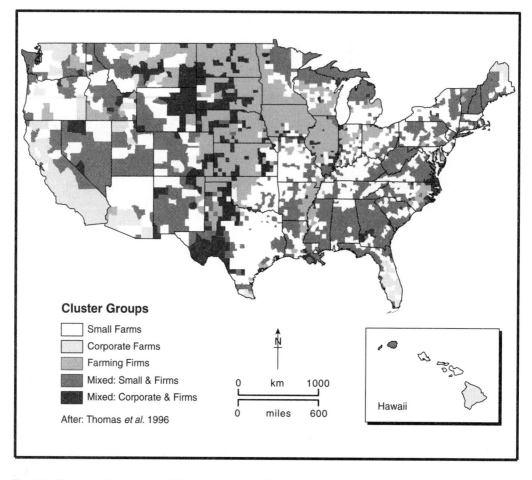

Cluster Groups

- ☐ Small Farms
- ☐ Corporate Farms
- ☐ Farming Firms
- ☐ Mixed: Small & Firms
- ☐ Mixed: Corporate & Firms

After: Thomas *et al.* 1996

N

0   km   1000

0   miles   600

Hawaii

Fig. 9.3 Structural typology of US agriculture, 1992

and either small farms or corporate firms. The largest number of mixed counties was in the former group. The small farm and farming-firm mix dominated the eastern United States and the interior West. The greatest concentration of corporate farm and farming-firm counties were located on the western side of the Great Plains, situated near the large expanse of farming-firm counties.

AGRICULTURAL REGIONS

The shift towards larger, more specialized farming operations has meant increased geographical specialization in farm output. Greater geographical concentration and

homogenization of the agricultural landscape have accompanied this transformation. In terms of regional agricultural activity, this translates into more intensive and well-defined production districts. In the context of this process, spatial agricultural demarcation has become more accurate and easier at county or sub-county levels (Fig. 9.4).

While there are a number of important agricultural production districts in the United States and Canada, two stand out for their economic importance. The Interior Plains represents the traditional agricultural heartland of North America. It has dominated the agricultural scene, in financial and output terms, since it was put to the plough. In many ways, however, it might be seen as

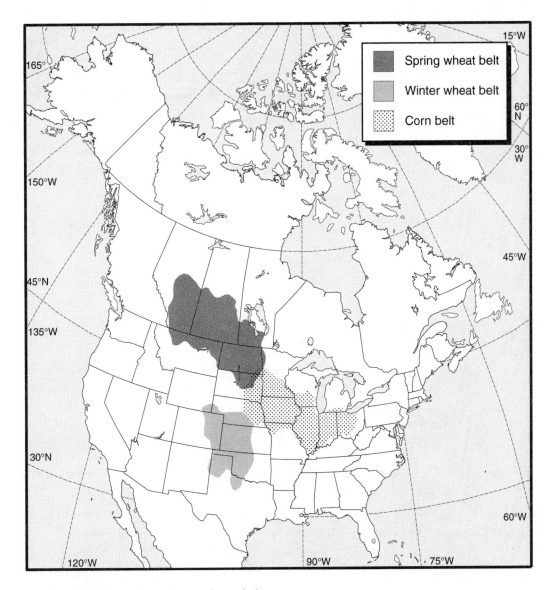

FIG. 9.4 North American wheat and corn belts

indicative of conventional pre-Second World War agriculture. The current revolution has changed the status and nature of this agricultural heartland (Fig. 9.5).

The second region reflects post-Second World War trends in North American agriculture. While encompassing only a single state, California's agriculture system is enormous in size and diversity. The issues facing California farmers mirror concerns in other parts of North America, but are often more advanced. As in other areas, California leads the way.

THE AGRICULTURAL HEARTLAND

The large Interior Plains of North America constitutes an agricultural heartland for the North American region. Blessed with the combination of rich soils, gentle terrain, and

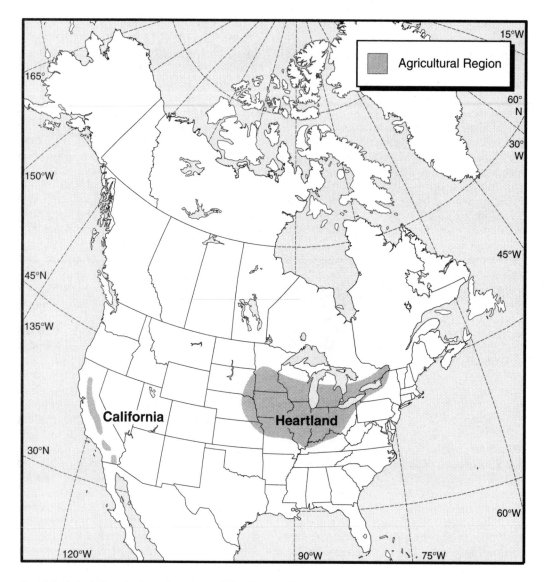

FIG. 9.5 Selected agricultural regions of North America

a favourable climate, it represents North America's largest area of highly-productive agricultural land. In turn, this region is the most important meat and grain producing district in the world. Nestled within the heartland and contributing to its significance are North America's most important corn, soya bean, small grains, dairying, and hog-producing districts.

The eastern portion of the region, commonly called the 'Midwest' in the United States and 'southern Ontario' in Canada, is a landscape dominated by mixed crop and livestock farms. Corn, soya beans, and other small grains are important income generators, while also providing rations for hog and beef herds on the farm. Family-operated farmsteads with highly mechanized and efficient production systems predominate. The popular public image of the modern yeoman farmer is constructed from this model.

The restructuring of North American agriculture has greatly impacted the character of the mixed crop and livestock farms. The number of family farms is dropping through consolidation. The larger corporate farm model is becoming more common. Similarly, more specialized production processes are accompanying this shift.

On the western side of the heartland, the mixed crop and livestock model gives way to specialized grain farming and feedlot operations. This is the drier side of the region and the climate is semi-arid. On this edge of the heartland is the North American wheat belt. Feedlots, especially beef operations, are widespread in Nebraska, Iowa and South Dakota. These operations are designed to fatten animals for market. They take advantage of low cost grain and the proximity of breeding herds further east in the region. The feedlots vary in size and complexity, from small family operated units fattening less than 100 animals annually to massive industrially-structured operations with thousands of animals in the feedlot at a single time. The latter scale increasingly represents the norm in the industry.

California

While total agricultural land use is greater in the heartland region, no single state can match the total value or variety of agricultural output from California. Endowed with fertile soils, climatic diversity, and supplies of irrigation water, California is a veritable cornucopia. The state leads the United States in the production of 75 different crops and commodities. With two per cent of the US farms, California agriculture generates nearly 10 per cent of the nation's gross farm receipts. In 1996 California agriculture generated $24.5 billion in farm receipts and income, including $12 billion in exports. The most important agricultural area is the Central Valley, producing 75 per cent of the farm output. Other farming districts fill particular niches in the agricultural complex. These include Sacramento Valley, San Joaquin Valley, Napa Valley, Salinas Valley, Oxnard Plain, and Imperial Valley. Eight of the ten top agricultural production counties in the US are in California.

Beyond the volume of production, California's farms produce an amazing diversity of agricultural commodities. Often California farming is thought of in terms of vegetables and fruits. Indeed, the state is the leading producer of fresh and processed vegetables and fruits, growing 55 per cent of the US's produce. However, California is also the nation's leading producer of milk and cattle, as well as a major producer of cotton, rice and small grains. The ascendancy of California to the status of agricultural powerhouse is closely tied to ambitious state and federal government irrigation schemes. Water resources in California are hydrologically mismatched; there are water surpluses in the north but water needs in the centre and south. Early private irrigation programmes stimulated agricultural development and created political pressure for publicly-funded irrigation. Today, the federal government's Central Valley Project and the state's California Water Project have reconfigured the distribution of California's water resources system to serve agricultural needs as well as the expansive thirsts of urban areas.

Into the twenty-first century, the growing urbanness of the state represents the greatest threat to California's agricultural economy. Expanding low-density urbanization has increasingly led to tighter controls of normal agricultural operations (e.g. spraying chemical materials, noise and smell complaints, movement of farm equipment on public roads), the alienation of farmland resources and the loss of irrigation water for higher-value urban users. The abundance of California farmland resources and the growing urban-related pressures on the resource base have made Central Valley the most threatened farmland in the United States according to the American Farmland Trust.

AGRI-BUSINESS

Post-Second World War North American agriculture has experienced two divergent transformative processes. Both have challenged traditional representations of the small, family farms nestled within a bucolic rural landscape that has come to represent

Canadian and American agriculture. And, both have raised serious public dialogue and public policy concerns about the future direction that agriculture will take in North America.

The first transformation arose from industrialization and globalization of food production and consumption. American agricultural interests lead the world in this new business-orientated configuration. This reshaping of agriculture followed earlier antecedent processes in North American manufacturing (the Fordist production model) and has become ubiquitous in other Western capitalized nations.

This restructuring of agricultural processes has given rise to an agro-food system in which individual farming units are bound together by a global network of institutions, technologies and products. The structural, technical and economic impacts of this reconfiguration are evidenced both on the farm and beyond the farm gate. Commodity chains, such as the California fresh produce industry, bundle and market agricultural products in the same fashion as manufactured goods. Under an agri-business model of agricultural reorganization, multinational corporate interests have become directly and indirectly involved in the farm production processes. In the former case, agro-business capital controls production from the fields and barns, through manufacturing, to the marketing of the final food product. In the latter case – indirect control – corporate direction is exerted through networks of markets contracts, credit arrangements and technical services to individual farmers or co-operatives. Farmers are told what to grow and how to do it. In return for giving up their independence, farmers are guaranteed a steady income stream.

Beyond the structural effects on the agricultural enterprise, the agri-business model has brought about significant changes in North American agricultural geography. Locational advantage within a food production chain causes shifts in food production, with concomitant losses in other production areas. As the importance of 'off-farm' sectors in the food production system increases, traditional production inputs – labour and land – become marginalized. The current explosion in biotechnologies heralds a further diminishment in traditional locational criteria, with 'super' plants and improved animal strains able to defy climatic and other production-related environmental constraints.

SUSTAINABLE AGRICULTURE

The second theme that has emerged among North American agricultural interests and consumers in the latter portion of the twentieth century is sustainable agriculture. Although consumer alarm over the safety of food products and the environmental impact associated with modern agriculture systems have been simmering issues for several decades, concerted efforts to develop sustainable approaches to agriculture are largely still in the beginning stages.

The North American concept of sustainable agriculture is closely tied to the sustainable development terminology and framework articulated by the World Commission on Environment and Development (1987) and the subsequent Rio Environmental Summit. Currently, both Agriculture Canada and the US Department of Agriculture (USDA) have adopted sustainable agriculture policy statements and initiated programmes to promote sustainably-based agricultural systems. Environmental critics, however, warn that the present policy shift is structurally inadequate with little real change in institutional focus or resource allocation away from the productionist model.

In the United States, the USDA defines sustainable agriculture as an integrated system of agricultural production that will over the long term: satisfy food and fibre needs; enhance environmental quality and the natural resources base; make the most efficient use of non-renewable resources; integrate natural biological cycles and controls into production processes; sustain the economic viability of farms and enhance the quality of life for farmers and the larger society. The achievement of these ambitious goals is framed around three broad research and educational programmes: the utilization of biological systems for pest control, the development of integrated management systems, and technology development and transfer.

Across North America, community activists, environmental organizations, and farmers have increasingly joined together to form research and educational centres and co-operative marketing apparatus promoting sustainable agriculture. Many of these initiatives are grass-roots orientated and evolved from the work of individuals or small groups. British Columbia's City Farmer, for example, is a non-profit society that promotes urban gardening and environmental conservation from a small office in downtown Vancouver. The group has an impressive array of technical reports and a sophisticated web site which foster organic and socially responsible agricultural practices. In the age of Internet communication, the power and capability to diffuse their message is magnified. In some cases, however, research-orientated universities have grafted sustainable agriculture research programmes on to their existing research framework. The Leopold Center for Sustainable Agriculture at Iowa State University and the University of California's Sustainable Agriculture Research and Education Program illustrate the latter cases.

While the total share of North American agriculture output that could qualify as sustainably produced remains low – less than five per cent, it is expanding rapidly and already represents a $3.6 billion segment of the economy. To date, consumer demand for environmentally-friendly (organic) foodstuffs has established a strong niche market in the North American food-processing and marketing industry. Market research indicates that North America's green consumers tend to be better educated, more affluent and female.

CONCLUSION

North American agriculture is a powerful food and commodity production system. As we move into the next millennium, Canadian and US farmers are comfortably positioned to maintain their global status as competitive producers of food products for the con-

sumers around the world. But farming as we know it is changing. Conventional structures and production processes are giving way to new norms. Traditional agricultural geographies are being reconfigured by the same processes.

In the course of change, contradictory movements toward homogenization and heterogeneity in agricultural production are evidenced. The trend towards increased homogenization results from the adaptation of large-scale industrial techniques in agriculture. Under this regime, increasing quantities of standardized foods are produced at decreasing costs to the agro-food industry. Heterogeneity arises from flexible production models. Growing customer and producer attention to issues of the quality of food and the impacts of agricultural production on the environment are linked to sustainable agriculture and organic agricultural practices. How strongly each of these paradigm shifts will affect Canada's and the US's agricultural future is unknown. However, they promise to continue to impact the farms and ranches of the continent.

KEY READINGS

Ilbery, B., Chiotti, Q. and Rickard, T. (eds) 1997: *Agricultural restructuring and sustainability, a geographical perspective*. Wallingford: CAB International.

Opie, J. 1994: *The law of the land: two hundred years of American farmland policy*. Lincoln: University of Nebraska Press.

Sorensen, A.A. and Greene, R.P. 1997: *Farming on the edge*. DeKalb: North Illinois University, American Farmland Trust and Center for Agriculture in the Environment.

10

MANUFACTURING IN NORTH AMERICA

IAIN WALLACE

INTRODUCTION

It used to be easy to describe the geography of manufacturing in North America[1]. Fifty years ago, over two thirds of the continent's manufacturing employment was concentrated in the compact, approximately rectangular region known as 'the Manufacturing Belt'. This stretched westwards from the industrial port cities of the East Coast of the United States (Boston, New York, Philadelphia and Baltimore) to the mid-continental manufacturing hub, Chicago, via large specialized centres such as Pittsburgh (steel) and Detroit (autos). The linear industrial heartland of Canada, stretching from Montréal, through Toronto, to Windsor formed the Manufacturing Belt's northern fringe. Even at the turn of the twenty-first century, that regional concentration of manufacturing capacity remains very evident. In Canada, southern Ontario and Québec account for over 70 per cent of employment in this sector, and the states of the US Manufacturing Belt still contain over 40 per cent of American manufacturing employment. But neither the geography of manufacturing, nor the forces shaping it, can be written as simply today as they could in the early 1950s. We will begin by examining why.

In the first place, there is a definitional problem. 'Manufacturing' is not as clear-cut a category as it once was, and most of the developments which lie behind this have been pioneered in North America. Since the Second World War, industrial corporations have increasingly devolved 'non-manufacturing' tasks (from product design to payroll administration) to specialist suppliers of 'services'. This has tended to exaggerate perceptions of the decline of the manufacturing sector, and it has raised important questions about the degree to which a dynamic 'producer service' sector is functionally dependent on the health of manufacturing firms. A more profound ambiguity has been introduced by the rise of the 'knowledge-based economy', for even manufacturing processes can no longer meaningfully be conceived as involving only the transformation of tangible materials. In particular, 'high technology' manufacturing involves a fuzzy overlap between hardware and software. When the value of a product lies in its embedded computerized 'intelligence', traditional definitions and data categories that were designed in an earlier technological era cease to be really informative. The emergence of biotechnology-based industries compounds this problem.

The definitional problem is of course a symptom of the fundamental processes of structural change in advanced economies. In North America, just as the share of total employment in the once-dominant primary industries has shrunk steadily to a residual 4 per cent, so the share of manufacturing employment has been in decline since the 1950s, as the tertiary and quaternary sectors have dominated job creation. However, this relative decline has taken place against an underlying dynamic of initially rising, and then, since the early 1970s, falling absolute employment in manufacturing. The forces behind this important shift have to do both with the ratio of labour to capital in North American industry – which has fallen as a result of continuous investment in improved technologies – and with the competitiveness of North American manufacturing production in an increasingly industrialized world. Hence the vexed issue of 'de-industrialization' (i.e., manufacturing job loss) needs to be carefully contextualized, the more so as trends in manufacturing employment have been regionally differentiated.

The international dimension of these structural changes is that North America's share of global manufacturing has been in retreat for 50 years. Immediately after the Second World War, the US's dominance of world industrial output was overwhelming. But with the economic revival of Western Europe and Japan during the 1950s and 1960s, then the emergence of the 'newly-industrializing countries' (NICs), particularly the export-orientated Asian economies, in the 1970s and 1980s, and the addition of China's growing industrial output in the 1990s, the supremacy of US manufacturing has been eroded. This process has involved not merely the virtual disappearance of some sectors of production from North America (such as the manufacture of television sets), but also in many instances a loss of US technological leadership to Japanese or European competitors. As a result, interpretations of, and responses to, the changing geography of manufacturing production in the US have invoked both domestic factors and the global scale interests of the American government and of American corporations. In Canada, the rising volume of post-war US investment in the manufacturing sector gave rise in the 1970s to a period of nationalist concern about the weaknesses of a technologically-dependent 'branch plant economy' (Britton and Gilmour, 1978). Since the mid 1980s, debate has centred around the impact of free trade with the US on the viability of manufacturing in Canada. With the inclusion of Mexico in a continental free-trade agreement since 1993, fears of manufacturing job loss to its lower-wage economy have been an issue both in Canada and the US.

This chapter proceeds to analyse forces that have influenced the location of North America's manufacturing industry, both historically and in recent years. It then focuses in more detail on change within specific sectors. The conclusion provides a brief summary of regional shifts.

CONVENTIONAL LOCATION FACTORS

The North American Manufacturing Belt attained its dominant position in the nineteenth century on the basis of a regional conjunction of all the traditional location factors. It supported a reasonably prosperous agricultural sector, it was well endowed with industrial raw materials and energy (notably coal), natural waterways provided cheap bulk transport, and its initial advantage as a region with rapidly growing urban populations (fed by transatlantic immigration) was consolidated into dominance of national markets. Only in California did a concentration of population, markets and resources begin to emerge in the early twentieth century as the basis for a second regional concentration of manufacturing. In Canada, the

nation-building tariff policy of the 1870s was designed to ensure that the manufacturing industries of Ontario and Québec achieved viability in the face of established British and American competitors and consolidated their hold over the domestic market from coast to coast.

Conventional location factors have not lost all their significance in explaining the contemporary geography of manufacturing in North America. Proximity to raw materials remains important for industries processing primary products. This is particularly evident in Canada in the location of new pulp and paper mills in northern Alberta; the expansion and diversification of the petrochemical sector, also in Alberta; and the addition of new aluminium smelters to the existing concentration in the Saguenay and middle St Lawrence valleys in Québec, where cheap hydro power is the attraction. However, the scale of resource consumption in North America and heightened concern over its environmental consequences has brought important changes to the nature and location of resource inputs. The geography of steel production in the United States is increasingly marked by a dispersion away from the traditional locations (and traditional blast furnace technology) of the Great Lakes Basin. This has been made possible by the availability of large quantities of scrap metal that can be processed into a broadening array of products in 'mini-mills' reliant on electric furnaces (see below). Similarly, the strong support (legislated in many jurisdictions) that has developed for newspaper recycling has made large metropolitan areas a major source of fibre for pulp mills, supporting new plants in formerly unlikely locations such as Arizona.

Markets also remain a major locational determinant for many industries supplying final consumers. In this respect, the decline in the dominance of the American Manufacturing Belt since the 1950s is simply a reflection of the steady demographic reshaping of the United States. Population growth in the South and West has outpaced that in the Northeast, creating volumes of demand that have reached thresholds justifying regional production of a widening array of products in growing numbers of metropolitan markets. Internal migration (from the old, 'Rustbelt' industrial regions of the Northeast and Midwest); changing flows of immigration (with Mexico and Asia replacing Europe and the Caribbean as major sources); and the greater locational freedom that many people, both employed and retired, have enjoyed in recent decades to seek places of high amenity are factors that have combined to favour the loosely-described 'Sunbelt'. Other forces, such as the regional distribution of defence and aerospace expenditures, have reinforced the Sunbelt's rising share of American manufacturing output, but the changing geography of purchasing power accounts for much of it. In Canada, although Alberta and British Columbia have been gradually increasing their share of the national population, regional markets are still too small to trigger a marked expansion of manufacturing capacity.

TOWARDS A GEOGRAPHY OF REGIONAL COMPETITIVENESS

In modern advanced economies, world-scale forces interact with national and local conditions in complex ways. The globalization of financial flows, the industrial dominance of transnational corporations, and the steady movement towards freer international trade (globally and continentally) have combined with structural shifts in the economy and society of nations to create new locational variables. For example, in a knowledge-based economy, labour can no longer be viewed simply as a cost to be minimized in favoured locations. Similarly, the characteristics of firms are extremely important. Porter (1990) has captured many features of this new environment in his concept of 'national competitive advantage'. This incorporates a broader spectrum of a nation's human geography than the concerns of traditional industrial location theory. His 'diamond' of four interlocking characteristics comprises:

- *factor conditions*, which include natural resource endowments, localized pools of

skilled labour, and the quality of infrastructure;
- *demand conditions*, particularly the presence of technologically sophisticated firms which challenge their suppliers to be innovative;
- *related and supporting industries*, in particular the existence of national clusters of linked industries which transmit innovations and 'best practice' among themselves; and
- *firm strategy, structure and rivalry*, which captures the importance of management culture, the regulatory environment, and the presence or absence of real competition in the domestic market.

Together with the nature of government policies, and the chance outcomes of historical events that have resulted in acquired advantages or disadvantages, Porter argues that these variables combine powerfully to promote or inhibit successful industrial performance at the national level. Moreover, he recognizes that geographical concentration tends to intensify interactions, accelerating the growth of favoured regions and cumulatively retarding economic growth in regions which are adversely endowed. These dynamics are often most apparent at the scale of the large metropolitan region.

Factor conditions

Compared to Western Europe and Japan, North America is much better endowed, per capita, with natural resources. This is pre-eminently true of Canada, whose economic history until the 1950s could largely be written in terms of its natural resource-based 'staple' industries, including pulp and paper production, metal refining, food processing, and wood products manufacturing (see Table 10.1). Although the United States has lost its former high level of domestic self-sufficiency in natural resources (most notably in oil), its farms, forests and mineral resources continue to support large and technologically-advanced processing industries. Regional shifts in employment primarily reflect changes in the political economy of production (labour–management relations, inter-

TABLE 10.1 Sectoral distribution of manufacturing employment in Canada and the United States, early 1960s and 1995

	Canada		United States	
	1961 %	**1995** %	**1995** %	**1963** %
Food and drink	15.5	12.4	8.1	10.1
Tobacco	0.8	0.3	0.2	0.5
Leather goods	2.4	0.7	0.5	2.0
Textiles	6.5	2.8	3.2	5.3
Clothing	6.9	4.6	5.1	7.9
Timber and wood products	6.1	6.9	4.0	3.5
Furniture and fixtures	2.5	2.9	2.7	2.3
Paper and allied products	7.3	6.1	3.4	3.6
Printing and publishing	5.6	7.3	8.2	5.6
Primary metals	6.6	5.0	3.7	6.9
Metal fabrications	7.5	8.9	7.8	7.0
Machinery	3.7	5.2	10.3	9.0
Transport equipment	7.3	13.4	8.1	9.5
Electical/electronic products	6.6	7.0	8.2	9.3
Scientific instruments	1.0	1.2	4.3	1.9
Stone, clay, glass products	3.2	2.5	2.7	3.5
Petroleum and coal products	1.2	0.7	0.6	0.9
Plastics	na	3.2	4.1	1.0
Chemicals	4.7	4.8	4.5	4.5
Rubber products	1.6	1.4	1.3	1.5
Miscellaneous manufactures	2.9	2.7	2.1	3.9
Total employees ('000)	1353	1715	18760	16232

Notes: na = not available, included elsewhere
Data are essentially, but not absolutely, comparable between years and between countries
The years are displayed in such a way that a direct country-to-country comparison can be made for 1995
Sources: Statistics Canada, *Manufacturing industries of Canada* (annual); US Department of Commerce, *Statistical abstract of the United States* (annual)

corporate competitive strategies) and in technological possibilities (see meat packing, below). Improvements in transport (notably air and road) and communications infrastructures have increased the relative accessibility of the Sunbelt to national markets.

The geography of labour as an industrial location factor can be analysed first in terms of the nature of North American capitalism (or, rather, -isms). The cultural and political climate of the United States has always been more supportive of the unconstrained freedom of private individuals, particularly as owners of property, than has that of Western Europe. Canada has tended to occupy an intermediate position, with higher levels of unionization than the US and a more favourable legislative environment towards the labour movement. In the post-war heyday of the Manufacturing Belt, a predominantly Fordist regime of industrial employment (of which the auto sector was the epitome) supported a powerful union movement that developed a stable relationship with the large oligopolistic firms that dominated the core industries. However, as US corporations began to feel intensified competitive pressures in the 1970s, one strategy they identified to lower costs and outflank the labour movement was to invest in new plants in southern states. Here, 'right to work' legislation and a limited history of industrial employment made it easy to recruit a cheaper and more pliable workforce, particularly for manufactured goods embodying established technologies and demanding a relatively unskilled workforce. One could argue that the industrial migration of the textile and clothing industry which took place from New England to the Carolinas in the early twentieth century came to be replicated on a wider sectoral and geographical scale. The contemporary geography of the US auto industry (see below) shows the degree to which these two 'labour regions' are now linked.

Demand conditions

North America's favourable population/resource ratio and its cultural openness to innovation has led to high average living standards and has fostered markets for technologically advanced products. Post-war suburbanization and the growing participation of women in the labour force stimulated a wide range of manufacturing industries associated with modern consumerism. Antitrust legislation ensured a reasonably competitive environment in US industry, but America's post-war economy, shaped by its Fordist production regime, became characterized by cosy oligopolies in each of the leading manufacturing sectors, replicated in the branch-plant economy of Canada. The comfortable dominance of domestic firms in the North American market was disturbed, however, in the 1970s by the penetration of strongly competitive products from European and Japanese corporations. The experience of the auto industry is recounted below, but similar developments occurred in the capital equipment and consumer products sectors. In response to the new global trading conditions, many US manufacturing corporations sought to reduce their costs by moving production to lower labour-cost settings, notably to Mexico and the Asian NICs. These job losses were disproportionately felt in the former Manufacturing Belt, turned Rustbelt. At the same time, the substantial military–industrial sector, which was protected from foreign competition on strategic grounds and protected internally from price competition by a culture of 'cost plus' contracting, prospered. This benefited primarily its southern and western Sunbelt strongholds (until the downturn in defence and space programme spending in the 1990s). Hence structurally-different demand conditions within the US's manufacturing sector can be seen to have direct geographical correlates.

Related and supporting industries

The size and affluence of the American market, and the global technological dominance of the United States in the 1950s and 1960s, supported clusters of linked industries in many sectors. Many of these were geographically regionalized (for instance, rubber tyres in Akron, Ohio), both within the Manufacturing Belt and beyond it, such as the aerospace complex of greater Los Angeles. The loss of technological leadership by American firms in many sectors since 1970 has 'hollowed out' many of these clusters, however. But the growth of foreign-owned manufacturing in the US economy has seen new clusters developing. This has been particularly evident as 'transplant' Japanese auto firms (see below) have fostered supply chains of leading-edge component manufacturers, regionally concentrated for easy delivery to their assembly plants. The Cold War 'military–industrial complex', which has shrunk but is far from dead in the post-Soviet era, has constituted a huge and highly-variegated cluster of industries, and in this respect has served as the primary agent of 'industrial policy' in a country that otherwise rejects the notion of activist government involvement in industry.

In Canada, as in Western Europe, political support for government intervention to promote the emergence of successful industrial clusters was strong throughout the post-war period until the mid 1980s. The scale of American ownership of Canadian manufacturing and the technological dependence it involved was seen to justify this. But in practice, both at the federal and provincial level, funds have been more often directed to prop up existing manufacturing operations. The political pressure to do so is all the greater when the plants involved are in isolated, single-industry, resource-based towns. Federal government defence spending, minuscule in comparison to that of the United States, has scattered contracts across the country rather than seeking to build up regional clusters of expertise (with contracts for naval frigates, for instance, split between shipyards in New Brunswick and Québec). But the geography of Canadian manufacturing capacity is such that southern Ontario and Québec are always the overall beneficiaries. Government funding has played a role in the emergence of the Ottawa high-tech complex (see below) and has assisted in the expansion of the Montréal-based aerospace industry.

Firm strategy, structure and rivalry

The US economy in the post-war era was the environment which gave substance to the elaboration of Fordism as a theoretical construct. Large firms, benefiting from economies of scale in serving mass markets and providing stable blue-collar employment, were the paradigmatic form of manufacturing industry. But they did not constitute the entire manufacturing sector. In particular, low value-added industries, such as clothing, textiles, leather goods and furniture, were dominated by small firms, operating with lower levels of, and less current, technology and were frequently located in small single-industry towns where unions were more easily kept at bay. This production regime was as typical of southern Appalachia as was the Fordism of the Great Lakes states. Inter-regional and metropolitan/small-town differences in industrial structure and the dominant culture of the workplace remain extremely important, and they are actively exploited by management as a strategy of control. Such activity is less evident (but not non-existent) in Canada, partly because of the stronger presence of the labour movement, and partly because, within a much more restricted national ecumene, there exist far fewer places that are locationally attractive to manufacturers.

Throughout the industrial history of the United States, waves of corporate mergers have tended to mark changes in the underlying geographical structure of the national economy. The period of heightened global competition since the 1980s has seen a new round of takeovers and corporate consolidation. This has been given particular spatial dimensions by the pursuit of continental free trade, involving first the US and Canada (from 1988), and including Mexico also since 1993. The reduction or elimination of tariffs tends to erase the Canada–US border as a locational variable (although the gradually widening gap between the value of the US and the Canadian dollar since the 1970s has had locational consequences: see the auto sector below). Corporate rationalization of manufacturing capacity in the 1990s has particularly hit Canadian branch plants of US firms, concentrated in Toronto and southwestern Ontario, which were typically smaller and often older than comparable operations south of the border.

Having reviewed some of the underlying forces that are shaping the behaviour of manufacturing firms in North America, we will move on to analyse their interactions in the context of specific industrial sectors.

INDUSTRIAL SECTORS

Beef packing

It was the mechanized slaughter houses of Chicago on which Henry Ford based his assembly line technology. And until the 1980s, the geography of the North American beef-packing industry continued to be dominated by large (and increasingly antiquated), unionized plants in major metropolitan areas in the Manufacturing Belt. A combination of factors have finally changed this pattern in recent decades. The suburbanized affluence of post-war Americans brought about major changes in food retailing, notably the rise of the supermarket and the demand for cooking convenience. This gradually undermined the role of the retail butcher, processing large cuts of meat from the packing house, in a distribution system that originated before domestic refrigeration became almost universal. The large oligopolistic firms which grew up in this framework began to be challenged in the 1970s by new competitors which developed a different geography. Their production system was built around large, modern plants located on the Great Plains (close to the beef herds), supplying fully dressed cuts of meat for supermarkets and catering industries (Chapter 9). Their non-unionized plants in small rural towns paid lower wages and enjoyed more flexible working practices than the traditional firms, whose workforce fought and gradually lost the battle to maintain their higher wages (Broadway and Ward, 1990).

In Canada, a similar evolution in the structure and location of the beef-packing industry took place in the 1990s, with an added twist. Canadian transport policy had long subsidized the movement of feedgrain from the Prairies to central and eastern Canada, where feedlots sent finished cattle to the metropolitan packing plants. The removal of these subsidies has made it all the more profitable to locate processing plants close to the livestock in Alberta, where the large new Cargill plant at High River is within a 250 km radius of 80 per cent of the province's beef cattle. The Ontario meat-packing industry has, in contrast, undergone a major downsizing (Chiotti, 1992).

The steel industry

The North American steel industry has undergone drastic restructuring in recent decades. US output in the early 1990s was almost 40 per cent lower than in the early 1970s, and employment over 60 per cent lower. In Canada, output remained relatively stable but it was achieved with a third less labour. Technological change in steel production, and in the markets for steel (the reduced weight of North American autos, for instance), together with shifts in corporate strategy and the competitive environment, account for much of the sector's new configuration.

Although particular locations represented different combinations of advantage (Pittsburgh compared to Chicago, for instance), the historic home of the North American steel industry, in the Great Lakes Basin, combined the advantages of proximity to raw materials and to markets. The huge capital investment demanded by large integrated steel mills and their ancillary infrastructures strongly promoted geographical inertia. Outside the Manufacturing Belt, the location of integrated US steel plants in the early 1970s reflected localized raw material cost advantages (as at Birmingham, Alabama) or the legacy of wartime investment in the 1940s to supply shipyards on the West Coast (as at Fontana, California and Geneva, Utah). In Canada, outside Hamilton, Ontario, home of the majority of domestic capacity, integrated mills existed at Sault Ste Marie, Ontario and Sydney, Nova Scotia (a chronically uneconomic plant in a high-unemployment region, sustained by government subsidy) (see also Chapter 19).

The emergence of scrap-fed electric furnaces ('mini-mills') as an alternative production technology began to change the geography of the steel industry in the 1960s. From being specialist suppliers to small market segments, mini-mills became increasingly versatile and competitive across a range of products. The firms that operated them were free to follow the growth of markets in the

American South and West, and at the same time free of the legacy of obsolescent plants and entrenched labour practices that handicapped the dominant integrated firms (US Steel, Bethlehem Steel, etc.). The mini-mill sector, which now accounts for over 40 per cent of US steel production, is characterized by non-unionized plants, predominantly outside the Manufacturing Belt, and leading firms (e.g. Nucor) that have been notably more profitable than their long-established competitors (Hogan, 1987).

Integrated firms still dominate certain major markets, notably the high-quality sheet steel used by the automobile industry. The major development in the 1990s in this sector has been the growing number of joint ventures between North American and Japanese steel firms, responding to the more stringent quality demands of the automobile industry, led by the Japanese transplants (see below). The location of investment in new facilities has been governed by the 'just-in-time' delivery schedules of the auto producers and hence has been concentrated in the western half of the Manufacturing Belt, in both the United States and Canada. The greater competitiveness of steel imports from Asia, together with the decline of auto production west of the Rockies, has resulted in a proportionately greater shrinkage of crude steel capacity in the American West than in the East.

The automobile sector

The modern geography of the North American auto industry has been shaped by three decisive events that have involved the close alignment of corporate and political interests (Holmes, 1992; Mair et al., 1988). First, the Auto Pact (1965), involved the elimination by the Canadian and US governments of cross-border tariffs on automobiles and original parts manufactured in North America. It also incorporated guarantees by the 'Big Four' North American auto manufacturers (General Motors, Ford, Chrysler, and American Motors) to maintain production levels in Canada proportionate to the volume of their sales in the Canadian market. As the Canadian automobile industry consisted almost

entirely of branch plants of the 'Big Four', the pact allowed the continental-scale rationalization of production of each model of car, and specifically eliminated the multiple high-cost, short production runs of the firms' Canadian plants. The industry remained centred on Detroit and adjacent parts of the Manufacturing Belt, but assembly plants were opened in the US South and West as their regional markets grew rapidly.

The second event, emerging in the early 1970s, was the arrival of Japanese imports as major threats to the market dominance and profitability of the US auto firms. The greater fuel economy and frequently better quality of Japanese products made them increasingly popular (at a time of increased cost and insecurity of US oil supplies), and they rapidly claimed a substantial share of the lower-priced car market. This threat to auto industry employment in North America, and to the US trade balance, created a powerful coalition of interest among governments, labour, and the US auto firms. The United States (with Canada as a minor partner) successfully pressured the Japanese government and auto makers to accept 'voluntary' export restraints. But as the Japanese responded by switching their attention to higher-margin luxury cars, this did not resolve the ongoing friction over trade. By the early 1980s, leading Japanese auto firms made the strategic decision to start manufacturing in North America, as the only long-term guarantee of maintaining a secure presence in the continental market.

In establishing their 'transplant' manufacturing operations, the Japanese aimed to implement the 'lean' and flexible production systems they had pioneered at home. This implied making a clean break from the strongly Fordist labour regime and generally adversarial climate of labour–management relations that typified the Detroit-based industry. Hence the Japanese firms chose assembly plant locations that allowed them to mould a labour force to their own specifications, meaning in localities without a tradition of unionized manufacturing and at some distance from the strongholds of the UAW (United Auto Workers). Smyrna, Tennessee, Georgetown, Kentucky, and Alliston, Ontario have these characteristics in common, but also that they are strategically located along freeway corridors focused on Detroit and within a few hours' driving time of most of their component suppliers (Fig. 10.1).

The third event has been the gradual incorporation of Mexico into 'North America', culminating in its partnership in the North American Free Trade Agreement (NAFTA) since 1993. For the automobile industry, Mexico has grown in importance as a site where labour costs are significantly lower than north of the Rio Grande, but where production of increasing skill-intensity can be undertaken. The *maquiladora* border industrial zones, established in the 1960s, have allowed US corporations to send components to Mexico to undergo labour-intensive operations and to pay duty only on the wage content of the returned products. So auto wiring assemblies, for instance, are now heavily sourced there. But by the early 1980s, partly in response to its debt crisis, the Mexi-

can government required all the major auto companies selling in Mexico to achieve a trade balance on their operations. This led to significant investment in vehicle assembly and parts plants by US, Japanese and European firms, including large-scale engine plants that proved to be a match in quality and efficiency with those in the US. With the greater access to US and Canadian markets that Mexico enjoys under NAFTA, it would not be surprising if the geography of auto assembly begins to shift. To date, lower labour costs in Canada (reflecting lower health insurance costs and a weaker dollar) have encouraged US auto makers to maintain assembly plants there, even as they have closed plants in the US (note the contrasted trends in auto industry employment in Table 10.1). And for the remaining big three American auto firms, the balance between Canadian and US operations is still governed by the Auto Pact. But the Japanese and European firms are not party to this agreement

□	Japanese or Japanese-joint-venture assembly point
•	Japanese or Japanese-joint-venture automotive supplier
⟍	Key interstate highways

0   km   250
0   miles   150

FIG. 10.1 North American automobile assembly and supplier manufacturing plants of Japanese firms, 1988

and have greater freedom of action. The new Volkswagen Beetle, for instance, is produced in Mexico for the North American market.

High-technology industries

By the close of the twentieth century, the term 'high-technology' has lost much of its definitional value. Increasingly, all manufacturing industries in North America employ advanced technologies. Laser scanners calculate the optimum cut of logs at sawmills in British Columbia; auto body parts are hydroformed (using very high-pressure water jets) in Ohio; and relatively ubiquitous large printing plants are full of computerized equipment. In all these examples, the software is the key to controlling the production process, and the critical labourforce skillsets require advanced levels of technological literacy, which is often hard to find (or to retain) outside metropolitan regions. However, a conventionally narrower definition of high-technology manufacturing encompasses the microelectronics and computer sectors, telecommunications equipment, machine and instrument making, and the aerospace sector. These are all industries that benefited heavily from US military spending during the Cold War.

Microelectronics and computers

The origins of this high-technology sector lie in the northeast United States, where the major electronics firms of the Second World War era were based. The first noticeable cluster of electronic and computing equipment manufacturers emerged in the western suburbs of Boston, along Route 128, in the 1950s. It drew on the expertise of the city's universities, especially the Massachusetts Institute of Technology (MIT), and on New England's long tradition of industrial innovation. By the mid 1970s, however, Route 128 had been overtaken in terms of high-tech employment by a second regional cluster of electronics/computing in Santa Clara county, 'Silicon Valley', near San Francisco. Its origins also had a strong university component, in the academic entrepreneurship of a Stanford engineering professor who, from the late 1930s, encouraged students to commercialize

their research. Among the earliest to do so were the founders of Hewlett-Packard.

The different growth patterns of high-tech industry in Route 128 and Silicon Valley have been analysed by Saxenian (1996) in terms of contrasted regional business cultures. Although engaged in innovative technologies, the firms of the Boston area tended to replicate the structure and practices of established East Coast firms. Traditions of corporate secrecy and self-sufficiency remained strong, as did tendencies towards hierarchical management. Relatively conservative community values tended to favour company loyalty and employment stability. As firms such as Digital Equipment grew in size, risk taking became less favoured. In contrast, the business environment in Silicon Valley, which until the 1940s was an agricultural area with no industrial tradition, was characterized from the outset by greater fluidity and openness. Most people in the high-tech sector were new to the region: without established ties, they collaborated professionally and networked socially. As new firms were established, it became standard practice for people to switch firms or to start their own. Risk taking was part of the regional culture, not least among venture capitalists, whose willingness to back innovative entrepreneurs contrasted with the much more conservative lending practices of the New England bankers.

The ascendancy of Silicon Valley as the US's premier region for computer-related technologies was consolidated in the early 1980s. The leading firms of the Boston area, having successfully developed the minicomputers of the 1970s, with proprietary technologies and software that tied their customers to them (following the pattern of IBM's mainframe computers), failed to recognize the revolutionary impact of the personal computer, both as a low-cost competitor and as the basis for an industry built around shared technological standards. Rather than relying on vertical integration, the computer firms of Silicon Valley relied on networks of component suppliers and were quick to take advantage of the lower costs of microprocessor assembly available in Asia. Innovation also diffused more quickly in this environ-

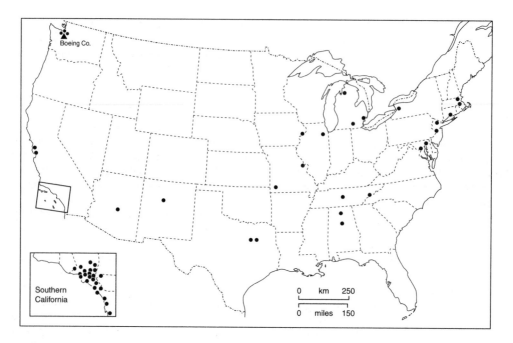

Fig. 10.2 First-tier subcontractors for the Boeing Company for the Saturn V Space programme, 1962–73
Source: After Scott (1993), Fig. 6.4

ment, and specialized small firms readily gained market access.

Other localizations of computer-related high technology employment have emerged in North America in the past 20 years, showing differing patterns of origin and development (Bathelt, 1991). Canada's 'Silicon Valley North' has grown in the Ottawa area out of the work of federal government laboratories and a major corporate laboratory (Nortel), both of which have acted as 'incubators' of new firm formation. North Carolina's 'Research Triangle' emerged in the 1960s from the collaborative efforts of state and local governments and regional universities to foster high-technology employment in a depressed region. Its initial dependence on attracting research facilities of large firms based elsewhere has been lessened, but the region has been less of an incubator than Boston or Ottawa. Microsoft, based near Seattle, is by far the largest of a number of computer firms located in cities (such as Boulder, Colorado and Orem, Utah) in the western United States marked by high amenity environments and the presence of significant university or defence-related research capacity.

Aerospace

The North American aerospace industry has been the principal beneficiary of US defence spending since 1945. This has powered its growth, but at the cost of vulnerability to changing strategic and budgetary priorities. Such significant events as the end of the Vietnam War, the collapse of state communism in Eastern Europe and Russia, and the cancellation of particular space programmes have resulted in declines in employment, plant closures and corporate rationalization. The geography of the industry was established during the Second World War, with massive growth in the aircraft industry in southern California and expansion of the aero-engine and related engineering industries of New England and other parts of the Manufacturing Belt. From the 1960s, the US space programme supported expansion in the West and across the South, from Texas to Florida. Major defence and space projects have involved complex patterns of inter-regional subcontracting by the prime contractor (Fig. 10.2). Canada's very much smaller industry is overwhelmingly concentrated in the Montréal and Toronto metropolitan areas.

CONCLUSION

The challenges of defining and measuring the importance of manufacturing in contemporary North America, which were analysed at the beginning of this chapter, return as one attempts to interpret Fig. 10.3. This county-scale map of earnings from manufacturing in the United States reveals many things and hides others. It shows how extensively an above-average dependence on industrial employment has deconcentrated away from the Manufacturing Belt. It is now found over a large and continuous area comprising most of the eastern and central US, including many rural counties quite distant from metropolitan areas. Among the areas of greatest reliance on manufacturing earnings there is considerable variation in industrial structure. The industrial heartland of Ohio, Michigan and Indiana retains its strength in engineering and 'metal-bashing', much of it linked to the auto industry; single-industry communities in Maine and the Pacific Northwest are primarily dependent on forest products ; a swathe of counties from the Carolinas to Arkansas rely on branch plants in a whole range of industrial sectors, attracted in large part by local and regional conditions that enhance management control of the workplace.

But the data mapped in Fig. 10.3 tell only part of the story. The relative importance of manufacturing has diminished with the growth of the service sector especially in large metropolitan areas. New York and Chicago are not prominent on this map, and California as a whole has few counties with an above-average share of earnings from manufacturing. Yet in 1994, California was

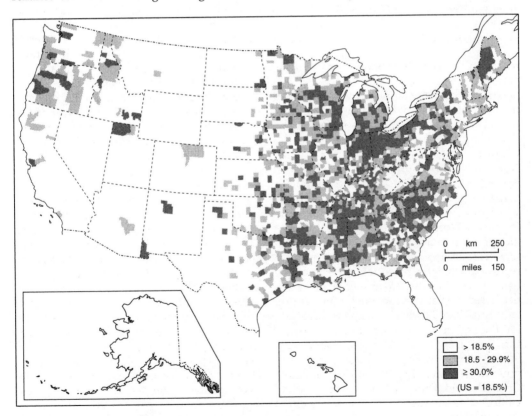

FIG. 10.3 Earnings from manufacturing by county, 1995

the US's leading manufacturing state by all measures, with over 10 per cent of national employment and value added, and almost 10 per cent of shipments. Moreover, as Scott (1993) argues, the southern California metropolis stretching from Santa Barbara, through Los Angeles, to San Diego has become the continent's largest and most dynamic industrial complex. Its growth has been powered not only by the defence and high-technology sectors noted above, but also by low-technology, sweat-shop industries. These have thrived on its influx of immigrants (including undocumented arrivals from Mexico), who have constituted a low-paid workforce for production that would otherwise have been sourced in developing countries (see Chapter 15). In sum, California is a reminder that even if North America is portrayed as a 'post-industrial' society, the state that is often taken to represent the leading edge of cultural and social change is home to a diversified and thriving manufacturing sector.

ENDNOTE

1. In this chapter, 'America' and 'the US' are used interchangeably, as are 'manufacturing' and 'industrial'. In both cases, the advantages of stylistic flexibility outweigh the slight possibility of definitional ambiguity. 'North America' excludes Mexico, except where it is included as a party to the 'North American Free Trade Agreement'.

KEY READINGS

Britton, J.N.H. (ed.) 1996: *Canada and the global economy: the geography of structural and technological change.* Montréal: McGill-Queen's University Press.

Holmes, J. 1992: The continental integration of the North American automobile industry: from the Auto Pact to the FTA and beyond. *Environment and Planning A* **24**, 95–119.

Noponen, H., Graham, J. and Markusen, A.R. 1993: *Trading industries, trading regions: international trade, American industry, and regional economic development.* New York: Guilford Press.

Norcliffe, G. 1994: Regional labour market adjustments in a period of structural transformation: an assessment of the Canadian case. *The Canadian Geographer* **43**, 2–17.

Storper, M. and Walker, R. 1989: *The capitalist imperative: territory, technology, and industrial growth.* New York: Blackwell.

11

THE SERVICE ECONOMY

WILLIAM B. BEYERS

INTRODUCTION

The economies of the United States and Canada are dominated by employment in services, and the majority of gross domestic product in both countries stems from service industries. The economies of these two countries have been dominated by services for over half a century, and the share of service employment continues to rise. This chapter documents recent trends in service employment. The first section develops a national perspective on change, it addresses reasons why service employment has expanded in relative importance in these economies. Then dynamics in the geographical distribution of services among states and provinces are described. This section is followed by a discussion of location factors and the basis of demand for services, including their role in inter-regional and international trade. The chapter concludes with comments about the future of services in Canada and the US.

Although the primary focus of this chapter is on service economies in Canada and the United States, the analysis is presented in the context of overall employment in the two countries. The reason why I have chosen to do this is to present a balanced perspective on the service economy within the larger economy into which it is embedded.

GROWTH AND CHANGE IN THE SERVICES

Service employment in the United States and Canada, as in other advanced economies, is composed of a broad array of industries, not a simple category labelled services, as documented by Illeris (1996). Each country measures the composition of the service economy in a somewhat different manner. However, we can broadly group services into categories according to their function in the economy, and around the markets they serve, as proposed by Browning and Singelmann (1975). Tables 11.1 and 11.2 provide an overview of trends in employment in the US and Canada over recent 20-year time periods. These tables provide us with several key per-

spectives on the role of services and the growth of service employment in the development of these two countries.

While the data in Tables 11.1 and 11.2 are not for identical years, the broad trends are similar. First, employment within goods producing sectors has remained almost constant over the time periods in Tables 11.1 and 11.2[1]. Second, employment in services has grown rapidly, accounting for 92 per cent of job growth in the United States and 95 per cent of job growth in Canada. Third, employment growth rates within the services have been differentiated, with sectors such as producer and health services growing rapidly, while other sectors such as transport and public utilities have experienced slower growth.

Before proceeding further, let us define the nature of these various service industries more clearly. Table 11.3 defines typical industries within the broad services categories contained in Tables 11.1 and 11. 2. Table 11.3 also describes in a generalized way the markets served by different lines of services, making it clear that the bases of demand for services vary among sectors. This table is based on the US definition of services, and does not match perfectly the Canadian definitions. The industrial components described in Table 11.3 are not exhaustive; rather they are intended to be examples of the types of industrial activity found in these industrial groupings. Readers may study Table 11.3 and discover for themselves the rich variety of industries included in the service economy. The complex set of industries and the diversity of firms within each of these industries calls attention to the importance of functions which have emerged in the modern service economy, and indicates the variety of market-types served.

Why have services expanded so rapidly?

The rapid increase in aggregate service employment has been the subject of considerable controversy, as wage levels in some sectors such as retailing are well below wage levels in manufacturing, while other services such as computers and security brokerages have high wages. This bifurcated pay struc-

TABLE 11.1 US employment change, 1975–95

Sector	1975 ('000)	1995 ('000)	Change	% change
Farm employment	3948.0	2984.0	−964.0	−24.4
Agricultural services, forestry and fishing	659.3	1821.9	1162.6	176.3
Mining	875.9	922.0	46.1	5.3
Construction	4664.3	7649.6	2985.3	64.0
Manufacturing	18654.7	19225.9	571.2	3.1
Transport and public utilities	4981.3	7079.7	2098.4	42.1
Wholesale trade	4871.0	6953.5	2082.5	42.8
Retail trade	15165.4	25181.3	10015.9	66.0
Finance, insurance and real estate	7629.2	11088.6	3459.4	45.3
Producer services	4665.4	13650.7	8985.3	192.6
Health services	6272.5	13803.3	7530.8	120.1
Other services	9304.2	17319.6	8015.4	86.1
Federal civilian	2912.0	2976.0	64.0	2.2
Military	2656.0	2234.0	−422.0	−15.9
State and local government	11940.0	16400.0	4460.0	37.4
Total employment	99199.2	149290.1	50090.9	50.5

Note: Service sectors shown in bold
Source: US Regional Economic Information System and US County Business Patterns

TABLE 11.2 Canadian employment change, 1977–96

Sector	1977 ('000)	1996 ('000)	Change	% change
Agriculture	473.3	453.3	−20.0	−4.2
Fishing	20.1	35.6	15.5	77.1
Logging and forestry	71.4	75.9	4.5	6.3
Mining	155.9	168.3	12.4	8.0
Utilities	111.8	147.0	35.2	31.5
Manufacturing	1951.0	2082.5	131.5	6.7
Construction	650.8	718.6	67.8	10.4
Transport services	737.1	872.6	135.5	18.4
Trade	1733.4	2361.2	627.8	36.2
Finance, insurance and real estate	554.8	799.9	245.1	44.2
Education	716.9	929.0	212.1	29.6
Health and social services	796.2	1425.7	629.5	79.1
Business and personal services	1281.7	2786.4	1504.7	117.4
Public administration	723.8	820.1	96.3	13.3
Total employment	9978.2	13676.1	3697.9	37.1

Note: Service sectors shown in bold
Source: Statistics Canada, Employment, Earnings and Hours (monthly)

TABLE 11.3 Components of the service economy

Service sector	Industrial components	Sectoral markets
Transport, communications and public utilities	Truck, rail, passenger, air and waterborne transport services; services to transport; telephone, radio and television; water, gas, sewer, and other utilities	Mixed consumer, government and intermediate markets
Wholesale trade	Merchant wholesalers specializing in commodity lines such as groceries or fabricated metal products	Retailers, construction and manufacturing, agriculture and service industries
Retail trade	Retail lines such as groceries, furniture, clothing, automobiles, as well as eating and drinking establishments	Primarily household consumers
Financial, real estate and insurance	Banks and non-depository financial institutions such as mortgage lenders and security brokers; insurance carriers and agents; real estate agents, holding companies	Mixed consumer, government and intermediate markets
Producer services	Advertising, computer, legal temporary help, engineering and architecture, management consulting and public relations, accounting, research and development	Primarily intermediate markets, but also some government and household markets
Health services	Hospitals, speciality clinics, offices of physicians and dentists	Households
Other services	Hotels and other accommodations; repair services; educational and entertainment; social services	Primarily households, but also business markets
Public services	Local and state/provincial government services including education, regulatory agencies, utilities, social service agencies; federal civilian and military agencies	Primarily households, but also business support

ture is not the only issue related to strong growth in services; there are concerns surrounding relative rates of productivity improvement (Baumol, 1967), the role of services in the economic base of communities (Thurow, 1996), and trends towards downsizing, re-engineering, and outsourcing discussed by Reich (1992), Tapscott (1996), and Rifkin (1995). The bases of growth for the service economy are multiple, and should be viewed along sectoral lines. Let us briefly summarize forces explaining the relative expansion of this sector in North American economies. First, part of the expansion is just proportional, as the aggregate economy has expanded there has been a proportionate expansion in services employment to serve growing populations. Sectors such as retailing and financial services exhibit growth rates similar to aggregate growth (see Tables 11.1 and 11.2), although within these lines of services there are segments which have grown faster or slower than the aggregate (such as the rapid growth of eating and drinking establishments or security brokerages). Second, while it is very difficult to measure productivity in services, there is a general feeling that services productivity has lagged that of the goods-producing sector. This is a matter over which there is disagreement, and it is easy to identify examples of service sectors with rapid productivity improvements (such as telecommunications or banking with the automation of many functions previously handled by human labour), as discussed by Kutscher (1988) and Tschetter (1987). Third, technological innovations have changed the way in which many services are produced, and have also led to the invention of new services. Thus, there has been an ongoing changing division of labour within the services, and simultaneously the marketplace adoption of new lines of services (Beyers and Lindahl, 1996a). The expanding demand for and the supply of computer software is a good example of an innovative service sector. The relatively rapid growth of the producer services has been explained by this combination of innovation on the supply side, and a growing demand for specialized information (Beyers and Lindahl, 1996a; Quinn, 1992; Tschetter, 1987).

While it is popular to regard much of the growth of the producer services as the result of outsourcing and downsizing on the part of clients, this force appears to be weak compared to the increasing need for ever more specialized expertise by clients spread across all industries, the public sector, and in some cases by households (Illeris 1996; Beyers and Lindahl, 1996a). The relatively rapid growth of health services is related to demographic shifts – a larger share of elderly population with relatively strong health care demands – as well as to improved knowledge of diseases and medical treatments. From a regional standpoint the growth of services employment may also be related to the trade of service activity to other regions. Examples include financial service capitals such as New York City, New York or Charlotte, North Carolina, entertainment centres such as Las Vegas, Nevada or Honolulu, Hawaii, government centres such as Washington, DC or Ottawa or the state and provincial capitals, specialized health care centres such as Rochester, Minnesota, and university-based towns such as Hanover, New Hampshire or Athens, Georgia.

Each of the factors just described is responsible for the collective growth of the services, as well as for the patterns of growth in particular regions. Before turning to the location of particular service industries, let us examine the aggregate pattern of employment in particular service industries among regions. One way of summarizing the distribution of industries across a set of regions is through the use of coefficients of industrial concentration. Tables 11.4 and 11.5 present these measures for the United States and Canada; in computing this index the distribution of a particular sector is compared to the overall distribution of employment.[2] If the industry is distributed in the same proportions among the regions as all industries, then the index will be zero. Indices which are relatively high are indicative of industries with a sharply uneven distribution across regions. States and provinces have been used to compute the indices in Tables 11.4 and 11.5. Employment in all industries is included, to provide comparative measures between services and other sectors. In both

TABLE 11.4 Coefficients of industrial concentration, United States, 1975 and 1995

Industry	1975	1995	Change
Farm employment	0.544	0.521	−0.023
Agricultural services, forestry, fishing	0.460	0.312	−0.148
Mining	0.916	0.959	0.042
Construction	0.176	0.136	−0.040
Manufacturing	0.264	0.265	0.000
Transport and public utilities	0.107	0.101	−0.006
Wholesale trade	0.124	0.091	−0.034
Retail trade	0.061	0.052	−0.010
Finance, insurance and real estate	0.151	0.154	0.002
Producer services	0.251	0.186	−0.065
Health services	0.150	0.134	−0.016
Other services	0.115	0.171	0.056
Federal civilian	0.377	0.335	−0.042
Military	0.502	0.450	−0.052
State and local	0.073	0.078	0.005

TABLE 11.5 Coefficients of industrial concentration, Canada, 1977 and 1996

Industry	1977	1996	Change
Agriculture	0.647	0.568	−0.080
Other primary industries	0.460	0.692	0.233
Utilities	0.149	0.158	0.008
Manufacturing	0.267	0.257	−0.010
Construction	0.139	0.149	0.011
Transport services	0.139	0.086	−0.052
Trade	0.042	0.017	−0.025
Finance, insurance and real estate	0.079	0.094	0.015
Education, health and social services	0.074	0.067	−0.007
Business and personal services	0.065	0.068	0.002
Public administration	0.048	0.102	0.054

Canada and the US the trade sector is the most evenly distributed, followed in the US by state and local government employment. Most US service sectors exhibit declining indices between 1975 and 1995, which means that most sectors became more dispersed in their pattern of location. Federal civilian and military employment are relatively unevenly distributed, compared to other services. The producer services were relatively concentrated in 1975, and the index declines considerably for 1995, indicating relatively rapid

growth in regions lacking this sector in 1975.

The sectoring plan used in Table 11.5 differs somewhat from Table 11.4, and producer services are not identified separately. The indices for the various services are somewhat lower for Canada than is the case in the United States, suggesting that their distribution is relatively even among the provinces. A major conclusion that can be drawn from these coefficients of industrial concentration is that primary and secondary sectors exhibit a very uneven distribution in both countries when compared to services.

An alternative perspective on the distribution of sectors among the regions of Canada and the United States may be obtained by focusing on the aggregate structure of each region compared to a national benchmark. Coefficients of regional specialization are statistics measuring these patterns. Fig. 11.1a presents coefficients of regional specialization for the states of the US, while Table 11.6 presents these measures for Canadian provinces. The lower the index the closer the structure of the region is to the nation.[3] The coefficients in Tables 11.6 show a general tendency to decline, which means that regional industrial structures of Canadian provinces have become more like the nation. Fig. 11.1b presents changes in coefficients of regional specialization for US states, and all but 10 states exhibit decreases in these coefficients over the 1975–95 time period. Why? It is because the growth in employment in services, which Tables 11.1 and 11.2 clearly document at the national scale, has been spatially extensive and broadly distributed among the states. Relatively small states or provinces (as measured by population) such as Wyoming or Newfoundland tend to have high indices, reflecting their relatively narrow industrial base, as is also the case for Washington, DC. In contrast the most populous states tend to have low indices, reflecting the diversity of their economies. The pervasive decrease in the coefficients of regional specialization indicates that service-dominated growth in the US and Canada has been geographically widespread.

Although service-dominated employment growth has been geographically widespread in the United States and Canada, the coefficients of industrial concentration presented in Table 11.4 for the US indicate relatively uneven distributions for several service industry categories, including the rapidly growing producer services. Let us examine the distribution of several of these sectors that are unevenly distributed with maps of location quotients. Location quotients are index numbers, which compare the share of employment in a state with the share of employment in the same industry in the nation. For example, in New York State, 14.4 per cent of employment is in finance, insurance, and real estate, while nationally this sector accounts for 7.4 per cent of jobs. Dividing New York's employment percentage by the national percentage yields a location quotient of 1.94, indicating that New York has roughly double the national percentage of employment in finance, insurance, and real estate. Figs 11.2a and 11.2b portray the location quotients for four relatively unevenly distributed service industries.

Fig. 11.2a presents location quotients for the producer services. Only 14 of the 50 states and the District of Columbia have values greater than 1.0. These are generally the nation's most populous states, and they reflect the concentration of this sector in

TABLE 11.6 Coefficients of regional specialization, Canadian provinces, 1977 and 1996

Province	1977	1996	Change
Newfoundland	0.341	0.383	0.042
Prince Edward Island	0.306	0.239	−0.067
Nova Scotia	0.195	0.164	−0.030
New Brunswick	0.214	0.158	−0.055
Québec	0.087	0.092	0.005
Ontario	0.119	0.089	−0.029
Manitoba	0.205	0.171	−0.034
Saskatchewan	0.413	0.315	−0.098
Alberta	0.258	0.212	−0.046
British Columbia	0.144	0.130	−0.014

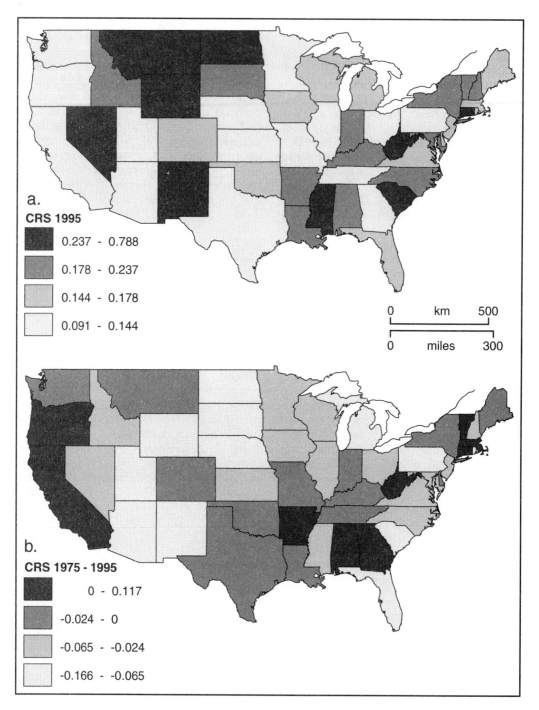

Fig. 11.1 (a) United States: coefficients of regional specialization, US states, 1995 (b) change in coefficients of regional specialization, United States, 1975–95

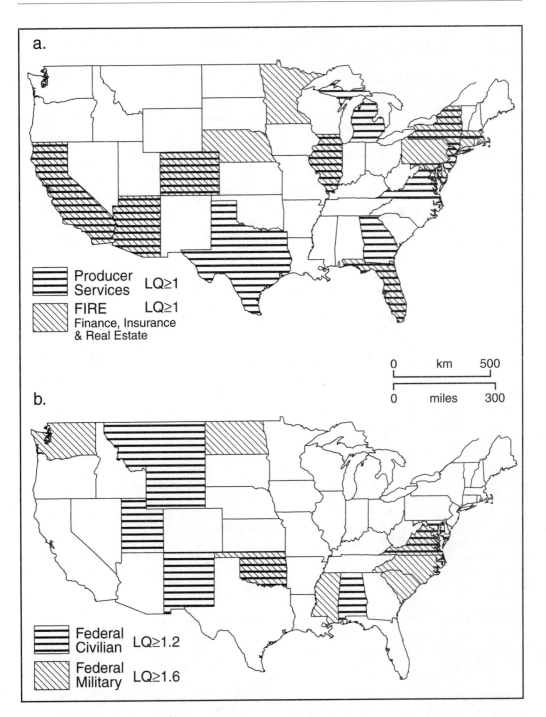

Fig. 11.2 (a) United States: location quotients, producer services and FIRE (finance, insurance and real estates) employment, US states, 1995 (b) United States: location quotients, federal civilian and federal military employment, US states, 1995

major metropolitan areas, as documented for an earlier time period by Beyers (1991, 1992). States with a strong agricultural economy (such as in the upper Midwest) tend to have a small share of employment in producer services. While the producer services are unevenly distributed, they are deconcentrating and experiencing relatively rapid growth in many smaller metropolitan areas, and their growth rate is as rapid in non-metropolitan areas as it is in metropolitan areas (Beyers, 1998).

Fig. 11.2a also illustrates the pattern of location quotients for financial services (FIRE). This pattern is similar to that for producer services, but several centres of insurance services put states into the group with high location quotients for finance, insurance, and real estate (Nebraska, Minnesota, Connecticut). Federal civilian employment is also relatively unevenly distributed, as mapped on Fig. 11.2b. This figure depicts two distinct concentrations of federal civilian employment down the Rocky Mountains and in the greater Washington, DC region. The concentration in the Rocky Mountain region reflects the large area of federally-managed land in these states, with a strong presence of employees in federal land management agencies in the Departments of Agriculture and Interior, as well as installations of the Department of Energy. Additionally, Fig. 11.2b illustrates the pattern for federal military, and here the historic concentration of this sector in the Southeast is evident, along with the perimeter states of North Dakota and Washington.

Dynamics in the location of services

The expansion of service employment in the United States and Canada has not been just proportional across the states and provinces, as some regions have experienced relatively rapid growth, and others have grown slower than national growth rates. A useful method for illustrating these differential trends is through the use of shift-share analysis. Shift-share analysis uses regions and industries to describe patterns of change, separating expected growth in each region and industry (share) from shifts which are the differences between expected and actual change. This technique adjusts the shares by industry-specific growth rates, to develop estimates of expected change which take into account differences in national industry growth rates. The difference between these expected changes and actual changes are referred to as competitive shifts, highlighting changes which are not expected (Stevens and Moore, 1980). An example of this methodology is contained in Table 11.7, which describes the shift-share components for producer services in the state of North Carolina. There were 65 143 producer service jobs in 1975 in North Carolina, and 301 481 producer service jobs in 1995. If growth occurred at the overall national rate, North Carolina would have added 32 894 producer services jobs (its share). However, this sector grew very rapidly nationally (it grew over and above the national growth rate by 142.1 per cent) leading to an expected additional number of jobs in North Carolina of 92 568 (industry-mix shift). However, producer services

TABLE 11.7 North Carolina producer services shift-share components

1975 Jobs	65 143	
1995 Jobs	301 481	
Share	32 894	50.5% = US overall growth rate × 1975 jobs
Industry-mix shift	92 568	142.1% = Producer services − US growth rate × 1975 jobs
Competitive shift	110 875	Residual = change − share − industry-mix shift
Change	236 338	

growth in North Carolina outstripped the national growth rate, gaining another 110 875 jobs (competitive shift). This North Carolina illustration for producer services could be repeated for each industry in North Carolina, and the sum of all the shift-share components across industries would yield the state's overall shift measures. Repeating these calculations for each state, we can develop a national view of relative growth, with some states growing faster than the average, and others lagging.

In the United States the aggregate competitive shift was +/− 10.142 million between 1975 and 1995, or approximately 20 per cent of total growth, while for Canada the figure was +/− 0.516 million, or about 14 per cent of total growth for the 1977–96 time period. As is commonly the case in shift-share analyses, the industry mix adjustment was relatively small compared to the competitive shift component, being +/− 2.068 million for the US, and +/− 0.033 million for Canada. The interpretation we can make of these data is that growth (or lack of it) in particular states *was not* driven primarily by the differential growth rates of industries, but rather by *shifts* in the geographical location of overall employment.

In Canada three provinces exhibited positive competitive shifts: Ontario (small positive), Alberta, and British Columbia, with the latter two provinces accounting for almost all of the positive shifts. On the negative side, Québec had the largest negative competitive shift, accounting for over two thirds of the negative competitive shift. The largest sectoral contributors to the negative competitive shift in Québec were business and personal services, followed by education, health and social services, and trade. In Alberta the principal contributors to the positive competitive shift were business and personal services, followed by manufacturing and other primary industries. In British Columbia the principal contributors to the positive competitive shift were business and personal services, education, health and social services, and then trade.

In the United States the pattern of competitive shifts is very clear, as illustrated in Fig. 11.3. The West, Texas, Tennessee, and the Atlantic coastal Southeast show strong positive shifts, with more modest positive shifts in upper New England, the Rocky Mountains (except Montana), and in Minnesota, South Dakota, Kentucky, and Delaware. Negative shifts are evident across much of the old Industrial Belt and the Great Plains, and down the Mississippi River valley. Table 11.8 indicates the principal sectoral contributors to the positive and negative net shifts for the states with the largest competitive shifts. The role of manufacturing as a leading sector in the economic base of these state economies is evident in Table 11.8; all but one of the big negative competitive shift states were led by manufacturing, meaning a decline in manufacturing employment. For example, New York State manufacturing employment fell from 1.45 million to 0.98 million jobs between 1975 and 1995. In each state with big negative competitive shifts retail was also a major component, linked via negative multiplier effects to manufacturing job losses. New York is the sole exception, where producer services had a larger negative shift than manufacturing – in this case only rising by 0.35 million jobs instead of the expected increase of 1.25 million jobs. Manufacturing and other services (which includes hotels, amusement and entertainment services, and motion pictures) were the leading sectors in the regions with large positive shifts. In the case of Nevada manufacturing is absent – the driving force has been the growth of gambling and entertainment/tourism. The leading role of other services in Colorado is also related to tourism, while in California it is linked to tourism and the multimedia/entertainment complex, including motion picture and television production industries, as discussed by Pollard and Storper (1996). While not reported in Table 11.8, it should be noted that in most cases, the other service sectors also had positive competitive shifts in the states with positive shifts listed in Table 11.8. While manufacturing has often been considered the leading sector in the economic base of regional economies in the US, the data in Table 11.8 suggest that services are now playing a larger role in the economic base. This role has been documented in a number of studies at the metropolitan scale (Beyers and

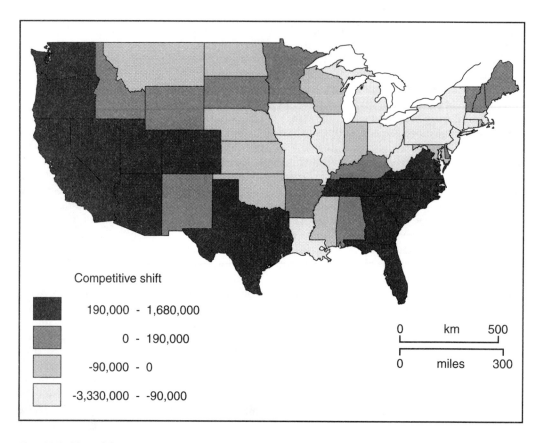

FIG. 11.3 United States: competitive shift, 1975–95, US states

Alvine, 1985; Harrington *et al.*, 1991; Illeris, 1996). This author has recently estimated that almost 80 per cent of the employment growth in Washington State over the 1963–87 time period can be attributed to the growth in trade in services (Beyers and Lindahl, 1998).

TRADE IN SERVICES AND LOCATION CONSIDERATIONS

Generalized markets of the various service sectors were identified in Table 11.3. This table suggested that retailing, health services and other services primarily served household markets, while wholesalers provided services to retailers as well as goods producing sectors, and producer services had primarily intermediate markets. For the sectors with low coefficients of industrial concentration, such as retailing (even when measured

at the gross level of spatial resolution of states used in this chapter), we can assume that markets are primarily localized. In contrast, for sectors such as producer and financial services, it is by now clear that there are specialist firms who have national or international markets for their services. Let us consider the patterns of trade and locational considerations for two groups of services: producer services, and services induced by goods movement.

Producer services

The rapidly-growing producer services sector has been the subject of considerable research in the United States and Canada in recent years. Most of the growth of this part of the economy appears to be associated with the creation of new establishments, not the growth in the average size of existing firms.

TABLE 11.8 Principal sectoral contributors to positive and negative competitive shifts, United States, 1975–95

Positive competitive shift		Largest sector	2nd largest sector	3rd largest sector
Arizona	+811	Retail	Other services	Manufacturing
California	+1050	Other services	Manufacturing	Retail
Colorado	+465	Other services	Retail	Producer
Florida	+1673	Producer	Retail	Health
Georgia	+783	Retail	Manufacturing	Other services
North Carolina	+652	Retail	Manufacturing	Health
Nevada	+417	Other services	Retail	Construction
Texas	+1526	Producer	Manufacturing and retail	Health
Washington	+615	Other services	Retail	Manufacturing

Negative competitive shift		Largest sector	2nd largest sector	3rd largest sector
Illinois	−1228	Manufacturing	Retail	Health and state and local
Massachusetts	−553	Manufacturing	Retail	Health
New Jersey	−478	Manufacturing	Retail	Finance, insurance and real estate
New York	−3325	Producer	Manufacturing	Retail
Ohio	−811	Manufacturing	Retail	Producer and other services
Pennsylvania	−1481	Manufacturing	Retail	State and local

The typical establishment is small, employing about a dozen people. However, the producer services also include globally-linked firms with hundreds of offices (such as Andersen Consulting), and giant corporations like Microsoft with hub-spoke industrial structures linked from the hub in Redmond, Washington, to spoke offices networked around the planet, as described by Markusen (1996). Research on producer services finds businesses with widespread geographical markets located alongside firms with all their clients located nearby. We have found that market orientation tends to be somewhat bimodal, and have labelled proprietors (one-person businesses) with strong export markets 'Lone Eagles', and firms with more than one employee and also

strong export markets as 'High Fliers' (Beyers and Lindahl, 1996b).

Many of the Lone Eagles and High Fliers are in lines of producer service work that requires travel to the client's location to produce the service, such as a management consultant needing to come to his client's office to explore the need for specialized assistance (Beyers and Lindahl, 1996b). In contrast, many of the locally-orientated producer service establishments do work in which the client travels to them, such as law firms. These contrasting situations are reflected in the factors surrounding locational choices. Lone Eagles and High Fliers have a degree of footloose-ness about them: if their clients are spatially dispersed they can be in the urban area of their choice, or can select a non-metropolitan location. In our recent research, we have found that many rural producer service firms were located in response to quality of life considerations, and this has also been an important locational factor for the urban Lone Eagles and High Fliers, who frequently locate their business near to the residence of the owner or founder (Beyers and Lindahl, 1996b). In contrast, firms with predominantly local markets indicate the need to be located in close proximity to clients – a profile similar to that which prevails for consumer-orientated services such as retailing.

Services induced by goods movement

A substantial segment of the service economy is associated with the movement of goods, whether it be functions such as apple warehouses which store fresh apples for up to a year for distribution into the 'fresh' apple market, or the typical merchant wholesaler stocking specialized lines of products such as nuts, bolts, and screws, or leather products. These enterprises are providing on the one hand a storage function between the grower or manufacturer and ultimate customer, and on the other hand are located centrally to either the commodity that they are gathering (as in the case of apple warehouses) or the market they are serving (as in the case of the nut, bolt, or screw wholesaler). These functions are also linked to the transport services sector, which must haul products from growers or manufacturers or mines, or from wholesalers to clients such as retail stores or construction contractors. Insurance services are often needed for goods in this chain of movement and storage, and service agents are needed to manage the flow of information surrounding this system of goods movement.

The geographical distribution of components of this service economy linked to the movement of goods is complex and multilayered. It involves fleets of trucks operating across the Interstate Highway System, as well as across international borders between the US and Canada and the US and Mexico. It has regional hubs in each major metropolitan region. It has speciality centres related to locations where particular crops are grown

subject to handling in the way described above for apples, but also globalized distribution systems for perishable crops like fresh cherries increasingly moved by air freight to spatially-dispersed markets. It is also being impacted strongly by computerization, which is reducing the ratio of inventory in wholesale channels to inventory in retailers, and moving a larger and larger fraction of product directly from manufacturers to retailers and other clients (OTA, 1995).

FORWARD TO THE INFORMATION SOCIETY

The shifting mix of business activity within the burgeoning services, and the changing nature of production in manufacturing, has been interpreted by many as a shift towards an information society or information economy (Gottman, 1979; Bell, 1973; Castells, 1996). Some have argued that we are in an era of deindustrialization, but it would be more accurate to say that we are in an era in which goods production remains strong, but it is not creating jobs at the national level in the United States or Canada (Rifkin, 1995; Beyers and Lindahl, 1998; Coffey and Shearmur, 1997). Some futurists argue that computers will displace human labour for many functions – as has been the case in telephone switching equipment (Rifkin, 1995; Tapscott, 1996). Visions of personal digital assistants have been created that will do everything from diagnose your diseases to optimally manage your financial investments (Harmon, 1996; Tapscott, 1996). In contrast, earlier in this chapter it was argued that this growth of the service economy especially the information-orientated producer services was not driven by downsizing and outsourcing but rather was a product of new technologies and innovative notions for services.

Each of the characteristics just described has been to some extent a source of angst about where labour markets in the United States and Canada are headed. Projections of employment change visualize a yet larger service sector, increasingly connected in cyberspace with new business innovations such as online WWW-based sales enterprises, virtual magazines and newspapers, virtual education, and free software. Where will this trend towards a more information-orientated economy end? Will we see the end of work as we now know it? Will there be massive job losses due to computer networks taking over many aspects of production, distribution, and even dramatically shaping consumption? Will the geographical consequences be centralization of power and the settlement system, or will dispersal prevail, and our existing congested, polluted, and high cost cities be abandoned for the bucolic countryside?

If there is any lesson to be learned from forecasting, it is that forecasts are always wrong. We cannot anticipate the discoveries that will reshape our economy tomorrow. If we put ourselves back in 1975 or 1976, the early years for data presented in this chapter, just think how different the service economy was. The PC had yet to be invented, much less the types of Internet connectivity which we now take for granted. Microsoft had yet to be founded. Online bookkeeping systems, such as are now common in retailing, had yet to be developed for inventory management and optimizing sales promotions. At universities people carried boxes of punch-cards to computer centres to have mechanical devices read them – and hopefully not mutilate them! Fibre-optic cable and TV cable networks were in their infancy. The computer technology was not with us to allow small package courier services which today link every part of the continent on an overnight basis by firms such as FedEx or UPS.

There can be no doubt but that we are moving further and further into the information economy, and its development will have unanticipated consequences on our society. Some are good, some may be judged to be not so good. We cannot yet determine what the impact of these changes will be on cities and regions, much less nations. However, to the extent that advances in the information economy release money for consumption and production, via decreases in the costs of goods or services, people and businesses will most likely behave like they have in the past: they will buy other goods and services. To the extent that the consumption of specialized producer services is driven by market conditions and technological innovations into new divisions of labour, we will continue to have the birthing of new industries creating wealth in new and old locations in space.

CONCLUSION

This chapter has described the aggregate power of the service economy in recent job growth in the US and Canada. While we cannot predict the outcome of, or the exact shape of economic changes coming, we can anticipate that the places currently favoured for growth by today's hot industries – such as gambling and entertainment in Las Vegas – will be eclipsed by some new set of industrial players. The complex of information-related industries which have been so powerful in reshaping our economy in recent years are likely to be key sectors in the near-term evolution of the North American settlement system. But even these industries are dynamic and can have rapidly-shifting geographies, as the history of computer equipment manufacturing and related hardware and software codes illustrates for the recent past. The conjunction of telecommunications and computer technologies for household, business, and public sector applications will undoubtedly create new divisions of labour in new industrial spaces.

ENDNOTES

1. Goods producing sectors in the US are defined here as farm; agricultural services, forestry, and fishing; mining, construction, and manufacturing. In Canada they are defined as agriculture, fishing, logging and forestry, mining, manufacturing, and construction.

2. The index is computed by taking the sum of the absolute values of the difference in the percentage of the national employment in a sector found in a particular region (in this case state or provinces) less the national share of employment in all

industries found in a region. For example, the index for producer services is calculated by summing the absolute values for individual states and the District of Columbia. For the state of Alabama the value entering the coefficient of industrial concentration for the year 1975 is 0.603. This figure is calculated by subtracting from Alabama's share of national employment in producer services (44 567/4 665 405) Alabama's overall share of national employment (1 545 807/99 199 200). This computation is repeated for each region, and all elements are summed to produce the values for each industry in Tables 11.4 and 11.5.

3. This index is calculated by taking the sum of the absolute values of the percentage of employment in a region in an industry less the national percentage in that industry. If the region has exactly the same share of employment as the nation in each industry, the index would be zero.

KEY READINGS

Beyers, W.B. and Lindahl, D.P. 1996a: Explaining the demand for producer services: is cost-driven externalization the major factor? *Papers in Regional Science* 75(3), 351–74.

Castells, M. 1996: *The information age: economy, society, and culture, Vol. I: The rise of the network society.* Cambridge, MA: Blackwell Publishers.

Coffey, W. and Shearmur, R.G. 1997: The growth and location of high order services in the Canadian urban system, 1971–1991. *The Professional Geographer* 49(4), 404–17.

Illeris, S. 1996: *The service economy: a geographical approach.* Chichester: Wiley.

Porter, M. 1998: *The competitive advantage of nations.* 2nd edn, London: Macmillan.

Quinn, J.B. 1992: *Intelligent enterprise: a knowledge and service based paradigm for industry.* New York: The Free Press.

Section E

The urban scene

12

THE NORTH AMERICAN URBAN SYSTEM: THE MACRO-GEOGRAPHY OF UNEVEN DEVELOPMENT

LARRY S. BOURNE

INTRODUCTION: SETTING THE STAGE

North America, as an urban society, is both old and new. It is old in the sense that Native settlements have existed for more than a millennium. Even European-imposed colonial settlements, in what is now the political territory of the United States and Canada, date from the sixteenth and early seventeenth centuries, first in the Spanish areas of the Southwest and Florida, then in the French settlements of Québec, and later in the English settlements along the East Coast. It is also a young society in the sense that the settlement system is still growing rapidly, becoming more urban, and evolving geographically – continuing a three-centuries long extension from the east to the west and south. The future state of that system and the living environments that its urban areas will provide, however, have yet to be defined.

This chapter sets the stage for more detailed discussions to follow of the internal structure of cities and the characteristics of individual regions and urban areas. Building on previous chapters, the text begins with a brief overview of post-Second World-War population and economic growth in the United States and Canada and then examines the outcomes of this growth in terms of the evolution of a system of metropolitan areas – or more broadly, the urban system – in the two countries. Particular emphasis is placed on accounting for variations in rates of growth among metropolitan areas and regions, and shifts in the functional hierarchy of urban places, whilst also identifying the most recent winners and losers in the competition for a share of national growth. Four themes are highlighted: (1) the rapidity of change in the urbanization processes affecting both countries; (2) the immense shifts among regions and metropolitan areas in population and economic activity, and thus in wealth and political power; (3) the apparent paradoxes of a continued concentration of population and production capacity in metropolitan areas combined with widespread decentralization, and the contrast between the increasing homogenization of urban economies (through globalization or continentalization) with the heightened importance of locality-specific characteristics; and (4) the dramatic effects of high rates of immigration in augmenting levels of social diversity and in accelerating a divergence in the social character of individual places. The concluding section speculates on future trends in urban and regional development and raises some of the issues that flow from those trends.

It is appropriate at the outset to clarify two concepts. First, the concept of the urban system is used simply to refer to a set of cities in a given territory or nation. The concept highlights the fact that individual cities, even large metropolitan areas, do not exist in isolation. Cities, instead, are interdependent parts of larger systems of cities, regions, and nations, with which they interact and from which they derive their characteristics and their growth prospects. The economies of developed nations are organized by and through their cities, particularly their metropolitan areas. Thus, our initial interest here is seeing how the two sets or systems of cities in the United States and Canada have changed in terms of their size, growth rates and attributes, the functions they perform, and their links with other places. This approach in effect constitutes the macro-geography of the urbanization process.

Second, all of the data used in this chapter refer not to cities as political entities but to extended or 'functional' urban areas. Typically, these areas combine a central city (or cities) and those adjacent suburban municipalities that are closely integrated with that city. The most common of such definitions is the Metropolitan Statistical Area (MSA), and the Consolidated Metropolitan Statistical Area (CMSA) used by the US Census, and the Census Metropolitan Area (CMA) as defined by Statistics Canada.[1] The rationale for using these statistical constructions is twofold: first, they offer a more consistent basis for comparing urban areas in different countries, and second they provide a more accurate description of the size and extent of urban development than do local political (e.g. municipal) jurisdictions. In most parts of North America urban development has spread far beyond the boundaries of the original central city, boundaries which have often not changed since the 1920s.

EVOLUTION OF THE SETTLEMENT SYSTEM

The current settlement system in North America has evolved over almost four centuries from its European-imposed base on the east coast and the St Lawrence Valley (Fig. 12.1). From that small base, population and economic development spread westwards, pushing back the frontier but still anchored to the initial urban centres in the East. It was not until the first decade of this century that the outline of the current settlement system, and the current hierarchy of urban centres, are clearly evident. Growth rates in this early period, as in other parts of the so-called new world, were phenomenal. New towns and rural settlements appeared almost overnight. Some later disappeared, most persisted, and a few grew to become the metropolises of today. By 1910 the US population was predominantly urban (i.e. more than 50 per cent urban); Canada followed in the early 1920s. In fact, by 1950 the US had became a predominantly 'metropolitan' nation, in that over 50 per cent of the population lived in MSAs; Canada similarly by 1970. It is this inherited settlement system which provides the backcloth for a discussion of urbanization in the recent post-war era.

Since 1950 there have been major changes in the spatial distribution of population, the structure of the economy, in technology and living standards, and in the relative importance of different regions, cities and metropolitan areas in both countries. Between 1950 and 1995 the population of the US increased by 73 per cent, from 152 million to almost 263 million; in Canada the national population grew by 105 per cent from 14 to almost 29 million.[2] Over the same period, the proportion of the national populations living in urban areas of all sizes rose from 62 to about 76 per cent in the US and from 61 to over 77 per cent in Canada.[3] Even these figures, however, underestimate the size of the urban population: in Canada, of the other 23

per cent only 3 per cent are classified as rural, the majority being rural non-farm residents. Under these conditions, and given the rapid diffusion of communications technologies, and the life-style homogeneity produced by mass consumption, the traditional urban–rural dichotomy may now have lost any meaning.

The rate of economic growth has been even greater over this period, despite a number of severe recessions. The Gross National Product (GNP) of the US increased almost 30-fold, from $288 billion in 1950 to over $7500 billion (in constant 1986 dollars), and from $1900 to over $28 000 per capita. Canada, at least until the latest recession in the early 1990s, grew somewhat faster over the post-war period, and its per capita GNP remains close to that of the US (allowing for variable exchange rates). This growth in income, wealth and production, in turn, provides the context for the continued redistribution of population and massive economic restructuring that has both underlain and generated substantial changes in the urban systems of the two countries.

The overall geographies of the two urban systems, and their reorganization over time, clearly mirror shifts in national and regional economies. Indeed, a description of the evolution of the urban system can be considered a 'map' of the changing economy. The dramatic sectoral transformation of the economies of both the United States and Canada have been outlined in previous chapters and need not be repeated here. It will suffice to note that, over the same post-war study period, the economies of both countries have shifted from resource based to manufacturing to service based, with immense consequences for both labour and for the macro-geography of urbanization. This chapter attempts to illustrate how these sectoral shifts have translated into a distinctive urban system – a metropolitan system – in each country and into different patterns of urban and regional growth (Fig. 12.2).

Metropolitan growth and concentration

The story of post-war urbanization in North

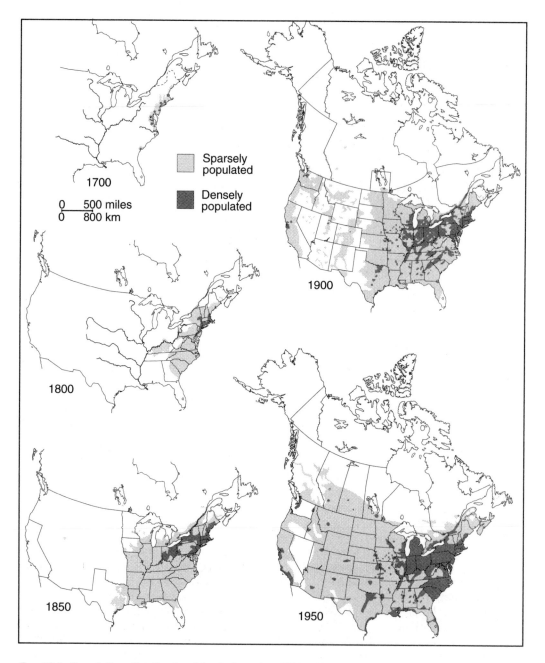

Fig. 12.1 Population distribution, North America, 1700–1950

America has been the emergence of national systems of metropolitan areas that have come to overwhelmingly dominate the economic and social geographies of both nations. In 1950 the US Census recorded 169 metropolitan areas (MSAs), housing 85 mil-lion people or 56 per cent of the national pop-ulation (Table 12.1). By 1995 there were 253 MSAs and 18 consolidated metropolitan areas (CMSAs); the latter included 73 MSAs that had been combined into larger CMSAs because of their high levels of economic and

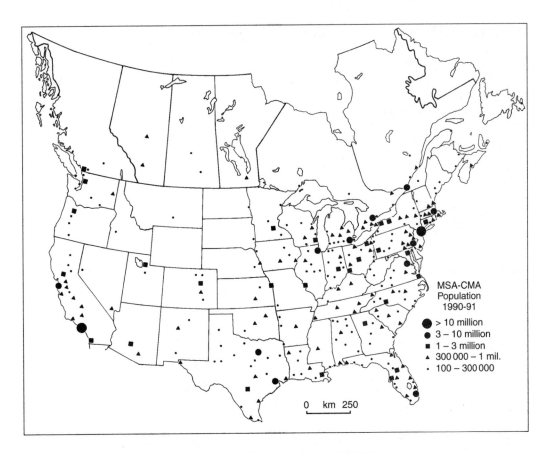

Fig. 12.2 Hierarchy of metropolitan areas, North America, 1990–91

social integration. These 271 places accommodated almost 208 million people, or 79.2 per cent of the nation's population.

The degree of concentration of both population and economic activity in metropolitan areas is even higher than these figures suggest. Table 12.1 also shows that from 1950 to 1995 the number of metropolitan areas with more than one million population increased from 14 to 40, and these were the places of residence for 50 per cent of the entire US population. Some of these consolidated metropolises are now of immense size and often cross several state boundaries: the New York–New Jersey–Long Island CMSA has a combined population of 19.7 million (in three states); the Los Angeles–Long Beach–Orange–Riverside CMSA over 15.3 million; the Chicago–Gary–Kenosha CMSA over 8.5 million. Even members of the second tier of

large metropolitan areas are now of mega-status: Washington–Baltimore (7.1 million), San Francisco (6.5 million), as are much younger centres such as Dallas (4.4 million) and Houston (4.1 million). Clearly the US is not abandoning large cities; in fact, it is even building new ones.

The primary measure of urbanization now used in the United States is not urban v. rural, but the differentiation between metropolitan and non-metropolitan areas.[4] The latter is a residual category, combining small cities and towns and all of the rest of the country outside of the metropolitan areas. Table 12.2 provides statistics on the rate of growth of metropolitan and non-metropolitan populations in the US from 1950. Overall rates of growth, as expected, have been declining but remain high by the standards of other developed countries, and particularly in compari-

TABLE 12.1 Evolution of the American urban system: number and population of US metropolitan areas, 1950–95

	1950	1960	1970	1980	1990	1995**
Number of metropolitan areas*						
All areas	169	212	243	284	284	271***
Areas with over 1 million						
population	14	22	33	35	39	40
Metropolitan area populations						
(in millions)						
All areas	84.8	112.9	139.5	172.6	192.7	207.7
In areas with over 1 million						
population	44.9	63.9	80.6	104.0	124.8	129.2
% US population in						
metropolitan areas						
All areas	56.0	60.0	68.6	76.2	77.5	79.0
In areas with over 1 million						
population	29.6	35.6	39.6	45.9	50.2	49.2

Notes: * Includes Metropolitan Statistical Areas (MSAs) and Consolidated MSAs (CMSAs)
      ** Estimated
     *** Number of separate MSAs reduced by amalgamation into larger CMSAs
Source: Bogue, D. (1953): Frey, W. (1995); *Statistical abstract of the United States*, 1991 and 1996, Government Printing Office, Washington DC

son to Europe. Note that in the 1950s and 1960s, and again in the 1980s, metropolitan growth rates far exceeded the growth rate of non-metropolitan areas. These were periods of further metropolitan population concentration, and with it a concentration of economic power and wealth.

In the 1970s, however, non-metropolitan growth exceeded that of metropolitan America, likely for the first time in US history. This period (and allowing for the difficulties of measurement), has been widely described as one of 'de-urbanization' or 'counterurbanization' (Berry, 1976), and as evidence of a rural revival, if not a renaissance. The trend, however, may have in part been a statistical artefact (e.g. a failure to extend MSA boundaries); it was also short-lived. Evidence for the 1990s, unfortunately, is as yet incomplete. In any case, these shifting growth rates provide the basis for a periodization of urban development introduced later.

Table 12.3 provides comparative statistics for the metropolitan systems in both Canada and the United States, using 1995 estimates for the latter and the 1996 Census of Canada. Given the ratio of national populations of roughly 10:1, we would expect the Canadian metropolitan system to be about a tenth the size. This is in fact the case. In 1996 Canada had 25 census metropolitan areas (CMAs) with over 100 000 population (and nine other census agglomerations – CAs – also over 100 000). These 34 urban areas combined had nearly 19 million people, or 65.6 per cent of the national population. The latter figure, a crude index of metropolitan concentration, is substantially lower than the 79 per cent estimated for the US. This difference reflects two conditions: first, the more restricted definition of the geographical size of metropolitan areas in the Canadian census; and second, the relatively larger role played by smaller urban centres (including resource-based communities in the north). In other ways the two

TABLE 12.2 Rates of change in metropolitan and non-metropolitan populations in the United States, 1950–95

% change	1950–60	1960–70	1970–80	1980–90	1990–95
Metropolitan	24.8	17.2	10.5	11.8	4.9
Non-metropolitan	4.9	2.6	14.3	3.9	3.8
TOTAL	18.9	13.4	11.4	9.8	4.7

Source: US Bureau of the Census, various years. Metropolitan areas as defined at each census

TABLE 12.3 Comparing metropolitan systems in the United States and Canada, 1995–96

	United States (a)	Canada (b)
Number of metro areas	271 (326)(c)	25 (34)(d)
Total metro population (in millions)	207.7	17.9 (18.9)
National population in metro areas (%)	79.0	61.9 (65.6)
Average size of metro areas ('000)	766.4	711.1 (556.9)
Number of metro areas over 1 million	40	4
Metro population in areas over 1 million (%)	62.2	58.4
Change in metro area populations 1985/86–1995/96 (%)	11.8	16.8
National population in 3 largest metro areas (%)(e)	16.6	32.6
Total national population (in millions)	262.8	28.8
Change in national population 1985/86–1995/96 (%)	9.7	11.1

Notes:
a) US metropolitan areas as defined 30 June 1995. Includes Metropolitan Statistical Areas (MSAs) and Consolidated Metropolitan Statistical Areas (CMSAs) in which two or more MSAs are combined.
b) Census Metropolitan Areas (CMAs) as defined in the 1996 Census of Canada.
c) Includes 25 MSAs with populations less than 100 000. The number in parentheses is the total number of MSAs.
d) The numbers in parentheses include nine Census Agglomerations (CAs) with total populations over 100 000 but not defined as CMAs in 1996.
e) Includes the New York–Northern New Jersey–Long Island CMSA (19.7 million), Los Angeles–Riverside–Orange CMSA (15.3 million), Chicago–Gary–Kenosha CMSA (8.5 million).

systems are rather similar; for example, in the proportion of the metropolitan population resident in areas of one million or more, and in the average size of metropolitan area.

One marked difference is in the level of concentration at the top of the urban hierarchy. As might be expected given the relatively small population of Canada, and the history of two founding peoples each with their own metropolis (see Chapter 16), the three largest CMAs – Toronto (4.2 million), Montréal (3.3 million), Vancouver (1.8 million) – held almost 33 per cent of the entire Canadian population, compared to 16.6 per cent for the three largest CMSAs in the US. As a result, not surprisingly one hears many complaints from other Canadians (at least English-speaking) about the degree to which the country's economy (e.g. banks), cultural institutions (e.g. television and the arts) and politics are dominated by a few urban centres, and especially by Toronto.

One of the most common ways to illustrate the importance, if not the dominance, of particular metropolitan areas in the urban system is to describe the hierarchy of places in that system. Since space does not permit us to examine or even list all 271 US metropolitan areas, or for that matter the 25 CMAs in Canada, Table 12.4 provides a simple ranking of the 20 largest MSAs and CMSAs in the US in 1990 and again in 1995 (the latter are

TABLE 12.4 US and Canadian metropolitan area rankings, 1995–96

UNITED STATES

Rank	Metropolitan area (CMSA/MSA)*	Population ('000)		% change	
		1990	1995	1980–90	1990–95
1	New York–New Jersey–Long Island CMSA	19 342	19 696	3.4	1.8
2	Los Angeles–Riverside–Orange CMSA	14 532	15 302	26.3	5.3
3	Chicago–Gary–Kenosha CMSA	8 240	8 527	1.5	3.5
4	Washington–Baltimore CMSA	6 727	7 051	16.1	4.8
5	San Francisco–Oakland–San Jose CMSA	6 253	6 513	15.4	4.2
6	Philadelphia CMSA	5 893	5 959	4.3	1.1
7	Boston CMSA	5 455	5 497	−2.0	0.8
8	Detroit CMSA	5 187	5 256	6.5	1.3
9	Dallas–Fort Worth CMSA	4 037	4 362	32.5	8.1
10	Houston CMSA	3 731	4 099	19.6	9.9
11	Miami CMSA	3 193	3 408	20.7	6.7
12	Atlanta MSA	2 960	3 331	33.0	12.6
13	Seattle–Tacoma CMSA	2 970	3 226	23.2	8.6
14	Cleveland–Akron CMSA	2 860	2 899	−2.6	1.4
15	Minneapolis–St Paul MSA	2 539	2 688	15.5	5.9
16	San Diego MSA	2 498	2 632	34.1	5.4
17	St Louis MSA	2 493	2 536	3.2	1.8
18	Phoenix MSA	2 238	2 473	40.0	10.5
19	Pittsburgh MSA	2 395	2 402	−6.8	0.3
20	Tampa–St Petersburg MSA	2 068	2 157	28.1	4.3

CANADA

Rank	Metropolitan area (CMAs consolidated)**	Population ('000)		% change
		1991	1996	1991–96
1	Toronto CMA + Oshawa CMA + Hamilton CMA + Barrie CA	4 858	5 303	9.1
2	Montréal CMA + adjacent CAs	3 461	3 581	3.5
3	Vancouver CMA + Fraser Valley	1 771	2 035	14.9
4	Ottawa–Hull CMA	942	1 010	7.3

Notes:
* A Consolidated Metropolitan Statistical Area (CMSA) includes two or more Primary Metropolitan Statistical Areas (PMSA) that are closely integrated, while a Metropolitan Statistical Area (MSA) is centred on one large city.
** CMAs = Census Metropolitan Areas. CMA boundaries have been extended and adjacent CMAs (or CAs) consolidated to more closely resemble the US definitions of MSAs and CMSAs.
Sources: US Bureau of the Census; Statistics Canada

estimates).[5] For comparative purposes we also list the four largest metropolitan areas in Canada, but here we have combined adjacent CMAs (and CAs) to produce Canadian figures more compatible with those for the US.

Perhaps the most valuable information that readers can take from the table is the massive size of these urban agglomerations, and for the United States in the 1980–90 period, the extreme variability of population growth rates. A number of these urban regions, particularly those in California, Arizona, Texas and Florida, as well as Atlanta (Chapter 14), Washington, DC, and Seattle, grew by rates of 25–40 per cent over one decade. Such high growth rates for entire metropolitan areas are seldom found in other developed countries. The lesson here, which is reiterated in the concluding section, is that urban growth is not only rapid but variable over time and regionally focused. The whys, wheres and consequences of this uneven pattern are examined below.

The demographic basis of urban growth

The post-Second World War transformation of the population of urban North America is the result of changes in three principal components: natural population increase; immigration; and the ebb-and-flow of internal (i.e. domestic) migration. Both the United States and Canada experienced a massive 'baby-boom' after the Second World War, particularly from 1948 to 1963. The birth rate rose dramatically, the average age at first marriage and at birth of first child dropped, and death rates declined. The result was a population boom, especially of young families, that fuelled rapid growth in the consumer economy, in the demand for jobs, new housing and social services, and thus an accelerating rate of suburbanization. This huge baby-boom cohort continues to send shock waves through both housing and labour markets, and will into the next century, with its effects exaggerated by the sharp drop in birth and marriage rates (the 'baby-bust') in the late 1960s and 1970s.

As rates of natural population increase

declined, and then stabilized across space, immigration and internal migration have become more important as determinants of uneven urban and regional growth. In any given year about 18 per cent of urban North Americans change their place of residence; over five years almost 50 per cent move, roughly half of whom are migrants (moving from one urban area to another). Such high mobility and migration rates suggest that the potential for a continuing spatial redistribution of population and economic activity is very high, and thus difficult to predict.

The other, and now perhaps the more significant component of social change in urban North America, is immigration. Both the United States and Canada have always been – aside from the Native populations – nations of immigrants. The numbers are staggering. Since 1820, for example, the US has accepted some 61 million immigrants; Canada over 14 million. Since 1960 the US has received 18.2 million immigrants, while Canada has received 4.1 million (roughly double the US rate). Although the rate of immigration was highest early in the century, and low during the depression and the Second World War, in the 1980s and 1990s it has increased once again (Chapter 6). In the last few years the US has been admitting just over one million immigrants annually, Canada over 220 000.[6]

Two properties of these recent immigration flows make their impacts on urban North America even more significant than in the past. The first is the shift in the source countries for immigrants, signalling a major transformation in the ethno-cultural characteristics of the immigrants themselves. Historically, almost all immigrants were from Europe, and in the immediate post-war period over 50 per cent were from that continent (although the countries varied). After 1965, however, the sources of immigrants changed, both as a result of policy decisions and as a reflection of improved economic opportunities in Europe. In the last three decades only 15 per cent of immigrants were from Europe; over 48 per cent came from elsewhere in the Americas and 37 per cent from Asia and Africa. For Canada, the overall proportion of immigrants from non-traditional sources is about the same, while the proportions from Asia (60 per cent), Africa (7 per cent), and the Caribbean (10 per cent), are substantially higher. These flows, in turn, have transformed the cultural, ethnic and racial make-up of the receiving cities.

Second, the flows of immigration are much more uneven than are those of internal migrants. Immigration tends to be intensely concentrated geographically, focusing on only a few 'gateway' centres. During the 1980s five metropolitan areas – New York, Los Angeles, San Francisco, Miami and Boston – received over 58 per cent of all immigrants to the United States. The two largest areas, New York and Los Angeles, alone received 41 per cent. Canada showed even higher levels of concentration; since 1991 almost 80 per cent went to the five largest centres, and over 60 per cent to Toronto and Vancouver. These gateways offer the combined attractions of diverse job opportunities and heterogeneous populations – indeed most immigrants follow the same paths and choose the same destinations as earlier migrants from the same country – a process called chain migration. As a result of this extreme concentration, the gateway metropolitan areas are beginning to look and feel very different from other metropolitan areas and those parts of both countries that receive few (if any) immigrants.

Patterns and timing of urban and regional growth

The scale and complexity of the changes in the economy, demography, governance and living environments in the North American urban system, unfolding as they are over a vast and diverse continent, defy simple generalization and easy explanation.[7] As a result most writers offer some form of periodization of the evolution of urban development in the United States (and in Canada) over the post-war period (Borchert, 1991; Knox, 1994; Frey, 1995). These periods, in turn, are often linked to regional shifts in both population and economic activity. Three periods are suggested here, covering the years 1950 to the late 1980s, with a fourth period (the 1990s and beyond) still taking shape. To illustrate the shifts taking place, Table 12.5 summarizes

the varying growth rates of US metropolitan areas classified by major census region (North, South, West), and by size category, for the three decades from 1960 to 1990 (Frey and Speare, 1992).

The first period, including the decades up to and including the 1960s may simply be described as one of metropolitan growth and concentration. Confirming the initial evidence in Table 12.2, metropolitan areas grew faster than non-metropolitan areas, and large metros grew more rapidly than small ones. Populations in largely agricultural areas continued to decline. This was also a period of rapid urban growth almost everywhere, reflecting the effects of the baby-boom, high levels of immigration, rising incomes and the expansion of employment in almost all economic sectors – manufacturing, services and government (including the military).

The second period, roughly from the early 1970s to the early 1980s was described above as a period of national population deconcentration and a non-metropolitan or rural revival. This revival of small towns and rural areas, reflected in a reversal in net migration flows which had previously favoured metropolitan areas, has been attributed to numerous factors including changes in life styles and in attitudes about where and how to live. Note in Table 12.5 that the growth rate of the North (the Northeast and Midwest regions) declined sharply (to 2.2 per cent) in the 1970–80 decade, and the growth of large metropolitan areas dropped to 8.1 per cent, well below that of smaller metropolitan areas (15.5 per cent), and even non-metropolitan areas (14.3 per cent).

This non-metropolitan turnaround, widely (but incorrectly) described as initiating a post-urban era, was the result of several factors acting in concert: declining rates of natural increase, an ageing population, and widespread restructuring in the economy, notably the loss of jobs in manufacturing. The older metropolitan areas in the northeast Manufacturing Belt (e.g. Detroit, Pittsburgh, Cleveland, Chicago) suffered the most from this process of de-industrialization (see Chapter 10). They also lost out in the competition for expanding industries and services. At the same time, the explosion in

TABLE 12.5 United States: rates of metropolitan and non-metropolitan population change by region and size category, 1960–90

Region and metropolitan size category	1990 populations (millions)	% change		
		1960–70	1970–80	1980–90
Northeast/Midwest				
Large metro	62.9	12.0	−0.9	2.8
Other metro	25.6	11.1	5.2	3.3
Non-metro	22.6	2.6	8.0	0.1
South				
Large metro	28.2	30.9	23.4	22.3
Other metro	31.9	15.5	20.9	13.4
Non-metro	24.9	1.1	16.3	4.6
West				
Large metro	33.8	29.1	20.0	24.2
Other metro	10.8	24.8	32.2	22.8
Non-metro	8.1	9.0	30.6	14.1
Totals by region (a):				
North	111.0	9.8	2.2	2.4
South	84.9	14.2	20.1	13.3
West	52.8	24.6	24.0	22.2
US totals:				
Large metro	124.8	18.5	8.1	12.1
Other metro	67.9	14.5	15.5	10.8
Non-metro	56.0	2.7	14.3	3.9

Note: (a) Where a metropolitan area overlaps regional boundaries, it has been allocated to the region containing the principal central city. Large metro refers to metropolitan areas over 1 million.
Sources: US Bureau of the Census, 1960, 1970, 1980, 1990; Frey and Speare (1992)

the prices of resources (e.g. oil), and the expansion of retirement and life-style migrations, favoured cities in non-manufacturing regions, especially in the South and West. During this period over 50 per cent of US population growth took place in just three states (California, Texas, Florida). In Canada, growth in this decade also shifted away from the older industrial heartland of southern Ontario and Québec toward the western provinces of Alberta and British Columbia. Those two provinces recorded 50 per cent of total Canadian population growth, and an even higher percentage of capital investment.

This decentralized pattern of growth, however, was not to last beyond the next business cycle. The 1980s brought yet another, and even more complex, pattern of growth in the urban system. This period witnessed a modest 'metropolitan revival', indeed a renewed concentration of growth in metropolitan areas. Table 12.5 indicates that large metropolitan areas grew more rapidly (12.1 per cent) than both smaller metropolitan areas (10.8 per cent) and non-metropolitan areas (3.9 per cent), much as they did in the 1960s – but conditions were not the same as in the 1960s. The reasons for this reversal, once again, are complex. What clearly did happen was that resource prices declined, undercutting growth in many metropolitan areas in the western United States (and in western Canada). Manufacturing enjoyed a modest resurgence (but with few new jobs), and the financial services and cultural industries expanded in those metropolitan areas at the top of the urban hierarchy (e.g. New York). Other directions of growth – for example in retirement and life style migrations – continued. The subsequent recession of the early 1990s, however, appeared to usher in yet another period of change in the urban system.

Given the complexity of these trends, there is a tendency in the literature on urban growth to describe these uneven shifts in the provocative language of 'Sunbelt' v. 'Frostbelt', or in more vivid terms contrasting the sun-and-gun (i.e. military) belts of the southern and western states v. the northern manufacturing Rustbelt. In Canada a similar dichotomy is frequently set up between heartland and periphery. While it is true that metropolitan areas in the US South and West have grown faster than those in the North and Midwest for several decades (see Table 12.5), it is an over-simplification to categorize all places simply according to their region. The same applies in Canada to the heartland–periphery divide. There are rapidly-growing areas in otherwise slow-growing regions, and declining communities in both the US Sunbelt and the Canadian heartland.[8]

The winners and losers in the competition for growth

Understanding recent growth patterns, and identifying those places in the urban system that are the winners and losers, requires a closer examination of the economic bases for metropolitan growth.[9] Table 12.6 displays population growth for 234 US metropolitan areas for the 1970–90 decades in which those areas are classified according to their dominant sector of employment (i.e. their economic base). The differences in economic performance among the seven functional categories are huge. The fastest-growing areas tend to be resorts, serving recreational and retirement functions (46.6 per cent in 1980–90), followed by government and military centres (17.5 per cent), medical and educational centres (12.2 per cent), and those whose economies are driven by business and financial services (14.5 per cent) (see Chapter 11). Manufacturing centres grew hardly at all (1.5 per cent in 1980–90), and resource-extraction communities actually declined in population over the decade (−2.4 per cent). The contrast is most obvious between those metropolitan areas plugged into the 'new' life-style and service economy and those dependent on older sunset industries and mines. Interestingly, many of the new growth centres are not strictly outcomes of private market decisions. At least four of the fastest growing types of metropolitan areas have economies based on functions largely anchored in the public sector.

Who are the specific winners and losers? Recent trends suggest that there is as much

TABLE 12.6 Growth rates for US metropolitan areas by economic base, 1970–90

Dominant economic base of metro area*	Population change		Number of metro areas	Population (millions)
	1970–80 %	1980–90 %		
Manufacturing	2.9	1.5	86	89.6
Business–Financial Services	10.9	14.5	48	107.6
Government–Military	20.0	17.5	22	6.6
Medical–Education	16.6	12.2	38	8.0
Recreational–Retirement	57.4	46.6	14	5.6
Resource–Extraction	12.6	−2.4	10	1.4
Diversified	14.1	10.6	16	24.4

Note: *Functional categories based on employment structures in 1980.
Source: Frey and Speare (1992)

variability in urban growth rates within the above functional categories as there is between them. Older, specialized manufacturing centres in the US Northeast and Midwest, for example, have generally not done well. Indeed, some have declined at a disturbing rate. Yet other manufacturing centres, typically those who have a slice of the new 'sunrise' industries in particular 'niche' markets (e.g. computers, transport, high-tech industries) have done better. Those centres, however, tend to be smaller, and to have more diversified economies which in turn are often linked to the presence of government agencies, research institutions, universities and colleges – and most are located outside the traditional Manufacturing Belt. Obvious examples include Seattle, Silicon Valley (San Francisco CMSA), Salt Lake City, Austin, Atlanta. Still others benefited from the explosion in recreation (e.g. Las Vegas), entertainment (e.g. Orlando) and retirement (e.g. West Palm Beach) pursuits, or the expansion of government (e.g. Washington DC), medical (e.g. Minneapolis–St Paul), educational and research activities (e.g. Boston) (see Chapter 17) – sectors that are the linchpins of the so-called 'knowledge' economy.

Table 12.7 provides a tally of the most rapidly growing and declining metropolitan

TABLE 12.7 Winners and losers in the American urban system: extremes of growth and decline, 1980–90

Metropolitan area* (state) (over 250 000 only)	Population ('000)		Change %	Economic basis of growth or decline
	1980	1990		
Most rapidly growing:				
Fort Pierce (FL)	151	251	66.1	R and Retirement
Fort Myers (FL)	205	335	63.1	R and Retirement
Las Vegas (NV)	463	741	60.1	Recreation
Orlando (FL)	700	1 073	53.3	R and Retirement
West Palm Beach (FL)	577	864	49.7	R and Retirement
Melbourne (FL)	273	399	46.2	R and Retirement
Austin (TX)	537	782	45.6	Gov't/High tech
Daytona Beach (FL)	259	371	43.3	R and Retirement
Phoenix (AZ)	1 509	2 122	40.6	Services/R and R
Modesto (CA)	266	371	39.3	Services
Most rapidly declining:				
Davenport (IA)	384	351	−8.8	Agric. Services
Pittsburgh (PA)	2 423	2 243	−7.4	Industrial
Peoria (IL)	356	339	−4.8	Industrial
Youngstown (OH)	531	493	−7.3	Industrial
Huntington (WV)	336	313	−7.1	Resources
Charleston (WV)	270	250	−7.1	Resources
Saginaw (MI)	422	399	−5.3	Industrial
Flint (MI)	450	430	−4.4	Industrial
Buffalo–Niagara Falls (NY)	1 243	1 189	−4.3	Industrial
Beaumont (TX)	373	361	−3.2	Resources

Notes: * Metropolitan Statistical Area (MSA).
R and R = Recreation and Retirement.
Source: US Bureau of the Census, 1995

areas in the United States in the 1980–90 decade. The range in 'growth rates' is enormous – from over 66 per cent (Fort Pierce, Florida) to nearly –10 per cent (Davenport, Iowa) in a single decade. Almost all of these places are small, and they divide rather clearly into two functional groups. The rapidly-growing places, as noted above, tend to be in the South and West, notably in California, Florida, Texas and Arizona, and most have economies based on retirement, recreation, entertainment and other leisure pursuits, and local service provision. Most are also located in regions with high amenity levels, and warmer climates, as well as lower living and production costs. The declining communities, in contrast, are either industrial (e.g. Pittsburgh; Flint, Michigan), resource-based (e.g. Huntington, West Virginia), or agricultural service centres (e.g. Davenport).[10]

The Canadian pattern of growth and decline is similar, but with fewer extremes.[11] The most rapidly-growing communities tend to be retirement communities, as in British Columbia (e.g. Victoria; Kelowna) and in the region around Toronto, or cities that have been able to combine control of the resource sectors with financial service and management functions (e.g. Vancouver, Calgary and Toronto). Declining communities in Canada are also numerous, but almost all are small places in the periphery dependent on older mining and refining industries, or they are agricultural communities that have shrinking rural service area populations; few of these places, given their extreme isolation and specialization, have any alternative economic development paths.

Among the larger North American metropolitan areas, those places that have shown the most robust economies tend to be closely linked to the new service economy and to the global economic system (Kresl, 1995; Wilson, 1997). Typically these are centres of business services and financial capital (Sassen, 1994). The most obvious examples of global financial centres are New York and Los Angeles, and to a lesser degree San Francisco, Dallas, Chicago and Toronto. These metropolitan areas serve increasingly as the 'command centres' of their respective economies and urban systems, and as the principal conduits for exchanges with overseas markets and corporations. Although difficult to measure, it is likely that the degree of control exerted by these metropolises has increased, in line with greater levels of corporate concentration, increasing global interdependence and the growth of financial markets and related services.

In these trends, then, we see evidence of the apparent contrast between greater concentration of wealth and power in a few metropolitan areas, especially in terms of high-order financial and business service functions, and the widespread decentralization of manufacturing and other lower-order economic functions to newer regions and smaller metropolitan areas. The landscape of economic growth has become more complex and volatile, and more uneven.

THE NORTH AMERICAN URBAN SYSTEM: A TYPOLOGY

To this point we have examined the urban systems of the United States and Canada in parallel, largely because basic urban definitions, data sources and institutions differ, and because of the obvious size difference in their respective populations. None the less, and despite the absence of firm evidence, we might ask: is there a single North American urban system – a common North American city? The answer to both questions is a qualified yes, as any visitor from Europe (or almost anywhere else) will attest. North American cities look distinctive, and similar. But national differences remain important. Despite continued globalization, national borders, institutions and government policies – not to mention differences in history and politics – do matter. The settlement patterns and the urban systems of both countries obviously share common ancestors and a more-or-less similar time path of development, but have evolved largely independently.

Yet cities in the two systems have become increasingly integrated over a long period of time. Recent policy initiatives, such as the Canada–US Auto Pact (1963), the Canada–US

Free Trade Agreement (FTA) signed in 1988 and its extension to Mexico in 1993 (NAFTA), have simply accelerated a process that was ongoing for decades. Cross-border trade flows have increased, in absolute terms and in the Canadian case relative to the traditional east–west flows between Canadian regions. The economies of a number of Canadian metropolitan areas in effect have become 'continentalized'; that is, more tightly interwoven with American regional and metropolitan economies.[12] Flows, increasingly, are north–south. This integration is perhaps most obvious in the case of southern Ontario and its durable goods manufacturing industries (Courchene and Telmer, 1998). The east–west economic linkages built up in Canada behind tariff walls, and legal and legislative barriers, have traditionally provided the basis for social transfers that have served as the glue of confederation. It is possible that this arrangement will become unglued as the Canadian urban system becomes increasingly part of a larger North American metropolitan system dominated by those places at the top of the US urban hierarchy.

How then might one classify places in the contemporary North American urban system? Most classifications of urban systems undertaken in the past, and there have been many, are limited to one country or to one or a few attributes. Here I suggest that within an urban system context, North American cities can be arrayed along a small number of dimensions:

(1) *size*: perhaps the basic, and certainly the most obvious contrast, is between large and small metropolitan areas, particularly differentiating between those metropolitan areas with two million or more residents and those with much smaller populations. Larger places tend to have higher order functions, higher densities, some level of transit service, a more diverse economy and housing stock, heterogeneous populations, and by definition more linkages with other places and the global economy.

(2) *the age and timing of development*: cities mirror the time of their construction. The primary contrast here is between old and new metropolitan areas, especially between those built up during the pre-war period and those which are largely post-war constructions (and thus automobile-dependent).Contrasting examples: Boston and Phoenix; Québec and Calgary.

(3) *dominant economic base*: as the evidence above suggests, the role, position and growth of cities within the larger urban system depend heavily on their dominant economic activity. Among the obvious contrasts are between older industrial (e.g. Buffalo, Youngstown) and new retirement communities (e.g. Orlando), and between government cities (e.g. Washington DC) and market-sector management cities (e.g. Houston).

(4) *the socio-cultural origins of the population*: within the North American urban system cities differ relatively widely in terms of the social-cultural and racial diversity of their populations. Contrast the French presence in cities in Québec and Louisiana with the Anglo-Irish populations of nearby Ontario and Mississippi; the Mexican populations of south Texas and California with Hispanics in New York, Cubans in Miami, and Asians in Los Angeles, Toronto and Vancouver, and most importantly, those with and without large black populations.

(5) *political culture and organization*: the history and cultural diversity of North American populations and regions is evident in differing levels of political fragmentation within cities, and more broadly in the cultures associated with politics, political institutions and everyday life.

(6) *the US–Canada border*: the differences engendered by the national border between the US and Canada persist, and to a degree are strongly correlated with the above attributes. But these differences may be declining. On balance the urban system in Canada is more decentralized, with stronger regional integration (e.g. within Québec and the West) and weaker inter-regional ties. Individual cities often have much stronger linkages to nearby American cities (e.g. Halifax to Boston,

Montréal and Toronto to New York, Winnipeg to Minneapolis and Chicago, and Vancouver to Seattle, San Francisco and Los Angeles), than American cities have in the other direction.

In other words, national differences constitute one, but only one, of the dimensions along which cities in the North American urban system are likely to differ. Viewed from outside, American and Canadian cities look remarkably similar, at least in physical appearance; viewed from inside they are equally remarkably diverse and unequal – and becoming more so (see Chapter 13).

LOOKING TO THE FUTURE

As we enter the new century what can we say about the future geography of urbanization, and about patterns of future growth and the continuing reorganization of the urban systems in North America? As both Canada and the United States emerged from the recession of the early 1990s, the dynamics of population change and urban economic growth remained as volatile as ever. It is unclear, however, whether the trends and patterns described above will continue into the first decades of the twenty-first century, or whether a new urban era will ensue. What is clear, however, is that with lower rates of national population growth the processes that influence the redistribution of population and economic activity within the urban system will assume even greater importance. Among the host of factors that contribute to this persistent uncertainty are immigration and ethnic pluralism, an ageing population, shifts in life-styles and consumer preferences, economic restructuring, technological innovation, trade liberalization and global competition.

First, consider the implications of an ageing society. The front-end of the massive baby-boom population in both countries will reach retirement age in 2010 and then the proportion over 65 will reach 22 per cent. Where will this huge population live? Will they continue to reside in their present environments – the vast majority now live in the older and larger metropolitan areas – or will they migrate to smaller cities and towns in less expensive and/or amenity-rich areas?

Second, immigration from abroad – unless it is curtailed by policy decisions – will continue to have large and very uneven impacts on particular metropolitan areas. The cities that serve as primary destinations or 'gateways' for immigrants will be transformed, at least in their social structure – largely in positive ways, but still not without tensions. Some of the bigger metropolitan areas, such as Los Angeles, San Francisco, New York, Miami, Vancouver and Toronto, have populations that are now over 40 per cent foreign born.[13] Since most other parts of the two countries – except along the Mexico border – receive few immigrants, a clear differentiation is emerging between the gateways and the rest of North America. Indeed, there is an ongoing debate in the United States as to whether overseas immigrants are replacing or displacing native-born populations, and in so doing leading to a 'balkanization' of American metropolitan areas (Frey, 1995; Frey and Liaw, 1998). All of the above metropolitan areas, even Miami, have recorded huge losses in terms of net internal (domestic) migration, while gaining foreign migrants.

Adding to this uncertainty is continuing economic restructuring, increasing global economic integration through trade liberalization policies (e.g. NAFTA) and transborder flows, and the hyper-mobility of global financial markets. In the United States and Canada, assuming no surprises, those places which will benefit most from NAFTA and the process of globalization are likely to be the same places that benefited from shifts in national and regional economies over the last two decades, notably those urban areas in the southern and western states and in the two western provinces in Canada. Cities with high-tech industries, or high-order service and financial functions, as well as those with retirement, recreation and entertainment functions, or educational and health-related economies, will continue to do well. Those with low-wage economies, rusting industries, a limited range of services and few environmental amenities – and those poorly located with respect to the urban system – will suffer most from intensified competition.

The urban system writ large will continue to evolve, gradually and in directions established over the last half century, but with more regional variation. The urban hierarchy will shift gradually to the South and West, but again with a more complex organization than in the past. The largest 20 US metropolitan areas listed in Table 12.4 will continue to dominate, but with population and power shifting to the newer metropolitan areas, notably Los Angeles, San Francisco, Dallas, Houston, Miami, Atlanta, and to an even larger number of soon-to-be very large metropolitan areas, such as Phoenix, Tampa–St Petersburg, Orlando, Austin. In Canada, both Vancouver and Calgary will continue to top the growth charts, yet Toronto's growth and dominance will likely be maintained because of its diverse economy and through immigration. For Canada, however, the basic question is whether the Canadian urban system will increasingly become a dependent 'regional' urban sub-system within North America.

The scale of urban growth and particularly the long-term implications of creating such huge metropolitan agglomerations should not be underestimated. In two provocative papers, Chinitz (1991) and Borchert (1991) lay out the conditions that might produce even higher levels of concentration of population and economic activity in a few mega-metropolitan areas. Borchert speculates that by the second decade of the twenty-first century the emerging conurbation in the United States will be in southern California (stretching from Santa Barbara to Los Angeles and San Diego) with 22 million people, and this will have passed the New York metropolitan region (at 19.5 million) to become the dominant American metropolis. The San Francisco–Oakland–San Jose–Sacramento conurbation, with 11.7 million will pass the Chicago–Gary–Milwaukee region (11.2 million) and move into third place. Other conurbations will also move up the hierarchy: an emerging central Florida metropolis (Tampa–Orlando–Daytona Beach) could reach 9 million, Baltimore–Washington over 8 million, and the existing south Florida metropolis (Miami–West Palm Beach) might have 6.5 million. In

Canada, the emerging metropolis around western Lake Ontario, centred on Toronto–Hamilton and extending from Oshawa to Kitchener and Niagara Falls, could have nearly 8 million people by 2011. The physical scale and complexity of these agglomerations, and the degree of concentration in urban life they reflect, are staggering. How does one manage, or even understand such agglomerations?

Or will the expanding 'knowledge' economy, and the electronic 'wired world' of telecommunication and the Internet, override the tendency to further spatial concentration? Such innovations, in theory, make everyone and everywhere in the urban system easily accessible. No doubt an increasing number of routine functions will physically decentralize to smaller communities and peripheral areas, or go 'online'. Yet at the same time, the nature of the telecommunications industry, and the aspirations of most individuals in terms of where and with whom they want to live and work, almost ensure a continued growth (and concentration) of high-order services, information-intense functions and jobs in large metropolitan command centres, especially those areas that are also rich in amenities and cultural activities.

The most obvious generalization that can be made about growth and change within the urban systems of North America is 'even more of the same'. But the patterns of growth will be considerably more complex than in the past and more variable over space. One implication, and a continuing challenge for policy makers, is how to reduce the negative consequences of this uneven profile of growth and prosperity and how to alleviate the resulting disparities in the quality of urban life.

ENDNOTES

1. In both the US and Canada, a metropolitan area is delimited by combining a central city (the urbanized core) with those adjacent suburban areas that are functionally integrated with the urbanized core through the operation of local housing and labour markets. The total population

must exceed 100 000 (except 75 000 in New England). Given differences in the spatial units used as building blocks between the two countries, MSAs tend to be larger than CMAs. Other differences make direct comparisons difficult. The US Census also defines Consolidated Metropolitan Statistical Areas (CMSAs) that combine individual MSAs (called Primary MSAs) that are adjacent and tightly integrated. (See Bunting and Filion, 2000, and Knox, 1994, for details.)

2. The US and Canadian censuses are taken at slightly different times: at the beginning of the decade in the US (i.e. 1950, 1960, etc.) and a year later in Canada (i.e. 1951, 1961, etc.). Canada also holds a mid-decade census (e.g. 1996), the US does not. Unless otherwise stated comparisons between the two countries refer to 1990/1991, and any post-1990 figures for the US are estimates.

3. Again, census definitions of what is urban differ somewhat between the US and Canada. In Canada any settlement with more than 1000 population that also meets a certain level of population density (400 per km^2), is classified as urban. In the US the minimum threshold is 2500 residents. In both cases, rural areas (and populations) are residuals – i.e. they are what is left after urban areas are defined.

4. One frequent source of confusion is to equate metropolitan and urban populations. Surprisingly, the proportion of the population that is classified as metropolitan (77.5%) is higher than the proportion urban (72.5%). How is this possible? The explanation is that some 26.5 million residents of metropolitan areas are classified as rural, while 20.9 million residents of non-metropolitan areas are classified as urban (i.e. they live in towns and cities of less than metropolitan size).

5. Sports fans, at least baseball fans, will immediately recognize this hierarchy of urban areas as the baseball league tables. Professional baseball has 30 teams, including two in Canada. All but one (Washington) of the 20 US MSAs in Table 12.4 have teams, and four have two teams: New York, Chicago, Los Angeles–Anaheim, San Francisco–Oakland. The evolution of baseball franchises over time provides another map of the evolution of the urban system.

6. These totals refer only to legal immigrants (and refugees).

7. Lemon (1996) provides a fascinating account of the historical evolution of five large North American cities and the urban systems of which they were/are a part – Philadelphia (in 1760), New York (in 1860), Chicago (in 1910), Los Angeles (in 1950), and Toronto (in 1970) – each selected to illustrate the conditions of that period.

8. The lowest income metropolitan areas in the US are predominantly in the South, the highest income areas in the Northeast.

9. A detailed analysis of the sources of differential metropolitan growth in the US is provided in Mills and McDonald (1994).

10. As one example, in 1970 Flint, Michigan had 446 000 people, almost exactly the same as Orlando's 453 000. By 1995, Flint had only 430 000 people, 16 000 fewer; Orlando had over 1 200 000.

11. Parallel shifts have taken place in the Canadian urban hierarchy. Toronto passed Montréal by 1971, completing a transition that had begun in earnest in the 1920s; Ottawa–Hull, and then Edmonton and Calgary (Canada's Dallas and Houston), passed Winnipeg (Canada's Chicago) in the 1980s. (See Frisken, 1994; Bourne and Olvet, 1995; Bourne and Flowers, 1996).

12. More than 75% of all Canadian foreign trade is with the US, and the proportion has been increasing.

13. The Greater Toronto Area, with 4.6 million people in 1996, had over 1.8 million immigrants (39.7%). In the City of Toronto (formerly Metro Toronto), with 2.4 million, over 47% were foreign born.

KEY READINGS

Downs, A. 1994: *New visions for metropolitan America*. Washington DC: The Brookings Institution.

Hart, J.F. (ed.) 1991: *Our changing cities*. Baltimore: Johns Hopkins University Press.

Knox, P. 1994: *Urbanization: an introduction to urban geography*. Englewood Cliffs, NJ: Prentice Hall.

13

NORTH AMERICAN CITIES: THE MICRO-GEOGRAPHY

JOHN MERCER

INTRODUCTION

Approaching the close of the twentieth century, the vast majority of Americans and Canadians live in metropolitan areas (almost 80 per cent in the United States and just under 66 per cent in Canada), the outcome of social and geographical processes that were established at mid century. But within the continent's metropolitan matrix, a notable number of great cities and many lesser ones have (demographically speaking) faded in the face of burgeoning suburban population growth. Generally speaking, this has been less the case in Canada than in the US. As of 1990, almost half of the US population lived in the suburbs (46 per cent) and within the dominant metropolitan centres, those residing in the outer city outnumber those in the central city (roughly three to two). In Canada, metropolitan suburbanites account for only just over a third of the national population and the ratio of outer to central city populations is about 1.3:1. Even allowing for the more extensive areal definition of the American Metropolitan Statistical Areas (MSAs) (see Chapter 12) and the fact that in proportional terms the central city in the Canadian context accounts for a higher proportion of the overall metropolitan population than is generally the case in the US, these data suggest that it is in the US where the suburbs are more dominant (in accounting for President Clinton's re-election in 1996, the suburban living, mini van driving, soccer supporting 'mom' took on near mythic proportions).

An increasing and remarkable complexity characterizes American and Canadian metropolitan systems, arguably greater in the former case. There is a staggering amount of variation that makes it extremely challenging to generalize about American cities or Canadian cities separately, far less sweep across the whole continent. But some attempt to convey this complexity is necessary. In America, for example, racialized residential segregation is common everywhere but it takes a particularly pronounced form in certain places; other places are key locales for the Latino population whereas elsewhere they

are relatively insignificant, and only a few large metropolitan agglomerations are the scenes of large scale immigrant settlement – they have been termed 'gateway' centres (see Chapter 12.) Likewise demographic change, encompassing population and household dynamics, is also highly variable. Easy generalizations about declining central cities, mushrooming suburbs and decentralization only serve to mask a complex reality that needs to be depicted in a more geographically-sensitive manner. In the Canadian case, the immigration experience has transformed a few leading metropolitan destinations (such as Toronto and Vancouver), touched others and almost completely bypassed yet others. While 'race' is not the social marker it is in the US, racialized minority populations ('visible minorities' is the official lexicon) are strongly related to the nature of immigration in the last three decades, save for historic populations in a few distinct locations (such as Halifax, Vancouver and Victoria) and the indigenous peoples found in increasing numbers in urban locations. There is no Canadian counterpart to the Latino situation in the US but language issues are the stuff of social tension and conflict in various cities, particularly in Montréal (see Chapters 8 and 16).

The cities of North America are in a remarkable state of economic flux. Put simply and at the risk of overgeneralization, the urban areas of North America largely grew into their present prominence on the back of a globally experienced industrialization, led by the manufacturing sector, spanning just over a century or so. But the common use of terms like 'post-industrial', and 'the informational city' suggest that a dramatic transformation of this continent's economy is underway. Thus, in the form and organization of the metropolitan areas we should expect to find manifestations of this rapid economic change. Also, as both economies are open to a common global system and are 'managed' by central institutions (like banks and markets) in a broadly similar tradition of a liberal political economy, we might not expect especially marked cross-national urban differences to emerge here; although the role of the public sector and the commitment to social support programmes has historically been greater in Canada.

Some other structural characteristics of metropolitan areas and their residents do appear to be notably enduring, even in the face of demographic and economic change. They provide a basis for drawing contrasts between the structure of American and Canadian cities. Most notably, in the United States the metropolitan territory is markedly fragmented into a myriad local governments – this is much less the case in Canada. The American central city, typically the site of the metropolitan CBD, is often portrayed as the loser in an ongoing struggle between the constituent parts of any given metropolis for investment, government funds, taxable property and residents, especially affluent ones. Such a portrait is generally less appropriate for Canadian cities but in a few larger metropolitan areas conditions now are closer to the US norm. Again, North American cities literally function on the backs of autos and trucks, and are thus both captive to the automobile while achieving great personal mobility for commuters and residents. This auto commitment (some say love affair) is particularly strong in the US, whereas urban Canadians are significantly greater users of public transport. One enduring trait of both sets of cities is a pronounced inequality of living and working conditions, an inequality that is unsurprising in a capitalist political economy, one intensified by the ideological strength of free market doctrines in the last decades of the twentieth century – the welfare state, never especially strong in North America compared to Europe for example, is in retreat; however, inequality may be somewhat lesser in Canada and its cities. Lastly, the housing landscapes of North American urban areas are dominated by the single family detached housing unit (mostly in an owner-occupied tenure); more compact, less dispersed Canadian cities are indicated by a generally lower proportion of such units in the metropolitan housing stocks and by urban transportation differences compared to US cities.

METROPOLITAN AND CITY POPULATION DYNAMICS

Despite published statements and popular perceptions about a general central city decline in American cities, the situation is more complex. Here, I use a descriptive five-fold typology which not only helps convey the diversity of experience but also allows a Canada–US comparison in terms of population shifts within urban areas.

- *Type 1*: central city decline, allied to metropolitan decline
- *Type 2*: central city decline, and metropolitan growth
- *Type 3*: central city stagnation (equal to or less than 5 per cent[1]) and metropolitan growth
- *Type 4*: solid central city growth (in the 5.1 – 19.9 per cent range[1]) and metropolitan growth
- *Type 5*: booming central city growth (20 per cent and over[1]) and metropolitan growth

For the 1980–90 decade, America's metropolitan areas exhibit a considerable diversity of contexts (Table 13.1). About a quarter fit the almost classic urban model of outer area growth and central decline, and close to a fifth (17 per cent) are in serious difficulty (the entire metropolitan area is losing population). However, the remainder exhibit a range of growth experience in their central cities, including over 40 per cent which had solid or booming growth. The Canadian experience in this decade was notably different (see again Table 13.1). No metropolitan area lost population and hence there are no *Type 1* areas; in addition, there are few (4) metropoles in the so-called classic model of *Type 2*. At the other end of the continuum, there are proportionately more American places in the fastest growing category (*Type 5*). In short, the American system exhibits more diversity (a conclusion that mirrors the findings of Goldberg and Mercer (1986) for an earlier decade).

The situation in the declining American places and those in the second group is exacerbated by the loss of households.

TABLE 13.1 Distribution of metropolitan areas by type: US and Canada, 1980–81 to 1990–91

	Type 1		Type 2		Type 3		Type 4		Type 5	
	n	%	n	%	n	%	n	%	n	%
US	39	17.0	54	23.5	40	17.4	60	26.1	37	16.1
Canada	0	—	4	16.0	8	32.0	11	44.0	2	8.0

Source: Author's calculations

Population loss in and of itself may not be especially disadvantageous, if, for example, suburban-bound families are replaced by smaller households, then population decline may occur but housing units are being occupied and the household as a key consumption unit is maintained. But to lose households is more damaging to a city's quality of life, resulting in a softening housing market, a likely decline in retail purchasing power, housing vacancies and even abandonment which may bring in its train arson and demolitions. In the last decade for which household change data are readily available (1980–90), some 60 central cities experienced a loss in households (over 20 per cent of the nation's central cities larger than 50 000). Such a loss is common in *Type 1* and 2 cities as 28 of 39 in the first group and 25 of 54 in the second suffered this critical erosion. Household loss is most pronounced in the inner parts of larger metropolitan areas posing a long-term threat to many major places if this decline is not reversed in the economically more prosperous 1990s. In some cities, such as Newark (–17.5 per cent), Detroit (–13.7 per cent) and Pittsburgh (–7.6 per cent), this is a continuation of declines evident in the 1970s and is indicative of a serious decay not easily reversed. The metropolitan areas which have the weakest central cities, as manifested by household decline in the 1980s, were strongly clustered in America's industrial core – over half (34 of 60) were in the contiguous states of New York, New Jersey, Pennsylvania, West Virginia, Ohio (the most with eight), Indiana and Illinois.

Of course population loss in American central cities is hardly a new phenomenon, as numerous cities peaked in 1950 (for example, Detroit, Syracuse and others) and have declined inexorably ever since. But their overall decline masked an important change in racialized identity – as white populations migrated outwards to the beckoning suburbs or left the region altogether their departure was partially offset by significant inflows of African Americans and, in certain locations, Latinos. Now, however, in places like Cleveland, Detroit and Washington, DC, to name a few, the minority population is also declining, portending a difficult future indeed, especially if such cities do not attract large numbers of new foreign born immigrants (at least the DC metro area has this positive growth prospect).

As noted in Chapter 12, there is no official census of population and housing in the United States in a mid-decade year as there is in Canada, so population estimates have to be used (and cautiously) for America to compare with the 1991–96 changes that can be determined for Canada (even regular census data need to be treated cautiously as they can convey a false sense of precision and accuracy). These data do indicate something of a convergence (Table 13.2); this is a likely reflection of the depth of the recession in Canada in the early 1990s on the one hand, and the general strength of the American economy in the current decade. In proportional terms, the Canadian system has a higher percentage of metropolitan areas in *Types 1* and 2 than does the US and in 1991–96 there were no Canadian areas in the most rapidly growing group. Comparing Tables 13.1 and 13.2 suggests that for the US there is a general distributional shift from the declining *Type 1* towards the booming *Type 5*, whereas in Canada the movement has been in the other direction. Furthermore, there are also detectable regional effects in both coun-

TABLE 13.2 Distribution of metropolitan areas by type: US and Canada, 1990–91 to 1996

	Type 1		Type 2		Type 3		Type 4		Type 5	
	n	%	n	%	n	%	n	%	n	%
US	15	6.6	53	23.3	31	13.7	85	37.0	43	18.9
Canada	2	8.0	8	32.0	5	20.0	10	40.0	0	—

Source: Author's calculations

tries. In the US, a number of metropolitan areas in New England, but not the Boston area, now belong in *Type 1* along with several from upstate New York, Ohio and Indiana which were in this category in the eighties. The Canadian metropolitan places in the two least desirable categories are all in Atlantic Canada, Québec (including Montréal), and the resource dependent region of northern Ontario, with one exception – Edmonton, where the central city declined by a minuscule –0.1 per cent in the 1991–96 period. Weak regional economies and political uncertainty are taking their toll on the attractiveness of these places for investment and adversely affecting their growth potential.

Racialized divisions and residential segregation

The depth of racial aversion and geographical separation in urban America is starkly depicted in the title of Massey and Denton's *American apartheid* (1993). The overall situation in the US is still dramatically different in terms of racial separation from that experienced in Canadian cities although in several of the latter racialized identities are now a significant part of everyday life. As a consequence of immigration since the 1960s, there are now numerous individuals and families of Afro-Caribbean, East Asian (especially Chinese) and South Asian descent or birth in places such as Vancouver, Toronto, and to a lesser but still important extent, Montréal. Refugees and immigrants from Spanish-speaking parts of Central and South America, and also from various African countries plus indigenous peoples who have moved into cities, all combine to make major Canadian cities more diverse in terms of

racialized identity than at any previous time.

But dramatic as these changes are, there is no counterpart to the way in which race marks the American city, although the specific geographical form of that marking does vary by region in terms of the patterning of the black ghetto.[2] First, there are 17 central cities where African Americans now constitute the numerical majority, with Detroit (76 per cent) and Washington, DC (66 per cent) being the largest of these, although the declining steel town of Gary, Indiana tops the list at just over 80 per cent. The overall distribution of the black population in central cities is fascinating for what it tells us about difference and variability in urban America. Perhaps surprisingly, a full two fifths of all cities had fewer than 10 per cent of their population black, while a fifth had proportions in excess of 33 per cent, including such notable places as St Louis, Cleveland, Chicago, and Philadelphia. It is highly likely that many of these cities which exceed this one-third threshold will become majority black within a few decades, although the cessation of the massive migrations out of the South in the 1950s and 1960s to the then growing industrial cities of the Northeast, upper Midwest and West will undoubtedly slow this transformation.

But regardless of the proportion of a city's population that is black, it is necessary to ask where do African Americans live and under what conditions in America's numerous metropoles. After an exhaustive survey of available evidence and further empirical analysis of their own, the urban sociologists Massey and Denton (1993, p. 109) conclude: 'First, residential segregation continues unabated in the nation's largest black communities, and this spatial isolation cannot be attributed to class.' They further conclude that while open housing is accepted by majority whites in principle, its practice is found wanting in most markets as whites, harbouring negative feelings towards blacks as co-residents, are willing to tolerate only a modest proportion of black neighbours in their residential areas (see Chapter 8). Lastly, 'discrimination against blacks is widespread and continues at very high levels in urban housing markets' (p. 109). As they acknowledge, however, racially-based segregation has declined in certain types of metropolitan areas, notably smaller areas in the South and West. But in many instances, the proportion of African Americans in such places is relatively low and this context greatly affects the probability of black–white contact or interaction.

An important distinction is drawn by Massey and Denton between a form of segregation that is exemplified by midwestern places like Buffalo and Milwaukee, where the minority black population is highly concentrated in one principal city ghetto and very few are to be found in the suburbs, and another form which demonstrates segregation of black residence in the central city, as well as in the suburbs, where black proportions are more significant overall. The latter tend to be in southern regions, and are exemplified by Birmingham, Alabama, Washington, DC, and Atlanta (Chapter 14). This again points to urban variability – in this instance in social pattern and ghetto form. But black suburban residence can mean social isolation and segregation, as in the central city – illustrated by the development of patterns in suburban Washington (Massey and Denton, 1993, p. 70).

Updating and extending this case is telling. The African American population in the Washington, DC metro area has grown to over one million (1990) from just over 700 000 in 1970, but given the overall growth of the Washington agglomeration the proportion of African Americans has remained relatively stable (1970: 25 per cent; 1980: 28 per cent; and 1990: 26.5 per cent). There has been a striking growth of the black population in the suburbs but in the city this population has

fallen each decade – from over half a million in 1970 to just under 400 000 in 1990. This is most clearly revealed in the declining share of the metropolitan region's black population living in Washington, DC itself – from 76 per cent in 1970, to 53 per cent in 1980, and then to 38 per cent by 1990. Within the suburbs, there have been some equally dramatic changes. The proportion of African Americans living in Prince George's County shot up from 13 per cent to 29 per cent and then again to 35.5 per cent over the same period. In 1970, the other counties that made up the metro area contained only 8 per cent of the region's black population but their spread throughout the region is indicated by the 26 per cent now (1990) living in the suburban counties. Another notable feature in the making of suburban segregation is that the proportions of black people in those rural Maryland counties brought within the Census Bureau's expanding MSA have declined as white populations have migrated there. At the local scale, racially-based transitions are equally striking; communities with minority proportions in the 8–16 per cent range in 1970 reached the 70–85 per cent level by 1990, by which time there were 11 places in Prince George's County with majority black populations. These places are as ghettoized as city neighbourhoods in terms of residential segregation, even if housing and socio-economic circumstances differ. Similar portraits of suburban segregation could be sketched for Atlanta, Memphis and Birmingham, for example.

There is absolutely nothing comparable in a Canadian context. As Brian Ray (1992) has shown, the Jamaican immigrant population in Toronto, the principal destination for people of Caribbean origin migrating to Canada, is widely scattered across the area although there is a local clustering in several locations, some associated with suburban concentrations of assisted housing. In Vancouver, the other immigrant gateway, the proportion of Asian-born residents in the central city has risen rapidly with both Chinese and Sikhs being most prominent. But this proportion is still less than half of the total population and only in a few, small areas, such as those in and around Chinatown, has the concentration reached the levels commonly associated with black inner-city ghettos in American cities. That said, Canadians are aware of racial issues and racial change, fuelled by disputes over housing unit size and street appearances, or the phenomenon of Asian youth gangs, or charges of police brutality towards minorities – most Canadians however experience these at second hand, filtered through the media. Comparatively speaking, then, the cities of Canada simply do not have ghettos in the American sense of the term. A thorough comparative analysis by Fong (1996) concludes that whereas African Americans experience extreme residential segregation in the major cities where they chiefly live, blacks in Canada experience minimal segregation, even when accounting for the historically lower proportions of blacks in the urban areas. He further finds that the local labour market is a key factor in understanding the high level of segregation in US cities; given the structural changes in work and employment opportunities, together with the poor quality of education for many African Americans, as well as discrimination in housing and labour markets, the prospects for dismantling the ghetto and reducing significantly black–white segregation are bleak.[3]

While black Americans consider themselves as indisputably American and have sought for decades to fully enter the mainstream of society and especially the economy, Latinos (people of Latin American origin or ancestry) are widely perceived as immigrant newcomers who do not speak English (even although there are numerous parts of the Southwest where Spanish speaking peoples have resided for decades, even centuries, and many are also bilingual). This population also has a particular urban geography both within and between cities that has no counterpart in Canada. There are 10 central cities where Latino people constitute an absolute majority, and 34 where they exceed a quarter of the total population. As one might expect, all of the first group are in Texas and California, save for Miami. Something of the national spread from the historic southwestern 'hearth' region can be observed in that several cities in the Northeast and one in the

Chicagoland region fall into the second group, with New York City and Chicago containing proportionately and absolutely very significant though culturally diverse Latino populations (the 1990 proportions are 24 per cent and 19 per cent respectively). However, while Latinos reside in every metropolitan area and central city, their relative significance is a different matter – in over half of America's central cities they represent less than five per cent of the total population. So, while this socially and politically important group adds to the diversity of American urban life, it is particularly concentrated in a key subset of metropolitan areas. The Latino presence in suburban parts of the metro areas is especially noteworthy.[4] As Frey and O'Hare (1993) observe, almost half of Hispanic Americans live in suburban areas, and there are marked income and educational differences between them and their more disadvantaged ethnic peers living in central cities.

In certain large- and medium-sized inner cities, Latino ghettos (barrios) do exist, in some cases, alongside black ones (Chicago's West Side). But the evidence suggests that residential segregation for Latinos (or Hispanics, to use the census term) is notably less than it is for African Americans (Massey and Denton, 1993, p. 67). As a socio-linguistically (or culturally) defined minority, the Latino population reflects a variety of racial identities; it is a telling fact that the darker the self-identification, the higher the residential segregation, with levels at the higher end approximating those for black–white segregation. Economic deprivation, severe disadvantage and decimated residential areas are not unusual in the barrios of Chicago, New York, Los Angeles, Denver, Newark and Camden, New Jersey (see Vergara, 1995).

Social patterns

The social geography of North American cities has for some decades now been characterized by the basic dimensions determined from statistical analysis of large data sets encompassing many socio-economic, educational, demographic, housing and racial/ethnic variables. Such studies have generally summarized these complex geographies in terms of three basic factors or constructs: social class or socio-economic status; stage in the life cycle: ethnic or racial segregation. More recently, American analysts have been emphasizing greater and smaller scale complexities, amidst larger societal changes in family and household structures, life-style differences and greater disparities in wealth and income, especially in the last two decades as market hegemony has been a strong and countervailing force to the redistributional programmes associated with welfare state orientations (Knox, 1994, pp. 217–31).

This complexity is emphasized by the findings of a factorial analysis of the entire Canadian metropolitan system (Davies and Murdie, 1993). Now, nine dimensions (or factors) are seen as necessary to capture the variation in the rich data set for all the metropolitan centres combined. The 'classic' social rank or class factor breaks down into two factors: economic status, and impoverishment, suggesting a sense of greater disparity and the falling behind of low-income and single-parent families. The family status or life cycle factor decomposes into five factors reflecting demographic change and fundamental shifts in household structure (families; non-families; early/late family; young adult; housing). Lastly, the segregation factor (typically ethnically based in Canadian research while racially based in US studies) splits into an ethnic factor (tapping into the non-British and non-French populations) and a factor that Davies and Murdie label 'migrant', linking local moves and apartment residences. Further emphasizing complexity and the need for a more geographically-sensitive focus, they argue that, for example, in considering high status areas in Canadian metropolitan areas there are four types that need to be acknowledged: inner-city areas where high status persists; new suburbs, possibly linked to some recreational amenity; newly-gentrified areas (commonly located in the inner city); and other central-city areas which have been completely redeveloped to high quality and costly residential uses. Even allowing for greater complexity, the emphasis on high

status in various central city locations is noteworthy – a scenario much less likely in most US central cities.

But there may be more durability in the three basic constructs than the proponents of greater complexity might wish to concede. For example, in an analysis of the social geography of the Wichita, Kansas metropolitan area (with a 1990 population of just under 500 000), Yeates (1998b) finds that the pattern of three principal factors broadly corresponds to the classic trinity, with the segregation factor picking out a clear area of concentration of African Americans (the term 'ghetto' would not be inappropriate in this instance) and those of Hispanic origin, although the latter are more dispersed. Again, in a study of social patterning in Toronto, Fabbro (1986, cited in Yeates (1998b)) established a three-factor model comprising (a) social status, (b) life course, and (c) an ethnic factor (principally non-Anglo Saxon, which, by 1981, includes racialized groups such as those of African descent from the Caribbean, as well as people of Chinese origin – either by descent or place of birth). Nevertheless, the preponderance of evidence is that there is more socio-geographical complexity in the cities of North America than at any other time in the post-Second World War era.

INEQUALITY

This overview of the social geography in North American metropolitan areas provides a context to discuss inequality and disparity within urban areas. There has been much public comment and debate about increasing inequality in the advanced democracies in recent decades. This is a complex matter with difficulties in comparability of data, and scholarly debate over the interpretation of data and statistics computed from various databases. The social economists Gottschalk and Smeeding (1997) indicate that, whereas there has been a significant rise in income inequality in the United States from the mid 1970s to the mid 1990s, in Canada there was a modest decrease from the mid 1970s to the mid 1980s and then no change through to the mid 1990s. Other national-level evidence points in the same direction. Hanratty (1992) reports family poverty rates for Canada at 7 per cent and 12 per cent in the US (1986 data); this clear cross-national difference is particularly acute for families headed by a single parent with a US rate at just over 40 per cent compared to just over 25 per cent in Canada. Consistent with the evidence of Gottschalk and Smeeding, she argues that the lower incidence of poverty in Canada is because the transfer system there is more generous and effective. At the metropolitan scale, we should not be surprised that the inequalities between the nations are mirrored in the huge urban agglomerations and smaller metropolitan centres. Based on analysis of 1970s data, Goldberg and Mercer (1986) showed that the income disparity between the central city and the metro area as a whole was higher in the US than in Canada.

This work in part prompted other more detailed and fine-grained studies of this matter, especially in Canada. Bourne (1993) rightly argues for going beyond the rather arbitrary central city–suburban distinction in urban analysis and does so most effectively in a study of the geography of income in Canadian cities, although considerable weight is placed on Toronto as a case study within the broader Canadian context. A key finding is that inner parts of the Canadian central city are characterized by continued, or even deepening impoverishment, by pronounced accumulation (and display) of wealth and growth of elite districts, and a remarkable degree of persistence in both the location and composition of well-established high and middle income areas. One would be hard pressed to fit this characterization to the inner city of most American urban areas (though some particular areas might fit – possibly in Seattle and Minneapolis, Dallas and Houston?). Bourne also demonstrates that the classic gradients from the Chicago School era still have descriptive relevance in Canadian urban areas (household status and income rises systematically with increasing distance from the city centre). But echoing the Davies and Murdie argument, he concludes that the ecology of income is 'significantly more complex' (p. 1313). Underlying

this complexity is the large-scale construction of luxury and more modestly priced condominiums in Canada's central cities (and the suburbs in such places as Victoria and Vancouver with their large populations of retirees) and the concomitant destruction of cheaper housing often in the rental sector, thus pushing the residents of the latter into public sector housing.

Although Bourne and others want to move beyond the oppositional city–suburban debate in a public policy context, it is difficult to see this disappearing from public discourse as it frames the debate over fiscal well-being and taxation, particularly in American cities where 'the city' has all too often become a code word employed by predominantly white suburbanites for a racialized 'other' they are anxious to avoid or have consciously migrated away from. This is not to say that urban geographers and other students of cities cannot go beyond this dichotomy but it will be a hard task to move the public and the media. A steady diet of nightly TV news in America portrays the city as a place of fires (likely arson related), violent crime, complaints over inadequate city services, and beset with fiscal problems and crises (Beauregard, 1993).

The income disparities and socio-geographical polarization addressed above have been expected to worsen in the urban areas of both countries. First, the de-industrialization thesis has been advanced as a cause of greater income differentials within cities as the middle of the income distribution is believed to be shrinking, many lower-waged workers, especially males, have been thrown on the industrial scrap-heap, and the incomes of the managerial and skilled professionals have grown more rapidly than the norm. Second, studies in a number of cities have demonstrated greater polarization, notably in world cities (New York, Chicago and Los Angeles, for example would be American agglomerations in such a listing) – see, for example, Sassen (1991) and Fainstein *et al.* (1992). Canadian cities typically are not considered to be primary world cities although Toronto is often included at a second level in such hierarchies. Although many new jobs have been created in a rapidly-expanding service sector to the point that employment in services now dominates many if not most metropolitan labour markets (Chapter 11), it has been widely observed that many of these new jobs are low wage in nature, with a high proportion of part-time employment and a related lack of benefits. This should also contribute to greater disparities, since well paid professionals abound in the managerial levels of the service sector, especially where knowledge-based and information-handling skills are highly rewarded.

INDUSTRIAL STRUCTURE AND CHANGE

It would be a serious mistake to think that because of the widespread use of terms like de-industrialization, post-industrial and the like, the cities of North America are replete with industrial decline and structural decay of industrial landscapes – to be sure, this does exist but there are also areas of industrial growth, even boom. As Bourne has shown in Chapter 12, manufacturing-based metropolitan areas have generally done poorly in the 'growth stakes', some doing exceedingly poorly. But considerable industrial growth has occurred in metropolitan areas that have diverse economies, that have growing populations and hence markets for industrial products, especially consumer durables (e.g., Los Angeles and Phoenix), and where new kinds of industries have taken root (Los Angeles again, the famous Silicon Valley example, or the Research Triangle in the Raleigh–Durham MSA in North Carolina).

In terms of the internal structure of urban areas and the location of manufacturing, one has to keep in mind the overall role or function of the area within the larger, metropolitan-based national economic system. Places like Detroit, with less economic diversity than most agglomerations of this size (4.4 million in 1990), have experienced industrial decline in the city and in those parts of inner suburbs where some of the first large-scale industrial plants were built to the point that

the city, with over a million people, has only 20 per cent of the metro area's manufacturing jobs; however, manufacturing has prospered in the outer suburbs so that almost a third of manufacturing employment is now located here – the remainder is to be found in the inner suburbs. Overall, however, as Pollard and Storper (1996) show, routine manufacturing in the Detroit Consolidated MSA contracted more than any other sector (from a 1977 index value of 100 to 78 in 1987).

Such a spatial resorting of manufacturing plants and changes in employment, the latter possibly reflecting gains in productivity, might be defended on market-based efficiency grounds. However, in the case of Detroit, Cleveland and numerous other American cities, this particular economic geography combines with the geography of 'race' described earlier such that the large minority populations of the cities are spatially and thus economically disconnected from manufacturing job opportunities in the inner and, even more so, the far distant outer suburbs. Kodras (1997) paints an alarming picture of the overall increases in poverty in the City of Detroit in the face of massive industrial decline, with a disproportionate burden being experienced by the African American population now with a poverty rate of 35 per cent compared to 22 per cent for whites (1990 data). Distressingly, half of the city's black children were classified as being in poverty condition. In smaller- or medium-sized industrial complexes (e.g., Akron and Syracuse) where manufacturing employment has systematically declined over the 1982–92 period, the pattern is one of suburban plant closure and job shedding, most manufacturing having long since abandoned central city locations. On a smaller scale, the same spatial mismatch ensues, leading to a continued economic disadvantage for many central city workers or potential workers, unable to find jobs in service work or insufficiently prepared for knowledge-based jobs.

In the Canadian context, a good deal less is known about the changing geography of manufacturing, and even employment, within the metropolitan areas. Using special tabulation data, Norcliffe (1996) has shown that in Toronto, as in numerous US cases, there is a decline, in proportional terms, in manufacturing and total employment in the central city area but rapid growth in the outer city from 1971 to 1981, and again from 1981 to 1991. Now, the greatest amount of manufacturing employment is in the outer suburbs. However, one must balance absolute change with proportional change. Central city manufacturing employment consistently declined over the three censal years but total employment in all industries in the central city increased over the two decades, as it did in the inner and outer suburbs, attesting to the overall dynamism in Toronto in this period. This again suggests that great care is needed when using such a term as de-industrialization. This clearly occurred in terms of manufacturing in the City of Toronto (as defined before 1998), consistent with continent-wide trends, but major investment and employment growth took place in other employment sectors in the City *per se*. Drawing on local knowledge and the paintings of Brian Kipping (echoing the photographic record made by Vergara – see endnote 2), Norcliffe provides a qualitative sketch of manufacturing decline in old industrial districts around the booming CBD and on the waterfront where a variety of uses mainly devoted to consumption have largely displaced the former industries.

For an earlier period (1975–85), Yeates (1998b) demonstrates that for Canada's 'million plus' metro areas (Toronto, Montréal and Vancouver) manufacturing employment declined in the central cities and grew significantly in the outer suburbs. William Coffey (1994) shows that in the ten largest Canadian CMAs, the share of the manufacturing sector relative to the total experienced labour force systematically declined from 21.4 per cent in 1971 to 18.8 per cent in 1981 and then to 14.7 per cent in 1991. Both Montréal and Toronto shifted downwards similar in magnitude to the national trend. However, a particularly important point to emerge from his closer analysis of Montréal is that the strong economic performance of the CBD is in sharp contrast to the remainder of the central city. While employment decline in manufacturing in Canadian inner-city locations poses severe

challenges to working low and moderate income families and a spatial mismatch can be detected behind Norcliffe's Toronto data, and even though a landscape of industrial decline does result, these challenges pale in comparison to the devastation wrought on large numbers of residents and neighbourhoods in many of America's central cities (see again Vergara, 1995).

THE CORE CBD AND THE OUTER SUBURBS: CHANGING FORMS

With the explosive growth of service related employment in recent decades as well as new forms of industrial production, the form and landscapes of North American cities have changed to accommodate the demands of these sectors for appropriate sites. The decline in US CBDs, much commented upon in the literature of the 1970s and 1980s, has been arrested in certain cases. For those major metropolitan centres that play important command and control functions in the national, continental and world economies, there has been a massive re-investment in CBDs to produce a new generation of structures to house the information-rich activities associated with these functions. In the principal metro areas, then, there are striking new office towers and dramatically new skylines. For some who toil there, inner-city living is seen as and has become a viable option to the long commute from far-flung suburbs. This has assisted gentrification and condominium redevelopment in the inner city. But even in these dominant centres, while there has been absolute increase in the amount of office floor space in core locations, the rapid growth of office space in outer-city locations (from generally negligible levels) has meant that relatively speaking the core has become less important within the overall metropolitan matrix.

Further down the US urban hierarchy, the office and information-based activities are less important and it has been more of a struggle to revive downtown. Most of these CBDs have now completely lost their major retail function to suburban malls and only niche or speciality retailing remains, sometimes tied to entertainment districts. Those cities with at least a regional service function, including centrally-located, large-scale health and educational institutions, have a much better chance to create a somewhat viable core, even if it will never regain its overall significance. Declining industrial centres lacking these roles and facilities are at greatest risk of having eroding and under invested cores.

Since the cores of Canadian cities never declined to the extent noted generally for the United States, the scenario sketched above is less relevant, though the difficulty of generalizing about a system as highly varied as that of the US is further evidenced when one has to acknowledge that CBDs in expanding US metroplexes (like San Diego, Seattle, Phoenix, and Orlando to name but a few) are likely to have been the sites of far more investment than, say, Montréal and Winnipeg. However, the distinctiveness of cities with major command and control functions *vis-à-vis* less well situated or highly specialized places further down the hierarchy in terms of CBD investment and consequent physical transformation is, none the less, still very evident, as a visual comparison of Toronto, Vancouver, and Calgary (even Montréal) as against Winnipeg, Hamilton or Sudbury would attest. In that sense, there is a strong cross-national similarity.

In the outer city, the most widely commented upon forms of urban development are (a) edge cities, (b) regional and super-regional malls, and (c) defended or gated suburban communities. Muller (1981) was one of the first to identify these mini-CBDs in the outer city and they leapt into national prominence in the United States with Garreau's 1991 book, *Edge city*. Their basic attributes include several million square feet (186 000 m^2) of office space, between 0.5 and one million square feet (46 000 and 93 000 m^2) of retail floor space in a mall-type environment, adjacent residential development at a higher density than the typical suburban subdivisions, including a mix of apartments and townhouses, and a reliance upon automobiles as the principal mode of local travel. The development of these massive complexes in most American places is the result of private sector initiative, including the planning of new communities (Irvine, in the vast southern California agglomeration and Reston, in the Washington, DC metro area are two such examples).

Such types of development tend to be more rare in Canadian urban areas but outer cores, or town centres, have been actively promoted in the Toronto and Vancouver areas, for example. The major difference lies in an explicit and public-planning ideology that undergirds these, seen as they are as rationally located town centres designed to take unacceptable development and environmental pressures off core/CBD areas. By providing employment nodes within walking distance of high-density residential clusters and by linking these into suburban transit systems, an attempt is made to reduce long commutes to the core (a 'social bad' in the eyes of many Canadian planners, it appears). The financial and physical development of these nodes is left to the private sector and so the outcome may not be radically different in certain respects from the American 'edge cities', but the smaller size, transit links, and clusters of high-density housing do tend to differentiate them, in form, from their American counterparts. It seems reasonable to conclude, given the strength and vitality of the CBDs in both Montréal and Toronto and the nature of their outer city developments, that the American edge city model is not yet appropriate in Canada, either as description or explanation (Coffey, 1994).

It is beyond the scope of this chapter to give a detailed comparison of the development of malls in North American cities (for more detail see Jones and Simmons, 1990). Suffice it to say that this form of retail development is widespread throughout North America, especially in the outer city, and there is not a great deal of difference evident in Canada as compared to the United States. In fact, the largest mall in the continent was built in inclement Edmonton which has long, severe winters; not long thereafter, the same Edmonton-based developers built a similar mega-mall in suburban Minneapolis, an apt illustration of the continental reach of this almost hegemonic retail form, one that had added additional entertainment functions and even 'public' spaces (e.g. libraries and walking areas for exercise prior to mall openings for regular business). One possibly distinctive feature of mall development is the success of a number of city governments in Canada in integrating mall-style developments into their downtown cores. Moreover in Toronto and in Montréal, there are also extensive networks of underground shopping that provide a controlled, quasi-mall setting.

The gated community or defended suburb, some of which have created local governments, is now better documented (Davis, 1990; Mackenzie, 1994; Blakely and Snyder, 1997). It is widely seen as the ultimate expression of the ideology of American individualism ('my right to live there and wall myself off from others') and a withdrawal from the larger metropolitan community, including an abdication of personal responsibility for the social condition of that larger community. It is also a reflection of the extraordinary concern of Americans over violent crime as well as a practical step to insulate oneself and the family from the threat of city-based violence. However, such communities are still the preserve of a small number compared to the massive numbers of 'ordinary suburbanites' living in non-gated communities. In the non-gated suburbs, many of the same motivations that impel people into gated communities are common but with personal security systems and private guards or police available on demand activation, exclusive housing removed from road views, and road layouts that would confuse the casual visitor or scout, most feel secure. Such defended or gated communities are far less common in the Canadian urban scene but exclusive suburbs (including estates) which afford privacy, comfort and a sense of personal well-being and security are more common. It is perhaps in this outer city that Canadian and American cities are most alike. But there is a dearth of careful, cross-national comparative analyses of these outer-city areas either using municipalities or groupings of census tracts as spatial units of analysis. This would be an immense undertaking but it is overdue.

HOUSING AND URBAN TRANSPORT

One of the persistent aspects of any metropolitan area's character is its housing stock. A common perception of North American cities is that the stock is dominated by owner-occupied, single detached housing units. Generally speaking, this is an accurate enough perception although there are numerous large cities in both Canada and the United States where other forms of housing predominate, especially in the older, central city areas (examples would include Montréal, Québec City and Toronto, and New York, San Francisco and Philadelphia). In both countries, the level of home ownership is remarkably similar, typically fluctuating between 60 and 65 per cent of all units; this is all the more notable as Canadians do not have access to the various tax advantages that are provided to homeowners in America in order to promote this particular form of home ownership, widely believed to be an essential underpinning of capitalist democracy and a particularly fine representation of American individualism and its commitment to private property. Somewhat relatedly, in neither country is there a strong commitment to a public sector/social housing tenure which is low compared for example to the UK and other European countries. But if the dominant tenure is broadly similar between the two countries, the form of housing is decidedly not.

Although the single detached unit accounts for the majority of all units in both metropolitan Canada and the United States, there was a clear cross-national difference evident in the Goldberg and Mercer (1986) analysis of 1970–71 data; in the US, on the average, single detached units were almost 75 per cent of all units in metropolitan America but only 57 per cent in Canada's metro areas. Significant differences also existed for the central city and the outer city, differences which persisted even when controls for city size and geopolitical structure were applied. Examining this key indicator for 1990–91 demonstrates that this cross-national difference persists. On average, the single detached unit accounted for 61 per cent of all housing units in metropolitan America but only 55 per cent in Canada. While the difference also persists for the central cities (where the average proportions are 45 per cent for Canadian areas and 52 per cent in the US), there is an interesting reversal in the outer city where for US metros the average is 68 per cent while it is 74 per cent in the Canadian areas. The strong demand for housing has increased outer area land values in America; more multiple forms of housing are thus being constructed and there is also increasing diversity in household structures. No longer is the suburb the exclusive domain of the family. Moreover, in numerous metro areas in the US, especially in the rapidly-growing southern and western regions, the demand for single detached units is met by mobile home units (not included in this analysis).

There is a strong relationship between urban form and urban access and movement, especially in commuting. Previous research (Goldberg and Mercer, 1986) has shown that while urban North Americans are much more reliant on automobiles than in other economically-advanced societies, metropolitan Canadians were significantly less likely to use automobiles in the daily commute (66 per cent for Canadians compared to 85 per cent for Americans in the 1970s). An equally significant difference also existed then for urban public transport when a quarter of Canadian commuters used mass transit but only an eighth of Americans did so. More recent data indicate that this public transport difference is still evident. In a ranking of metropolitan areas using per capita public transport trips (1993–95), Canadian metro areas occupied 8 of the first 10 ranks and 12 of the top 20; since public transport use tends to increase as metro size increases, it is remarkable that medium-sized Canadian metro areas (under 500 000) have per capita values equivalent to American agglomerations numbering in the several millions. Public transport in American metropolitan areas is used by only 6.5 per cent of all workers (1990); even in the higher density central cities this proportion is just under 12 per cent. These are still very low levels of transit

use, consistent with the notion of a very loosely-structured city with overall low densities and dispersed employment opportunities.

Cross-national differences in transit use are also evident in comparisons of ridership trends from 1950 to 1990 which show a strong recovery on Canadian transit systems in the 1970s and 1980s, compared to very modest gains in US ridership (Perl and Pucher, 1995). Vigorous urban and suburban growth in Canada as against suburban expansion and urban depopulation in America are the principal explanatory factors, although the characterization of urban depopulation is an erroneous representation of what has actually happened in terms of urban population change, as Bourne (Chapter 12) has shown. However, Perl and Pucher report that in the 1990–94 period, Canadian ridership has steadily declined and at a faster rate than in the US, accounted for by changing demographic trends (a decline in the heavy transit use cohort, aged 15–24; declining household size in areas best served by transit), falling auto prices (in inflation adjusted dollars) and fiscal austerity in federal and provincial public sectors. They argue that unless there are policy interventions a transit system decline will result with an obvious increase in auto dependency.

METROPOLITAN GOVERNANCE AND PLANNING

Compared to most European countries, urban planning systems in North America have been historically weak and ineffective. This has been exacerbated by a notoriously fragmented local government system, especially in the United States. Local government systems are very stable in America where there is strong commitment to the concept of local autonomy. Some two decades ago, an index of municipal fragmentation in metropolitan areas was roughly 2.5 times higher in metropolitan America than in Canada (Goldberg and Mercer, 1986, p. 214). There is little to suggest that this situation has changed since then. Fragmentation has increased in

US metropolitan areas from 1977 to 1987, with the number of local governments per metro area rising from 93 to 113 (Rothblatt, 1994). In Canada, however, there have been bold reductions of local government numbers in two metropolitan areas in the last two years, in order to achieve greater efficiencies and to realize anticipated cost savings. In both cases the principle of local autonomy was little in evidence as provincial governments introduced the required legislation and ensured its passage regardless of the nature of local sentiment. In Atlantic Canada, the local governments in the region's primary metropolitan area of Halifax were merged into a single metro-wide unit, eliminating the cities of Halifax, Dartmouth and some smaller suburbs. Even more importantly, given the size of the populations affected, the City of Toronto and the five boroughs that made up the Municipality of Metropolitan Toronto were amalgamated into a new and larger City of Toronto.

Having studied regional planning systems in five paired comparisons involving 10 large American and Canadian metropolitan areas (from 1989–91), Rothblatt (1994) concluded that the metropolitan management system is more highly developed, with more authority and fiscal capability, in the Canadian areas than in their American counterparts. That said, however, the 'global' process of decentralization is weakening the stronger metropolitan systems as more and more of the rapid, outer city growth is beyond their territorial jurisdiction. Increasing complexity and tensions within and between an increasingly diverse set of municipal governments and quasi-governmental institutions adds to the pressures on the existing metro management systems.

CONCLUSION

There is little doubt that there is increasing variability and complexity evident within the metropolitan areas of North America where the vast majority of people reside. This means that research and teaching on cities will need to be more fine grained and geographically sensitive to the local, both within and between cities. Broad generalizations and simple models will likely be more difficult to sustain. However, as long as capitalism is the fundamental mode of production and consumption in North America, and representative democracy the basis for civic life, then marked social and economic inequalities will endure in its urban places, as will small local governments with more autonomy than in most other parts of the world.

The increase in variability and complexity is principally due to major social and economic changes that are global in nature, thereby affecting the open systems that are the metropolitan agglomerations of both countries. Chief among these are the general decreasing importance of manufacturing as the principal economic motor for urban growth and its replacement with a service-based metropolitan economy (although manufacturing is not everywhere in decline) (Chapters 10 and 11), the dramatic increases in immigration in the last two or three decades with the source regions being mainly non-European resulting in complex social changes in both inner cities and suburban areas where immigrants also now settle (Chapter 6), and the greater range of family and household arrangements evident and being transmitted through urban housing markets. Given the large scale of these transformations, it is not surprising that the changes they produce are being more or less equally felt in the urban places of both Canada and the United States. This points towards potentially greater similarities between the two sets of cities and some of the recent empirical evidence reviewed here is in that direction – leading to what some would call the convergence thesis.

A second major theme of this chapter has been cross-national differences between Canadian and American cities. These have arisen as one society and its people have made different choices from the other over centuries of colonialism (then independence), economic development, territorial expansion and massive immigration. These conscious choices and the urban landscapes that they created occurred in the face of common, even global processes. Is there the will amongst Canadians to continue to differentiate themselves from their massive neighbour, with whom they are so closely integrated economically, a neighbour that serves as their primary reference and point of comparison? Inherited urban landscapes, the political culture and institutional structures of government, including metropolitan governance, planning and intergovernmental relations, and the social mores that govern interpersonal and intergroup relations have all proven to be remarkably resistant to continental homogenization. This has resulted in detectable cross-national urban differences in both compositional attributes of cities and urban form and transportation.

Although these differences do not arise from public expenditures as an expression of political and popular will, there is little doubt that the state (especially at the provincial and local level) has played a key role in the life of Canadians, more so than in the United States (although it is greater there than many commentators recognize). As both the Canadian federal and provincial governments strive to eliminate deficits and balance budgets, the impact of expenditure cuts, a lesser influence of the public sector and a corresponding growth in market influence may all push in the direction of convergence, so that Canadian cities will become slowly more like American ones. But to the extent that the urban differences are caused by more than public actions and public dollars, and to the extent that the cultural and political will to differentiate survives within North America, these cross-national urban differences will remain – even if smaller, they will be 'small differences that matter' (Card and Freeman, 1993).

ENDNOTES

1. Numbers are rates of population change over a decade.

2. In this chapter I use the terms 'black' and 'African American' interchangeably; there is still much variation in the use of these terms although the latter's use has increased substantially.

3. Trained as a sociologist and an experienced street photographer, Vergara (1995) has produced a remarkable record of physical and social transformations in a number of American inner city ghettos through photography, supplemented with fine commentary including interview testimony. Beginning in 1977, he developed an archive of over 9000 slides which forms the basis for the book's photographs.

4. This presence is probably inflated somewhat by the use of the county as the basic areal unit in defining Metropolitan Statistical Areas so that in California and Texas, for example, a rural Mexican origin population is counted into the metro totals.

KEY READINGS

Bourne, L.S. and Ley, D. (eds) 1993: *The changing social geography of Canadian cities.* Montréal and Kingston: McGill-Queen's University Press.

Bunting, T. and Filion, P. (eds) 1991: *Canadian cities in transition.* Toronto: Oxford University Press.

Davis, M. 1990: *City of quartz: excavating the future in Los Angeles.* New York: Verso.

Goldberg, M. and Mercer, J. 1986: *The myth of the North American city.* Vancouver: University of British Columbia Press.

Knox, P. 1994: *Urbanization: an introduction to urban geography.* Englewood Cliffs, NJ: Prentice Hall.

Yeates, M. 1998: *The North American city.* New York: Addison Wesley Longman.

ATLANTA: METROPOLIS OF THE NEW SOUTH

DAVID M. SMITH

ATLANTA: THE PUBLIC SPACE

The 1996 Olympic Games focused world attention on the city of Atlanta, as the booming commercial capital of the so-called new South. In recent years, metropolitan growth of frenetic urgency has replaced the more leisurely pace of the old South, epitomized by the cotton plantation and its graceful antebellum mansion. The small regional centre sacked by General Sherman's soldiers during the Civil War and celebrated in Margaret Mitchell's *Gone with the wind* (1936) now makes claims to the status of international city. More than any other contemporary American metropolis, perhaps, Atlanta exemplifies the capacity to create and recreate itself in the American tradition of civic 'boosterism', with the distinction between image and reality not always clear (Allen, 1996; Rutheiser, 1996).

While the Olympic Games are Atlanta's most recent claim to fame, the city has attracted attention, also, for the manner of its emergence as a major commercial centre. Behind the economic motor of urban expansion over the past few decades has been a distinctive political regime, which enabled restricted and highly privileged groups to control the growth process and take major advantage of its benefits (Hunter, 1953, 1980; Stone, 1989). At first, city hall was monopolized by local business leaders – exclusively white – whose power enabled Atlanta to undertake a massive programme of urban renewal (removing large areas of inner-city slum housing and its poor, largely black residents) and associated freeway construction. This protected businesses downtown and saved this area from the decline experienced in many other American cities during the 1960s (Stone, 1976). Then, from the early 1970s, the regime was flexible enough to incorporate representatives of the African American population, hitherto excluded from both formal and informal political power. Meanwhile, economic power in terms of the ownership and control of the business corporations concentrated in Atlanta has remained largely in white hands.

Race is a major dimension of Atlanta's recent history and geography (Bayor, 1996). The city was the home of the Revd Martin Luther King Jr, closely associated with the early civil rights movement, whose old neighbourhood is preserved by the National Park Service as an official Historic District. Atlanta was the first major city in the South with an African American mayor, Maynard Jackson, originally elected in 1973, subsequently re-elected, and influential in the city's bid for the Olympic Games. There are also associations with the white-supremacist Ku Klux Klan, reputed to meet at nearby Stone Mountain. Atlanta has seen important struggles for racial equality and social justice, in a region which many Americans elsewhere would regard as backward in these respects.

The officially-defined Atlanta Metropolitan Statistical Area vies for a place in America's top 10 in terms of number of inhabitants. The City of Atlanta as a political jurisdiction has a population of about 400 000. Two thirds of these are African Americans. Outside the city in this sense, whites comprise four fifths of the suburban population of a wider metropolis housing about three million people today. This inner-black and outer-white distinction is, of course, typical of the US city in general, with its strong racial segregation intersecting with socio-economic or class differentiation as the housing market finds its spatial expression in neighbourhoods of varying quality.

Atlanta's recent growth has been phenomenal. Its population increased by a third during the 1980s. As its residential subdivisions, suburban shopping malls and office parks stretch ever outwards into the surrounding countryside, new sky-scraper hotels and office blocks compete for prominence downtown. There are futuristic monorail lines forming part of a mass-transit system struggling to keep commuters off the roads, and an airport of suitably gigantic proportions. There is the headquarters of international news service CNN, added to such long-established local businesses as Coca Cola and Southern Bell. There are distinguished higher education institutions and a sporting prowess led by Atlanta Braves baseball. The city epitomizes the economic resurgence of the so-called 'Sunbelt', often contrasted with the relative decline of the 'Snowbelt' or 'Rust-belt' of the Northeast. Atlanta is a confident, self-assertive city, proud of its prosperity and of its role on the international stage consolidated by the Olympic Games.

ANOTHER ATLANTA

However, there is another side to Atlanta, largely hidden behind the civic hype and 'imagineering'. This is the city of poverty and social deprivation, exemplified by the remnants of the black ghetto, parts of which are almost literally in the shadow of the Olympic Stadium. Just as its predecessor, Fulton County Stadium, was built on land formerly occupied by low-income black neighbourhoods, so the new stadium stands on what were once streets of small wood-framed homes typical of an inner-city poverty area. Indeed, it was the clearance of ghetto housing during the earlier era of urban renewal

that made the site for the stadium available so close to downtown.

The geographical pattern of inequality is illustrated in Fig. 14.1, which shows variations in median family income across the metropolitan area. The contrast between the affluent northern suburbs and the poor inner city is clear. The areas with largely black population indicate some association between poverty and race, though there are affluent suburbs in the mainly black southern part of the metropolis. The contrast between rich and poor Atlanta is revealed by the fact that the highest median family income in any of the subdivisions shown (about $150 000) is 30 times that of the lowest (about $5 000). And this pattern is reflected in other conditions, such as housing, education, health and crime.

The American city is frequently described as polarized, by race and geographical space. This is illustrated in Atlanta in Fig. 14.2. The distributions of predominantly black and

TABLE 14.1 Median family income ($) by census tracts in the Atlanta metropolitan area, 1960–1990

Tracts	1960	1970	1980	1990
All tracts: mean	6 023	9 907	17 877	35 216
maximum	18 548	29 793	54 081	150 001
minimum	2 261	2 273	2 907	4 999
ratio maximum/minimum	8.2	13.1	18.6	30.0
coefficient of variation	45.2	43.1	53.0	65.0
White tracts: mean (w)	7 121	12 001	25 928	62 423
Black tracts: mean (b)	2 973	5 709	9 667	17 024
White ratio of advantage (w/b)	2.40	2.10	2.68	3.67
White tracts: coefficient of variation	38.9	35.1	33.0	40.9
Black tracts: coefficient of variation	16.7	29.2	48.0	52.8

Note: White tracts are those with over 90 per cent of their population white; black tracts have over 90 per cent black.
Source: Smith (1994) data from US Census of Population and Housing, 1990

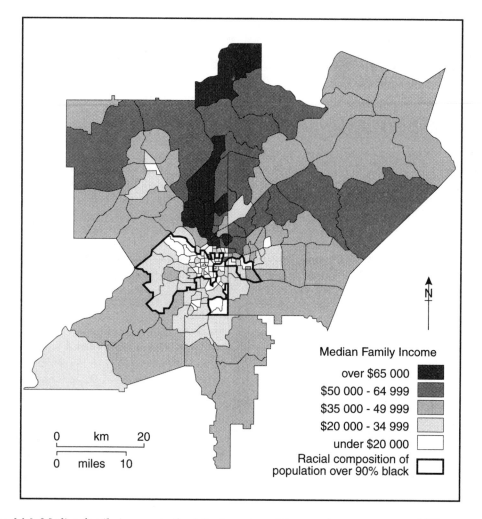

FIG. 14.1 Median family income in the Atlanta metropolitan area by census tracts, 1990

predominantly white tracts are plotted according to median family income (converted into standard scores in which the metropolitan mean is 0 and standard deviation 1.0). Despite the existence of some predominantly black tracts with incomes above the metropolitan average, this analysis identifies very little overlap between the two distributions.

Trends in income inequality over time are summarized in Table 14.1. This is based on census data which have been recompiled for identical subdivisions of the metropolis (tracts or combinations of tracts), for the four years indicated. The ratios of maximum to minimum values for median family income

show a steady increase in inequality, from 8.2 in 1960 to 30.0 in 1990. This trend is repeated, except for a slight decrease from 1960 to 1970, in a more comprehensive measure of inequality, the coefficient of variation (the average of median family income for all subdivisions divided by the standard deviation).

The ratio of advantage of white tracts over black (identified by the over 90 per cent criterion) has increased since 1970, indicating growing racial inequality after a reduction in the 1960s. And, although the coefficient of variation shows inequality among white tracts little changed in 1990 compared with 1960, there has been an increase for the black tracts year by year. This rising inequality

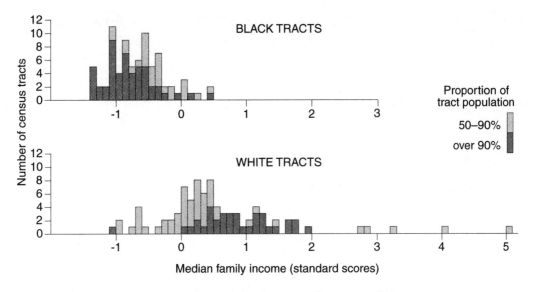

Fig. 14.2 Race–space polarization in the Atlantic metropolitan area, 1990

among black tracts is accounted for by the emergence of a relatively affluent black suburban middle class, increasingly differentiated, and distanced, from the inner-city poor trapped in the remnants of the ghetto (see Chapter 8).

The overriding experience of Atlanta is thus of disequalization (Smith, 1994). Against the expectation of gaps narrowing with continued economic growth and its supposed 'trickle-down' effect is the evidence of increasing inequality with a strong spatial expression. Inequality in the US city has its origins in a particular form of capitalist society, fiercely competitive and with very limited provision for those whose failure is so conspicuously associated with poor local environments and living conditions. In the South, the long history of racial discrimination going back to slavery has generated special disadvantages for African Americans, only partially overcome since the civil rights movement. Now, the globalization of economic relations, along with erosion of some of the limited social welfare provisions for the poor, has increased the vulnerability of local people to those market forces which drive corporate restructuring. The result is a

growing segment of the population (sometimes referred to as the 'underclass') whose exclusion from regular employment is compounded by the related problems of poor housing, education and health care, and by the criminalization of neighbourhoods in which drugs dominate both culture and economy.

The general prosperity of metropolitan Atlanta affords some protection from the extremes and scale of social deprivation found in some other cities. However, the City of Atlanta and its suburban jurisdictions still function for the affluent majority, with little more than token gestures towards neighbourhood renewal in the poverty areas. It is a reflection of the power of the image, and of the image makers, that so little of the underside of Atlanta reaches the attention of the wider world.

ACKNOWLEDGEMENTS

Parts of this chapter are based on Smith, D. M. 1996: Atlanta: inequality and social justice in the Olympic city. *Geography Review* **18**(1), 2–5. The research summarized here was supported by the Economic and Social Research Council and was assisted by Sandy Bederman, Tom Boswell and Steve Pile.

KEY READINGS

Allen, F. 1996: *Atlanta rising: the invention of an international city.* Atlanta: Longstreet Press.

Bayor, R.H. 1996: *Race and the shaping of twentieth century Atlanta.* Chapel Hill: University of North Carolina Press.

Rutheiser, C. 1996: *Imagineering Atlanta: the politics of place in the city of dreams.* London: Verso.

15

LOS ANGELES: THE AMERICAN DREAM?

JAMES P. ALLEN

LA. Is it the City of Angels, as we 'Angelenos' often call it, or is it America's corner of the Third World? Is it mile after mile of look-alike tract homes and shopping centres choked by freeway gridlock and smog, or do the warm climate and easy free-way access to an almost limitless range of work and recreation opportunities really lib-erate people and make them happier? Is it the new global city linking Asia and Mexico to America, the creative home of Spielberg and Disney, the high-tech city that built space vehicles, or a city so polarized by race and class that police brutality is commonplace and minority riots are triggered almost as regularly as earthquakes?

Even though such characterizations of Los Angeles may be inconsistent and overdone, each contains some bit of truth. The images of Los Angeles held by people around the world may be especially powerful because they are the cumulative product of a half cen-tury of Hollywood films and television shows designed to entertain and dramatize; of news reporting on real riots, killings, earthquakes, and fires; and of the often con-descending observations of writers from older more traditional centres of culture such as New York, San Francisco, and London who are offended by LA's rise to prominence. In reality, much of Los Angeles is either com-monplace or a much more subdued version of the easy caricature.

Because it is impossible to cover contem-porary Los Angeles in this brief sketch, I focus on a set of closely-related themes: cul-ture change, immigration, ethnicity, and the economy. The name 'Los Angeles' can refer to either of two different areas (Fig. 15.1). It can mean the city itself, second largest in the United States. Or it can be Los Angeles County, which includes Los Angeles City and 87 other cities, with a total population of 9.8 million in 1999. But even Los Angeles County is just the central part of a massive five-county functional region (not including San Diego) of 15 million people, referred to here as southern California.

Nearly all of what has been written about Los Angeles over the last half century has portrayed it as unique. On the other hand, geographers are especially likely to recognize that many of any place's characteristics are shared by other places. Although such a generic perspective is not stressed in this sketch, the reader should keep in mind that Los Angeles should also be seen as one of many large, highly-decentralized metropoli-tan areas in the United States.

PEOPLES AND CULTURES

Originally home to Indians, Los Angeles was founded in 1781 by the Spanish, and southern California evolved into a set of small settlements on the northern fringe of Mexico. By defeating Mexico in war, the United States took over California in 1848, but southern California remained mostly Spanish-speaking for another twenty years or so.

Massive in-migration and overpowering cultural change began after railroads made the connection to the Midwest in the 1880s. Advertising promoted southern California as the land of sunshine, and the first of millions of English-speaking American families arrived from 'back East' to make their new homes here.

White American settlers dominant before 1970

For the next half century southern California attracted people from the eastern half of the United States, some for its health benefits and warm climate and others to seek their fortune through work or real estate specula-tion. After the Second World War the region captured the hearts of young Americans thirsting for personal and social freedom, for the physical exhilaration of a life in the sun and the surf, and for opportunities in film and television. It beckoned to family men and women who heard about work in the many aircraft plants and were eager to trade winter coats and snow shovels for barbecues and swimming pools.

Because these new southern Californians were particularly likely to have come from farming areas and smaller towns, a conserv-ative Protestant Christian culture was unusu-ally widespread in southern California

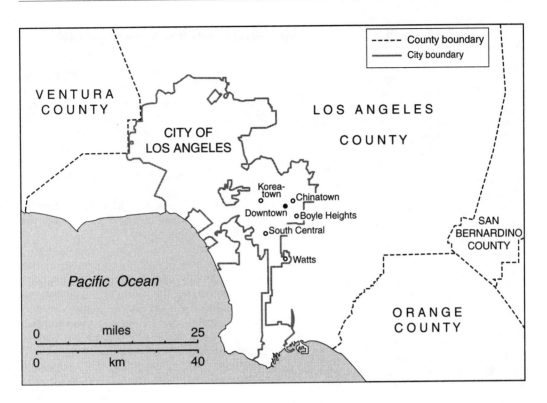

Fig. 15.1 The Los Angeles area

before the 1970s. At the same time, the city of Los Angeles attracted enough ambitious writers and would-be actors, social rebels, Jews, Catholics, and minority populations that the seeds of more individualistic and cosmopolitan values were planted. The net effect is that Americans in Los Angeles County have long been able to find others to share their specific lifestyle or organization, no matter how traditional or radically countercultural.

Although US-born whites represented 80 per cent of Los Angeles County's population in 1960, people of many different origins and races had long been attracted to the area. The largest minority group was Mexicans, including Mexican Americans – those who were ethnically Mexican but born in the United States. Some were descendants of pre-1848 inhabitants, but most had parents or grandparents who had come to California in the early twentieth century for farm and railroad work. Next in numbers were blacks, mostly US-born, who had migrated like all others to Los Angeles with hopes that hard work would be rewarded with good incomes and fair treatment. White discrimination against blacks and Mexicans in education, employment, and housing was taken for granted until the late 1960s or later. Much smaller in number were Japanese and other Asians, against whom most discrimination was eliminated by the 1960s.

Immigration since 1965

Huge differences in living conditions and economic opportunities have developed since the Second World War between the United States, with its political freedom, stability, high wages, and prosperity, and most less-developed countries. This growing contrast has been a major underlying reason why many people around the world have hoped to emigrate to the US. New immigration and refugee laws, beginning in 1965, made it possible for millions of people born elsewhere to become legal residents here (see Chapter 6).

These economic disparities and legal changes have resulted in a large and increasing immigration to the United States. Southern California's warm climate and the abundant economic opportunities generated by its large population have attracted immigrants from a great many different countries. In addition to new legal residents, thousands of others cross the US border illegally or overstay their temporary tourist or student visas to end up living here illegally. The net effect has been that the five-county region of southern California is now home to four million immigrants (people born in other countries). This region has a fifth of all the foreign-born in the US and far more than in greater New York, the next largest concentration.

The great range of countries and cultures of immigrant origin means that Los Angeles may be the most ethnically diverse place on the planet. Most newcomers, however, have come from Latin America and Asia. If immigrants and US-born from the same country of origin are combined into an ethnic population, Chinese, Filipinos, Japanese, Koreans, and Vietnamese are most numerous among the highly varied Asian groups. Although many countries in Central and South America are represented, Mexico has been by far the largest single source of all immigrants.

The characteristics of immigrants vary substantially by their countries of origin. Although the majority of immigrants to southern California – particularly those from Mexico – bring only a willingness to work hard, many from Asian countries are advanced students, professionals, or people with business skills and international contacts. These represent a 'brain drain' from their countries. A few immigrants even bring in sufficient money to buy homes immediately in the more affluent suburbs.

The internationalization of Los Angeles has been profound. It is hard to imagine anyone in the world whose beliefs and tastes in religion, politics, food, sports, music, art, the theatre, or night-clubs could not be shared and satisfied in the Los Angeles of the 1990s.

White out-migration since 1965

Another major component of the ethnic transformation of Los Angeles County has been the net out-migration of whites, usually to suburbs in the four outlying counties. This movement continues the suburbanization begun a century ago, as newer and often less expensive housing away from crime and poorer people in the central areas are still considered by many to be well worth a much longer commute. This selective out-movement has left the older, more central areas with higher proportions of poorer people and minorities.

At the same time, many blacks, Latinos, and Asians have been able to move to the suburbs. This has made several older suburban cities among the most ethnically diverse in the entire United States. However, the more expensive and newer suburbs still tend to have very high proportions of whites together with especially affluent Asians.

Because it has been the less educated whites who have felt most the effects of economic restructuring and the competition from immigrant workers, whites who are not college graduates have been particularly likely to migrate out of southern California. Whites with better education continue to move toward southern California because the surplus of low-wage workers has helped keep down general living costs. Thus, the net migration differential has exacerbated the contrasts between more affluent whites and Mexican immigrants.

An ethnic and cultural transformation

As of 1999, whites comprised only a third of Los Angeles County's population. Thus, there has been a major ethnic shift in the city of Los Angeles and, to a lesser extent, in the surrounding areas. Although many of the older residents and their descendants still reflect that earlier, small-town conservative Protestant heritage, most people of this background have died, moved away, or been changed to some extent by the events and people around them. The traditional, inward- and eastward-looking conservative

American perspective no longer dominates as it once did. At the same time, Los Angeles has been refashioned demographically by its many immigrants who have arrived with their very different cultures and international perspectives. The future of the city will depend on how well the new and the old accept each other and the degree to which the different groups come together socially and culturally.

THE ECONOMY

Southern California's warmth and its winter rains supplemented with underground and imported water made possible the countless orange groves and irrigated vegetable fields that, in turn, made Los Angeles County the richest agricultural county in the United States in the late 1940s. But soon the pace of urbanization became overwhelming, swiftly transforming old fields and orchards into homes, stores, offices, and parking lots.

Industries have played an important role in southern California's economy. The best known is entertainment. Movies were produced here by Jewish immigrants prior to 1920, and since the Second World War the LA area has remained the country's major centre of film and television production. Other industries here have included oil drilling and refining, automobile design and assembly plants, aerospace, and the making of various high-tech products. Apparel design and manufacture, specializing in sportswear and women's clothing, has also thrived in Los Angeles.

Economic changes since 1970

The ethnic transformation already discussed has been paralleled by major regional economic changes. These involve both shifts in types of employment in the region and the rapid expansion of trade and financial links to other countries, especially those across the Pacific Ocean. These changes also affected large metropolitan areas elsewhere in the United States, but they probably had a greater effect in Los Angeles County than in most other places.

The employment restructuring experienced in southern California in the 1970s and early 1980s had three components: (1) the loss of good jobs for less educated workers in heavy industry as steel, aluminium, and auto factories closed; (2) the growth of low-wage, low-skill work in services and in a few manufacturing sectors, notably furniture and apparel; and (3) the expansion of high-wage, skilled work in the upper end of the service and manufacturing sectors.

Many of the low-wage jobs have involved retail sales, but there are also more menial jobs – in restaurant kitchens and car washes, as construction helpers, lawn maintenance workers, and maids. Work in apparel manufacturing has also increased. The established presence in Los Angeles of apparel design and the growing number of Mexican immigrant women able to sew clothing rapidly and skillfully have made it worthwhile for owners to keep many of their factories here.

Some of the thousands of new jobs for better educated people have been in high-tech research and manufacturing, often involving computers and their peripheral devices, or aerospace sectors like space vehicles and guidance instruments. There has also been a large growth of high-end service-sector jobs. The robust southern California economy, technological advances, and the affluence of many southern Californians have created much new work in communications, finance and banking, investment brokerage, insurance, real estate, and legal and business services. In general, there has been a shortage of sufficiently skilled high-tech and advanced service-sector employees, so that pay scales have often risen substantially.

The net effect of these three trends in employment restructuring has been an increased polarization between the poorly educated and those with advanced education and training. Because whites and many Asians have been much more likely than blacks and Latinos to have the advanced education required for the higher pay, the polarization has become increasingly ethnic, as well as economic. In southern California, the social class divide is especially pronounced between the two largest groups: the more affluent whites and Mexican immigrants.

Landscape expression of the economic and ethnic polarization

Since the mid 1980s, the wealth of the rejuvenated downtown of Los Angeles and its noticeably white workforce have contrasted dramatically with the poverty and minority immigrant character of most of the surrounding area. Although all large American cities show this pattern to some extent, the sharp contrasts over short distances in Los Angeles are especially striking.

Downtown's revitalization has been led by a new financial district on Bunker Hill, south and west of the older civic centre and central business district. Many people, including leaders in both the private and government sectors, commute to jobs in downtown office buildings. Nearly all the people on the streets are well dressed and white. The architecturally-varied postmodern office buildings on Bunker Hill brighten and focus views from many miles away, and the spacious new central library, renovated exteriors of older hotels and office buildings, and newly constructed walkways make downtown Los Angeles an exciting place to work and visit. Close to sunset on a clear day, when office workers and tourists fill the sidewalks beneath the glittering lights of windows high above, the views are stunning and the ambience enchanting.

Surrounding the downtown, however, are abandoned railroad yards and decaying factories, warehouses, and apartment buildings that were built before 1920. Now these are either empty or are used as apparel industry sweatshops or flop houses for the poor. A mile or two from the downtown begin various residential neighbourhoods of older single-family houses, which represented the suburban periphery 90 years ago. These homes, once owned by whites, are now almost all owned or rented by blacks and low-income immigrants. Chinatown, Koreatown, Boyle Heights, and other centres for the poorer among recent immigrants are found in this more central section of Los Angeles. The largest such residential area is known as South Central, which extends from downtown south through Watts. This area includes the focus of the 1965 Watts uprising by poverty-stricken blacks and the area within which local residents rioted in 1992 after white policemen were acquitted in the beating of a black man, Rodney King.

Internationalization since 1970

As part of the general globalization of the world's economy, southern California has developed increasing trade and financial ties to Mexico and Asian countries. In the 1980s the Japanese bought many prestigious office buildings in downtown Los Angeles, several Japanese automobile companies have located their design studies here, and Sony now controls a major movie studio. Latin American pop singers and television programmes find audiences here, and vegetables grown in Mexico and trousers and semiconductor chips assembled in that country are widely sold across southern California.

Together, the adjacent ports of Los Angeles and Long Beach ship more cargo than any other port in the United States. Television sets, VCRs, machinery, shirts, shoes, and automobiles are loaded aboard containers in Asia for shipment to southern California, which in turn exports aircraft and electrical parts, plastics, California cotton, and Utah coal to China and Japan.

Wealthy immigrants from Taiwan, Hong Kong, and elsewhere have invested much money in southern California, and the networks of friendship and trust among Chinese in many countries help expedite the flow of capital into new southern California ventures. At the same time, the millions of dollars sent by immigrants in southern California back to relatives in their countries of origin play important roles in the economies of those countries.

CONCLUSION

Forty years ago most people in Los Angeles oriented themselves eastwards, to the states of their origin. Nowadays, however, residents are more apt to look outwards, especially to Asia and Latin America, as Los Angeles becomes more and more a hub of international flows of people, goods, and ideas.

There is a certain excitement to living in Los Angeles. The warm, dry climate remains attractive to almost everyone; but the easy, laid-back life in the sun of earlier decades has been replaced by a high level of energy that comes from the region's wide-open economic opportunities, from the frequency of shifts in political control, and from the relatively few restraints on individual freedom. Altogether, these attributes give residents of Los Angeles an exhilarating sense of living 'on the edge'.

Despite so much negative publicity over the last few decades, individuals who are innovative and ambitious still flock to Los Angeles for business opportunities and for work in entertainment and other high technology industries. The region's variety of employment, higher education, and recreational opportunities make it an especially favourable place for young adults. However, the labour surplus among less educated workers means that the chances of really prospering are much less for those without advanced training. This produces tensions, of course, in addition to opportunities.

As is true elsewhere in the United States, there is also a substantial degree of competition among the various ethnic groups for preferred jobs and higher incomes. Although political power has been diffusing among minority ethnic groups, most economic power in Los Angeles is still held by white men.

All this leads to the greatest problem facing Los Angeles today: the large social and income divides among its residents. The more affluent people – whites and some Asians, blacks, and Latinos – live in very different worlds from the poor, who are most commonly black or recent immigrants from Mexico.

Regardless of these tensions and difficulties, people still move to Los Angeles to achieve their dreams. Because their origins have generally shifted from the states 'back East' to the villages and cities of Asia and Latin America, there have been great difficulties of adjustment. Nevertheless, I suspect that those who have arrived since about 1970, regardless of where they came from, have been increasingly likely to be ambitious and hard working compared to earlier waves of migrants. In the long run that may be more important than their new countries of origin. Yes, Los Angeles still represents the American dream.

KEY READINGS

Allen, J.P. and Turner, E. 1997: *The ethnic quilt: population diversity in Southern California*. Northridge, CA: California State University, Northridge, Center for Geographical Studies.

Davis, M. 1990: *City of quartz: excavating the future in Los Angeles*. New York: Verso

Dear, M.J., Schockman, H.E., and Hise, G. (eds) 1996: *Rethinking Los Angeles*. Thousand Oaks, CA: Sage Publications.

Ong, P., Bonacich, E., and Cheng, L. (eds) 1994: *The new Asian immigration in Los Angeles and global restructuring*. Philadelphia: Temple University Press.

Pitt, L. and Pitt, D. 1997: *Los Angeles a to z: an encyclopedia of the city and county*. Berkeley: University of California Press.

Scott, A.J. and Soja, E.W. (eds) 1996: *The city: Los Angeles and urban theory at the end of the twentieth century*. Berkeley: University of California Press.

16

MONTRÉAL AND TORONTO: CHANGING PLACES?

FREDERICK W. BOAL

BACKGROUND

It is unusual for a country to be characterized by more than one dominant urban centre. With Canada, Montréal and Toronto fulfil this role – Vancouver providing a third, rising star on the Pacific. Montréal and Toronto encapsulate between them many of the key attributes one associates with Canada. They function as standard-bearers for the two 'founding peoples' – the British and the French (leaving aside consideration of the Native people) – they stand for the rivalry between them, and one – Montréal – actually incorporates this symbiotic rivalry within its own streets. In addition, the two cities demonstrate the increasing importance of the multicultural strand in Canadian society, seen in the varied, large-scale and complex immigrant presence.

From a geographical perspective the *raisons d'être* of the two cities provide a sharp contrast. Montréal's locational underpinnings are well defined, Toronto's much less so. It has been claimed that Montréal's destiny was 'engraved on the map' (*un destin inscrit sur la carte*) (Sénécal and Manzagol, 1993). The same cannot be said for Toronto. Indeed, in one sense, Toronto's early locational advantage stemmed not so much from where its site was located, but where it was not located – it was *not* too near the US border!

Montréal developed at a well-defined interface between ocean and continental interior, an interface between a maritime environment and one characterized by river routes. The St Lawrence estuary opened to the Atlantic while upstream the same river provided a routeway to the Great Lakes Basin and the vast heartland of North America. The Ottawa River, which joins the St Lawrence just upstream from the site of Montréal, offered a further passageway inland. Montréal Island, located astride the great routeway, provided a further incentive for settlement development, due to the rapids on the St Lawrence at that point – disrupting the flow of river traffic and stimulating the development of a crucial transhipment location.

Toronto also offered advantages of site and location, but to nowhere near the extent of Montréal. There was a reasonably sheltered harbour on Lake Ontario, a river and land route to Lake Huron started there (the Toronto Passage), and, very importantly in the early days, the site was some distance from the northern border of the United States, thus reducing the likelihood of a surprise attack from that quarter. The dramatic quality of Montréal's location was in sharp contrast to the almost nondescript attributes of Toronto's geography. Little in the comparison would provide a hint about the situation today, where Toronto has significantly outstripped its 'rival' in population size and in many aspects of urban function.

Montréal had its beginning as a missionary colony, founded in 1642. It also served as main operational base for the fur trade and as principal market for the furs themselves. By dint of its routeway connections, by 1700 Montréal was the dominant urban centre in North America, though its population barely reached 12 000 (Nader, 1972). Saving souls and fur trading lay in the hands of the French until the British capture of Montréal in 1760. However, by the early 1820s Montréal's continental role was over, though the city continued to grow as a regional service centre for the St Lawrence Valley. Although Montréal's transport function suffered from the opening of the Erie Canal in 1825 (thus bypassing the St Lawrence as the principal water routeway from the East Coast to the continental interior), the city began to regain a wide hinterland with the introduction of the steamship and the development of a trans-Canada rail network. Following Canadian Confederation (1867) Montréal emerged as Canada's metropolis – dominant in manufacturing, finance, transportation and just about every other aspect of economic life.

Toronto, some 500 km further into the continental interior than Montréal, developed later. In the eighteenth century the location had an unimpressive history as a French trading post and fort, burnt in 1759 to stop it falling into the hands of the British. By 1800, the settlement, at the time named York, had a population of 800 and had all the attributes of a frontier village. As southern Ontario began to fill with agricultural settlement, so

Toronto (as it became in 1834) could tap an increasingly rich market. By 1867 the city had been named as the capital of Ontario, its functional significance reinforced by the development of an extensive rail network. Major mineral discoveries in the Ontario section of the Canadian Shield towards the end of the nineteenth century gave a further boost to the city, not least because it began to function as the financial centre for mining – a prelude to Toronto's twentieth-century Canadian dominance in stock market trading.

THE TWENTIETH CENTURY SWITCH-OVER

Throughout the nineteenth century and for at least two thirds of the twentieth Montréal maintained its position as the largest city in Canada. In 1871, for instance, Montréal's population was twice that of Toronto, and by the turn of the century it was still one and a half times as big. As the twentieth century progressed the Montréal–Toronto size gap continued to close. It was not until the mid 1970s, however, that the population rank of the two cities reversed. None the less the underlying functional attributes of the two metropolises had been providing strong signals of impending change, long before it actually occurred. For instance, as Gad (1995) has noted, one component after another of Montréal's financial sector was overtaken or equalled by its Toronto counterpart: stock exchange activities in the 1930s, insurance in the 1940s and banking in the 1950s. Moreover the number of head offices of non-financial corporations had been in decline in Montréal since 1931, and, equally important, had been on the rise in Toronto. In addition, manufacturing has suffered a long-term relative decay in Montréal (Polèse, 1990).

Perhaps the simplest indicator of the relative positions of Montréal and Toronto[1] is provided by the data in Fig. 16.1, which shows the population sizes of the two metropolitan areas from 1931 to 1996. Until the early 1970s the two metropolises displayed vigorous growth (with the exception of the depression dominated 1930s), with Toronto continuing as the smaller of the two. However, between 1971 and 1981, Montréal's growth slowed, while Toronto's was maintained at a much higher rate. In consequence, by 1981, Toronto became, for the first time, bigger than Montréal. Subsequently Toronto has continued to sustain a higher growth rate, with the result that, by 1996, the Toronto metropolitan area housed almost a million more people than Montréal.

Although the switch-over of the two cities in the Canadian size hierarchy can be dramatized as a sudden event occurring in the mid 1970s, Fig. 16.1 shows that the cross-over was only a passing moment in a much longer term trend. None the less the change in the relative positions of Montréal and Toronto has been dramatized in publications such as Benjamin Higgins' *The rise and fall? of Montréal*. In so far as Montréal has 'declined' or 'fallen' it is in purely relative terms, and that relativity is almost always posed in terms of Toronto. Indeed Sénécal and Manzagol (1993) have referred to 'the obsessive *tête-à-tête*' of the two cities. Many cities vary over time in their relative growth rates, but in the case of Montréal and Toronto, attention to such change is greatly sharpened because of the symbolic significance, in Canadian terms, of the two cities concerned – as earlier noted they are seen as the standard-bearers of the French/francophone[2] presence on the one hand, and the British/anglophone on the other.

Explaining the 'switch-over'

It will be instructive to examine briefly the factors that have been put forward to explain the relative change in the size and functional significance of Montréal and Toronto. It is useful to heed Gunter Gad's advice at this point – 'the changing position of Montréal and Toronto in the Canadian urban hierarchy is due to a very gradual, multi-causal process, rather than a unicausal, revolutionary one' (Gad, 1995). Claims for a single cause – the up-welling of Québec nationalism and the drive to *francisise* everyday life in Montréal – have probably been motivated, in part at least, by opposition to those very enterprises. In addition, the fact that the

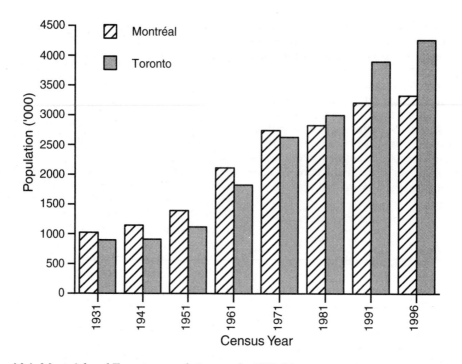

Fig. 16.1 Montréal and Toronto: population totals, 1931–96

changeover in the ranks of Montréal and Toronto coincided with the coming to power of the Québec nationalist party, the *Parti Québécois*, tended to create an association between language issues and nationalism, on the one hand, and a perceived decline of Montréal, on the other. Certainly, increased anglophone out-migration has been attributed to *francisation*[3] and to separatism concerns.

The trend in favour of Toronto has more broadly been explained by reference to a wide array of factors. The westward shift of the centre of gravity of the Canadian population and the development of the midwestern industrial heartland of the United States favoured Toronto's location. In addition Toronto's immediate hinterland (southern Ontario) provided a richer agricultural backdrop than that in proximity to Montréal. It also contained a constellation of sizeable other urban centres, something absent in the case of Montréal. Furthermore, acting against Montréal was the opening of the St Lawrence Seaway, which enabled large ocean freighters to bypass the port. The growth of air trans-

port further eroded Montréal's position as the centre of land- and sea-based routeways, again swinging activity towards Toronto and its more central 'hub' location. Finally Montréal was disadvantaged in that much of its traditional industry, in particular textiles and clothing, was subjected to severe foreign competition, leading to a fundamental restructuring of the city's manufacturing base.

It would be wrong, however, to ignore or attempt to dismiss matters of language and nationalism. This is particularly the case with financial services and also the control functions of head offices. Undoubtedly the slide away from Montréal to Toronto was underway long before the 1970s, but arguably the shift in business control was sharpened by events and perceived trends in Québec itself. The financial activities where the shift to Toronto was at its sharpest were those most clearly related to the financing of enterprises, and particularly of new enterprises: the stock exchange, the head offices and trading desks of the banks, the investment dealers, the 'money market' and the bond market

(Higgins, 1986, p. 61). In addition, 23 insurance companies moved their head offices from Montréal to Toronto between 1941 and 1961, while the move of Sun Life in 1977, given its temporal coincidence with an upwelling of language conflict in Québec, is frequently offered as an example of the impact of the changing politico-linguistic environment in Montréal. Mario Polèse has argued persuasively (1990) that in the financial sector, head-office location decisions are greatly influenced by cost factors, in particular the cost of specialized labour and the cost of communication associated with buying and selling. Changes in Québec from the 1960s onwards created a working environment increasingly dominated by French, while French speakers were vigorously recruited and promoted in the business world. This environmental shift, both in the workplace and in everyday life, increased the costs of recruiting anglophones, many of whom were migrating from Montréal in the 1970s, ostensibly because of the *francisation* policy of the Québec government. While most Montréal francophones could operate in English, Polèse claims that fluent use of the language of the client, together with what he calls 'cultural affinity', tended to advantage financial and other service organizations operating out of Toronto or other English-language environments. After all, as he points out, the Canadian market is 75 per cent anglophone, while the North American market as a whole is almost exclusively so (excepting Québec and a growing use of Spanish in the United States). Polèse concludes his analysis by claiming that 'to find the causes of the inversion at the top of the Canadian urban hierarchy, it is necessary to look at the linguistic transformation in Québec, and not at the traditional factors of economic geography' (Polèse, 1990; see also Coffey, 1994, p. 34).

In consequence of all this, Toronto has emerged as the metropolis for English-speaking Canada (most of the country, that is) while Montréal has 'declined' to be the regional metropolis of Québec. Québec nationalists, as will be argued below, would not see things quite that way.

ETHNICITY AND LANGUAGE

To further increase our understanding of the context of Montréal's 'decline' relative to Toronto we will now explore the ethnic and linguistic profiles of the two cities.

First Toronto. The population of Toronto in terms of the perceived ethnic origins[4] of the people, was overwhelmingly 'British'[5] until the last third of the twentieth century. Nineteenth-century Toronto was 90 per cent British and, even as late as 1971, 60 per cent of the population fell into that ethnic category. However, as Fig. 16.2 makes clear, from 1941 onwards the proportion of the population claiming British background was in steady decline, with the percentage neither British nor French increasing, until the British and 'Other' categories reached parity in 1981. By 1991 the British section of the population was in a minority. Throughout the whole period of the city's growth, the French origin section of the population formed only a very small percentage of the total (between 2 and 4 per cent). Thus, in ethnic terms, Toronto historically was a British city, but has now become increasingly, and indeed predominantly, multicultural. Throughout this transition, while the ethnic composition changed, there was no ambiguity whatsoever about the overwhelming predominance of English as the functional language of the city. Non-English speaking newcomers became and become anglophones. The linguistic hegemony of English in Toronto is just that – unchallenged.

Montréal provides a very different story, both ethnically and linguistically. In the early part of the nineteenth century there was a French ethnic predominance, with a large British minority. For several decades round mid century there was even a British majority. From the 1860s onwards, however, people of French ethnic origin again formed over half the population, and the French origin proportion has stayed above 60 per cent until the present day. As Fig. 16.3 shows, the British proportion has steadily declined over many decades, falling from 25 per cent in 1941 to about 6 per cent in 1991. At the same time the 'Other' (non-British/non-French

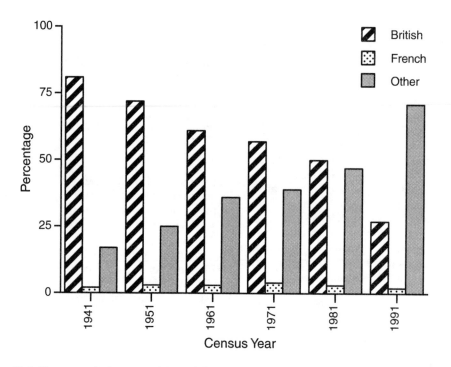

Fig. 16.2 Toronto: ethnic composition of the population, 1941–91 – British, French and 'Other'

category) has steadily increased, forming about a third of metropolitan Montréal's population by the 1990s.

Thus both Montréal and Toronto have experienced major shifts in their ethnic compositions – Toronto moving from an overwhelmingly British urban complex to a highly multicultural one, Montréal from a mid-nineteenth century British–French numerical equality to a city with a clear, though not by any means overwhelming, French majority, a much-contracted British group and a burgeoning multicultural component. As noted above, English has been unchallenged in Toronto. Montréal, on the other hand, has been a linguistic battleground. English for a long time served as the dominant language of commerce and higher-level business activity. This was true not only when there was a British ethnic majority, but also for more than 100 years subsequently, despite a French majority in the metropolitan population. However a central theme of recent decades in Montréal has been 'The Reconquest' (Levine, 1990) or *La Reconquête* (Levine, 1997). An anglophone-dominated

economy and a dominance of the English language in key areas of life has been overturned as part of radical social change in the province of Québec as a whole. This wide change has been called The Quiet Revolution (*La Revolution Tranquille*). Legislation passed by the Québec provincial parliament (*L'Assemblée Nationale du Québec*), in particular Bill 101 of 1977, mandated the use of French in the workplace, ordered the display of French-only signs in public places and required newcomers to Québec (and therefore immigrants entering Montréal) to enrol their children in French language schools. This last requirement derived from a concern that new immigrants, who were of neither French nor English mother tongue were tending to favour the use of English, seeing it as having more utility than French for their own social mobility and, even more, for that of their children. Increased opportunity for French speakers in the Montréal economy began to change this perception to some extent, but the schooling requirements of Bill 101 ensured that immigrant adaptation would be much more vigorously orientated

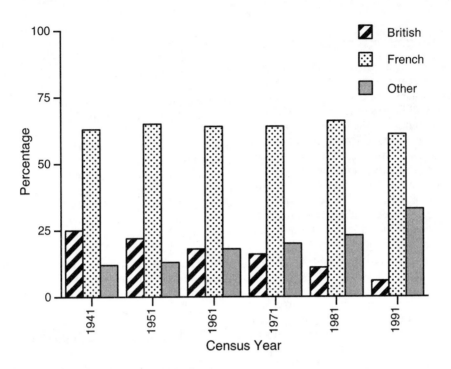

Fig. 16.3 Montréal: ethnic composition of the population, 1941–91 – British, French and 'Other'

towards the French community than hitherto. This impact can be seen in the patterns of school enrolment. In 1973–74 students from language backgrounds neither French nor English (the allophones) overwhelmingly registered in English language schools within the Montréal public system (89 per cent English; 11 per cent French). By 1987–88 66 per cent of the allophone students were enrolled in French schools, 34 per cent in English.

Thus, to a very significant extent, Montréal was reconquered for French within a few decades. Francophone Québecers became dominant players in much of the Montréal economy, the flow of immigrants was drawn much more into the French ambit and the visual environment (*le visage linguistique*) was being transformed into one that was unilingually French.

There were several other transformative effects of all this. Québec nationalism came much more to the fore, particularly with the election of the *Parti Québécois* to office in 1976. Referenda on the possible separation

of Québec from the rest of Canada were held in 1980 and again in 1995. In the latter case the anti-separatists had the narrowest of victories. Most significantly, a major component of the vote against Québec separation was centred on the western two thirds of Montréal Island – the area of concentration of the immigrant population and of most of the anglophones. This division, in turn, spawned two quite dramatic responses. In one, the leader of the *Parti Québécois* at the time of the referendum declared that 'money and the ethnic vote' had frustrated the aspirations of Québec nationalists. In this there was a suggestion that the immigrants were not real Québecers and were standing in the way of those who were. The other response came from some of those opposed to Québec separatism. Here emerged calls for the partition of Québec, and specifically for the partition of metropolitan Montréal – a division that would enable the western half of Montréal Island to remain in Canada in the event of the declaration of an independent Québec.

CONCLUSION

The changes and uncertainties swirling around Montréal for the past three decades stand in contrast to the dynamic certainties of Toronto. It will be obvious how the changes and the instabilities of the Québec metropolis could impinge unfavourably on economic performance. In such a context, Toronto offered a nearby, attractive alternative – an alternative location for business activity, and an alternative base for those sections of anglophone Montréal feeling threatened by the *francisation* of Montréal and the prospect of Québec independence.

From one perspective it can be argued that, for a variety of reasons, Montréal has contracted from being an urban centre of continental significance to become a regional metropolis for the province of Québec. From a different perspective, one could claim that Montréal has been adjusting from a pan-Canadian role towards becoming the metropolitan centre of a newly emerging country on the world map – Québec. Only the future will disclose where the truth lies. Interestingly, the very *francisation* of Montréal, which may have militated against the continuation of the city's continental role, served to provide an element of protection for the functions oriented to Québec itself. The socio-linguistic transformation that contributed to the relative decline of Montréal also set a limit to that decline (Polèse, 1990; Coffey and Polèse, 1991).

Toronto and Montréal, both still by far the largest urban centres in Canada, have, in concert, shown great change. They have expanded greatly during the twentieth century, and in the process they have traded places in the Canadian urban hierarchy. They have transformed ethnically from cities dominated, in the case of Toronto by one ethnic block (the British) and in the case of Montréal by two (French and British), to become multi-ethnic population concentrations. At the same time, while English has maintained its cultural and functional hegemony in Toronto, an English dominance in Montréal has been overturned to the advantage of French. Both cities have drawn in large vol-

umes of immigrants, particularly since the end of the Second World War. In this Toronto stands out, having attracted a much greater flow than Montréal. It has also received them into a context that is linguistically English. With Montréal, on the other hand, the immigrants initially were made not all that welcome by the French majority, and they were seen as a threat to the survival of French, given their propensity to add to the numerical and functional strength of the anglophone community. More recently, mainly due to changes in the ground rules for immigrant entry into Québec society, Francophone Montréal has become more welcoming. Thus, just as Toronto has been transformed into a stunningly varied multicultural complex, so francophone Montréal is now undergoing a similar, if more modest, change.

ACKNOWLEDGEMENTS

The Canadian High Commission in London provided financial support through its Faculty Research Programme. In Montréal the following were very informative: Mario Polèse and Damaris Rose at INRS-Urbanisation; Brian Ray and Sherry Olson at McGill University; Claude Marois at Université de Montréal; Robert Aiken, Pierre Deslauriers and Brian Slack at Concordia University. In Toronto similar help was provided by: Larry Bourne, Jim Lemon, Gunter Gad and Carlos Teixeira at the University of Toronto; Robert Murdie at York University.

None of the above bears any responsibility for the interpretations I offer in this chapter.

ENDNOTES

1. For the purposes of this discussion, the terms 'Montréal' and 'Toronto' do not refer to the central cities alone, but to the more extensive urbanized areas. Much of the data employed here refers to the Census Metropolitan Areas (CMAs). The general concept of a CMA is defined by Statistics Canada as being one 'of a very large urban core, together with adjacent

urban and rural areas which have a high degree of economic and social integration with that core' (Statistics Canada, 1989). To complicate matters, until 1 January 1998 the term 'Metropolitan Toronto' referred to a governmental unit, comprised of the City of Toronto and five contiguous municipalities.

2. 'Francophone' refers to those using French as their 'home language', 'anglophone' to those using English and 'allophone' to those using a language other than English or French. 'Home language' is defined as the language spoken at home by the individual concerned. If more than one language was employed, then that spoken most often at home was to be recorded in the census (Statistics Canada, 1989, xxvi).

3. According to The Oxford Hachette French Dictionary, 'francisation' translates into English as 'Frenchification'.

4. 'Ethnic origin' – refers to the ethnic or cultural group(s) to which the respondent or the respondent's ancestors belong. Ethnic or cultural group refers to the 'roots' or ancestral origin of the population and should not be confused with citizenship or nationality. Prior to 1981 only the respondent's paternal ancestry was to be reported, and only one ancestry recorded. From 1981 multiple ancestry could be claimed and could be traced through both maternal and paternal linkages. The data for 1981 onwards used here refers only to

single origin responses (Statistics Canada, 1989, xxv–xxvi). This is likely to bias the proportions in favour of those groups who are the most recent entrants into Canada – there has been less time for ethnic 'mixing' to occur.

5. Under the term 'British' the Census of Canada includes people of English, Irish, Scottish and Welsh origins.

KEY READINGS

Careless, J.M.S. 1984: *Toronto to 1918: an illustrated history*. Toronto: James Lorimer & Co.

Coffey, W.J. and Polèse, M. 1993: Le déclin de l'empire Montréalais: regard sur l'économie d'une métropole en mutation. *Recherches sociographiques* **34**(3), 417–37.

Higgins, B. 1986: *The rise and fall? of Montréal*. Moncton: Institut Canadien de Recherche sur le Développement Régional.

Lemon, J.T. 1985: *Toronto since 1918: an illustrated history*. Toronto: James Lorimer & Co.

Levine, M.V. 1997: *La reconquête de Montréal*. Montréal: VLB Editeur.

Linteau, P-A. 1992: *Histoire de Montréal depuis la Confédération*. Montréal: Les Editions du Boréal.

Polèse, M. 1996: Montréal: a metropolis in search of a country. *Options politiques/Policy Options* **17**(2), 26–9.

17

BOSTON: HIGH TECH
AND HERITAGE

STEPHEN J. HORNSBY

HISTORY

Founded in 1630, Boston is the second-oldest major American city after New York, and the largest and most vibrant city in New England. With a population of 3228000 in the metropolitan area and 574000 in the city (1990 census), Boston is a centre for state government (it is the capital of Massachusetts), business, high-tech industry, transport, health care, culture, education, and intellectual life. Moreover, its striking geographical setting and well-preserved urban heritage make it one of the most visually attractive cities in the United States.

Fronting island-studded Boston Harbor, which opens on to Massachusetts Bay and the North Atlantic, Boston is situated at a significant break in the coastal geomorphology of the eastern seaboard. To the north, jagged headlands jut out into the Gulf of Maine, while to the south, a low, sandy shoreline begins its sweep along much of the East Coast. Rocky Cape Ann and the sandy, northern hook of Cape Cod form the outer limits of Massachusetts Bay and frame the seaward entrance to the city. Boston itself is easily read. The focal point is the cluster of skyscrapers covering the Shawmut Peninsula, the original site of the city and now the central business district. To seaward, land has been reclaimed from Boston Harbor for a container port and Logan international airport; to the north, the Charles River forms a natural boundary between the cities of Boston and Cambridge; to the west and south, the city sprawls over a considerable area, at first densely in the Back Bay and South Cove areas, and then at increasingly lower densities through the old streetcar suburbs of Roxbury and Dorchester until it reaches the automobile suburbs and 'edge cities' that have sprung up along the perimeter highways. What is striking about much of this outer suburban development is that it is interspersed with mixed woodland and old glacial ponds rather than cleared farmland, a reflection of the poor agricultural hinterland of the city. From the air, Boston seems surrounded either by sea or heavily-wooded land, a juxtaposition more common in

Atlantic Canada than in the eastern United States.

During its first 200 years, Boston was dominated by merchant capital and long-distance trade. With a poor agricultural hinterland but rich fisheries offshore (symbolized by a wooden cod hanging in the Massachusetts State House), shipbuilding, and the coasting and Caribbean trades, Boston's merchants created a dynamic commercial economy and the largest city in the British North American colonies, until it was superseded by Philadelphia in the 1750s. After the American Revolution, the old colonial trades were supplemented by a profitable trade to China. Yet restrictions on maritime business in the early 1800s prompted Boston merchants to switch their capital from commerce to industry. Following the example of industrializing Britain, some of the city's leading merchants – the Boston Associates – invested in large, integrated textile mills on the periphery of the city. The first such mill at Waltham on the Charles River was followed by mills farther away at Lowell, Nashua, Lawrence, and Manchester on the Merrimack River. By mid century, the boot and shoe, railroad, and machine tool industries were also established in the region. With these developments, Boston was launched into an era of industrial capitalism and benefited enormously. Profits from the textile industry were ploughed into other sectors of the city's economy, including banking, insurance, and railroads. As a result, Boston maintained its position as the undisputed financial and service centre in New England and a major business centre in the country. Yet, by the early twentieth century, the economic foundations of the city were weakening. In the textile industry, the introduction of standardized machinery and increased automation after the end of the Civil War in 1865, as well as rising wages for New England labour, encouraged manufacturers to relocate to the southeastern states, where cheap, non-unionized, and unskilled labour was plentiful. In the boot and shoe industry, factories were moving out of eastern Massachusetts to the periphery of New England, again in search of cheap labour. As factories in the Northeast began to close, the

Great Depression in the early 1930s caused further economic dislocation. Yet the Second World War, considered the 'most momentous event in the modern history of New England' by economist Bennett Harrison (Harrison, 1984) laid the foundations for economic renewal. Massive wartime investment by the federal government in new plant and equipment gave a tremendous fillip to the arms and aero engine industries in southern New England, as well as to the fledgling computer industry. At the end of the war, much of this plant was transferred to the private sector, the basis for some of the high-tech industries of modern New England. Boston was particularly well-placed to benefit from these new industries. With the greatest concentration of colleges and universities in the country, the city was producing the highly skilled labour needed by the emerging high-tech sector. Just as Stanford University generated growth in Silicon Valley, so the Massachusetts Institute of Technology played a key role, through spin-off companies, in the development of research labs and electronic industries in and around Boston. By the mid 1990s, the Boston area was second only to northern California as a centre of high-tech industry. During this economic transformation, Boston maintained its leading role in New England as a provider of financial and commercial services, and has developed an international reputation, through its many medical schools and hospitals, for health care. Despite the economic difficulties associated with restructuring, Boston has moved from a maritime-based commercial economy, through the industrial era, and into the high-tech and knowledge-based industries of today. Many seaports and textile towns in New England have not been so successful.

SOCIAL GEOGRAPHY

Founded by English Puritans but best known today for its large Irish Catholic population, Boston's social geography has undergone considerable change over the past 350 years. During the colonial and early national eras, the city's population was among the least diverse of any major American city. Much of the population derived from the English Puritan and Scots-Irish migrations of 1630–60. By the early nineteenth century, an old-stock Yankee population dominated the city. Yet in the late 1840s, massive Irish Catholic immigration, caused by the great Irish Famine, altered the city's ethnic complexion. By 1850, Irish Catholics were the largest ethnic group, comprising a quarter of the city's population. During the late nineteenth century, there were large influxes of Maritime Canadians ('bluenoses' in Boston parlance), Russian Jews, and Italians. While the Maritimers integrated relatively easily into the Yankee city, the Irish, Italians, and Russian Jews created their own ethnic neighbourhoods, concentrated in the inner city. Although the quota acts of the 1920s choked off much of this Canadian and European influx, Boston continued to receive migrants from other parts of the United States, especially blacks moving north from the southern states, as well as Chinese filtering east from the Pacific Coast. Since the 1970s, the city has also attracted immigrants from the Caribbean, Latin America, and Asia.

The economic restructuring and changing social geography of Boston has had a marked effect on the city's physical landscape. The old Yankee mercantile city, concentrated on the original Shawmut Peninsula, is still readily apparent, despite considerable urban renewal. Most obvious is the irregular street layout of downtown Boston, so different to the grid-iron of urban America, and much of it dating to the seventeenth century. Overlaying this street pattern are remnants of old mercantile land uses. Along the waterfront, the economic heart of the mercantile city, are shipping wharves and warehouses, many now converted to residential and commercial uses. Inland, the Custom House district, comprising the federal Custom House, old mercantile exchanges, and offices, gives way to the financial district, where bank skyscrapers dwarf the Old State House, once the institutional apex of the mercantile city. On the northern fringe of the business district lie the former produce markets of Quincy Market and Faneuil Hall, both recycled recently to cater to office workers in the financial district and tourists. At the outer rim of the old mercantile city are the remains of the early elite residential areas. Little survives from the colonial era, largely because the North End, once the haunt of wealthy merchants and government officials, was abandoned by the elite after the American Revolution, leaving behind little more than the episcopal Christ Church (the Old North Church) and Copps Hill burial ground. Yet during the late eighteenth and early nineteenth centuries, the city's mercantile elite created in Beacon Hill one of the finest groupings of federal and Classical revival townhouses in the United States. For many Bostonians, Beacon Hill represents the city at the height of its early nineteenth-century commercial prosperity and Yankee 'Brahmin' dominance.

As the city industrialized and immigrants arrived during the 1840s, the city's landscape began to change. Although the textile industry was confined to peripheral villages, such as Waltham, where large brick mills were built, the inner city was transformed. With a burgeoning financial and commercial sector, the central business district expanded, displacing areas of slum housing. Recognizable financial, wholesaling, and retailing areas emerged. The former elite neighbourhood of North End became densely packed with brick 'dumbbell' tenements, housing first Irish and then Italian immigrants. Close by, the West End accommodated many Russian Jews and blacks, while other Irish immigrants moved into South Cove. The elite, meanwhile, shifted further out. The Back Bay mudflats were reclaimed from the Charles River and a grid of tree-lined boulevards and elegant townhouses modelled on Haussmann's Paris was laid out. As urban transport improved (Boston was a pioneer in electric mass-transit), streetcar lines were pushed into the rural hinterland, encouraging the suburbanization of the city. Streetcar suburbs comprising three-decker wooden tenements were built southwest of the city in rural Roxbury, Dorchester, and Jamaica Plain, areas which soon housed Irish and Russian Jews moving up the socio-economic ladder and out of the inner-city immigrant ghetto. Elite suburbs were also developing in Brookline, an area of well-treed, curving drives, and Tudor-revival homes. Among its residents was Frederick Law Olmsted, the master of picturesque landscaping in the United States and the designer of the 'emerald necklace' of parks that surrounds Boston's built-up inner city.

After the Second World War, Boston was marked by increasing suburbanization as well as urban renewal. The building of interstate highways by the federal government, federal mortgage insurance, and a massive increase in automobile ownership encouraged suburban expansion around many American cities. In Boston, the building of Route 128/Interstate 95 and the Massachusetts Turnpike stimulated suburbanization westwards, with many high-tech companies developing greenfield sites along the highways. In the 1960s, the growth of out-of-town shopping malls and, more recently, office and professional parks, led to further expansion of the 'urban realm' or what journalist Joel Garreau has called 'edge cities' (Garreau, 1991) (Fig. 17.1). At the same time, inner-city Boston was being transformed. Widespread urban clearance destroyed derelict areas as well as immigrant communities, most notoriously in the city's West End. Urban freeways, such as the elevated John F. Fitzgerald Expressway, carved through other neighbourhoods, while the building of Government Center in the downtown area created several concrete behemoths. In Dorchester and other parts of the city, massive public housing projects were started. Such upheaval provoked reaction. The congested North End was celebrated by planning critic Jane Jacobs for its vibrant street life and variety (Jacobs, 1965, p. 43), and by the 1970s parts of inner-city residential areas were undergoing 'gentrification'. City government, too, was recycling, rather than destroying, the historic market area. After decades of clearance, the city's heritage of buildings was being preserved.

CONCLUSION

Perhaps more than any other American city, Boston is a city of the mind. For MIT professor Kevin Lynch, Boston was literally a laboratory in which to chart people's perceptions

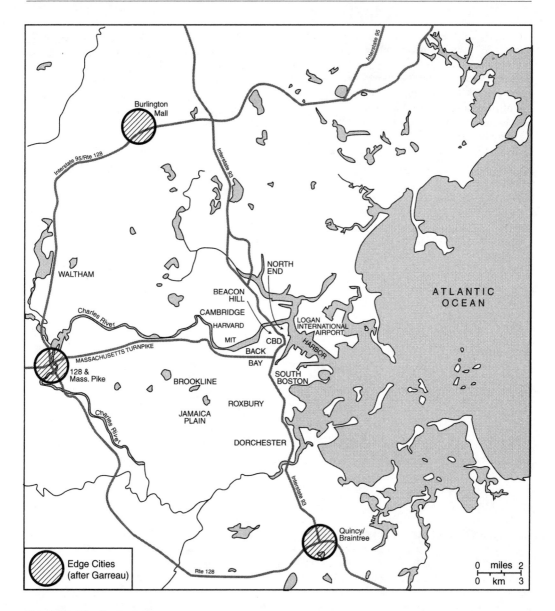

FIG. 17.1 The Boston area

or mental 'image of the city' (Lynch, 1960, pp. 16–25). For literary and cultural critics, Boston has occupied a central place in the American imagination. Ever since Winthrop's injunction to the Puritans to build 'a city upon a hill', writers and intellectuals have given the city a powerful sense of identity. Although Boston's cultural preeminence may have slipped since its heyday in the nineteenth century, when writer Oliver Wendell Holmes described it as the 'hub of the solar system' (cited in O'Connell, 1990, p. 4), the city still evokes a spirit of place rivalled by few other North American cities.

KEY READINGS

Conzen, M.P. and Lewis, G.K. 1976: *Boston: a geographical portrait*. Cambridge, MA: Ballinger Publishing Co.

O'Connell, S. 1990: *Imagining Boston: a literary landscape*. Boston: Beacon Press.

Whitehill, W.M. 1968: *Boston: a topographical history*. Cambridge, MA: Harvard University Press.

Wilkie, R.W. and Tager, J. (eds) 1991: *Historical atlas of Massachusetts*. Amherst: University of Massachusetts Press.

Section F

Regional vignettes

18

THE PACIFIC NORTHWEST: RURAL BACKWATER TO URBAN ECOTOPIA

LORETTA LEES

I wanted to be herbal, flannel, on-line and in-line. I wanted boots with the guts for mountain biking and day hiking. ... I wanted to dwell among people who wear socks with sandals and eat peanut butter so pure it has to be stomped with bare feet. ... I wanted to read magazines with stories like 'White lies about white flour', 'Are free-range chickens really cruelty free?' and 'What we learned from Biosphere 2' (Dowd, 1996)

The Pacific Northwest is 'naturally northwest!'

There are no commonly agreed upon boundaries for the region known as the Pacific Northwest. I include British Columbia, Washington, and Oregon under the umbrella; other authors might include western Montana, Idaho, northern California, and, less often, Alberta and southeastern Alaska (Fig. 18.1). Indeed, the identity of the Pacific Northwest is not determined by boundaries (except perhaps physically by the various mountain ranges and the Pacific Ocean), but by the transgression of boundaries – both physical and social. The Pacific Northwest crosses the US–Canada border and US state borders and is orientated towards the Pacific Rim by virtue of its position on the western edge of a large continent. The Pacific Northwest's geographical setting, its physical environment and natural resources are the special ingredients that have influenced not just its economic development but also its social and cultural development, and more generally its sense of place. Situated on the edge of the continent, the Pacific Northwest was relatively isolated from political control by Ottawa and Washington, DC. Distance facilitated a unique political culture common to both sides of the border, characterized by rugged individualism, liberalism, and resentment of eastern authority.

The historical development of the Pacific Northwest was based upon manipulation of the region's natural resources. Nature in the Pacific Northwest has been harnessed by Native Canadians and Native Americans, exploited by white colonizers and their resource industries and, more recently,

enjoyed through outdoor recreation and hyped through the West Coast rhetoric of Ecotopia.[1] Even before European settlers arrived Native Canadians and Native Americans had established trade networks bartering natural resources such as fish and roots. In the late eighteenth century the commercial fur trade was facilitated by capitalism and colonialism, linking the Pacific Northwest with markets in Asia, Europe and eastern North America. In fact before the Canadian Pacific Railroad connected British Columbia with eastern Canada, British Columbia's main exports – salmon, lumber, coal and gold – went south to California or west across the Pacific to Asia. This special Pacific Northwest relationship between Canada and the United States was to continue.

Throughout the nineteenth and into the twentieth century resource industries such as logging, mining and fishing characterized and dominated economic development in the region. Regional life featured boom and bust cycles of economic frenzy followed by depression, more often than not caused by external economic factors, by speculators, foreign markets and competitors. Despite the rapid development of cities in the region such as Seattle and Vancouver, the Pacific Northwest was to remain a provincial, rural backwater until the 1930s.

Urbanization in the Pacific Northwest took off during the Second World War as industrialization became linked to the Allied war effort. With aluminium smelters processing local ores with cheap electricity from the massive dams along the Columbia River, ship building and aircraft manufacture (the Boeing headquarters are in Seattle) became mainstays of the new urban industrial Pacific Northwest. This urbanization accelerated in the first two decades after the Second World War when the forest products industry was at its most successful, partly as a result of the building boom associated with rapid suburbanization (Schwantes, 1996, p. 441). In the 1950s, the economy was influenced by the principles of Fordism as the forest products industry became dominated by larger firms created through mergers and take-overs which saw the Pacific Northwest's economy continue to be dominated by the export of

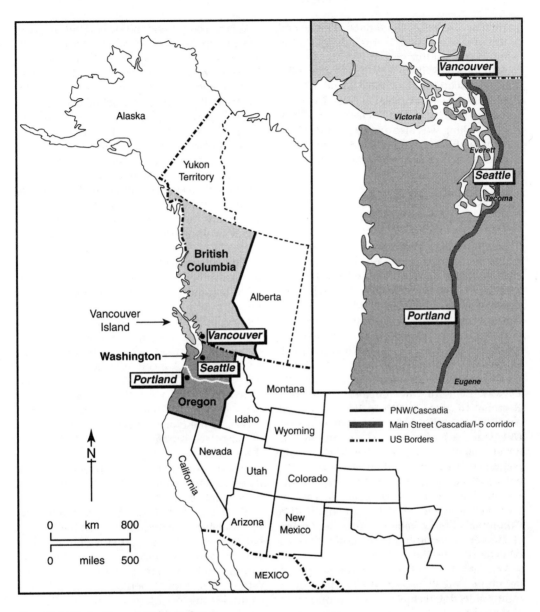

Fig. 18.1 Cascadia and its Main Street

natural resources. This high point of Modernism in the Pacific Northwest was celebrated in 1962 when Seattle held the 'Century 21' world's fair which featured the Space Needle and a monorail. But the boom was not to last. In the 1970s the region entered a period of industrial decline. Resource industry jobs were cut as new technology and the labour market flexibility of

the present post-Fordist industrial regime has meant that white collar job opportunities have outrun blue collar job opportunities.

As the Pacific Northwest was dealing with the decline of its resource-based industries, it was also negotiating the reformulation of its identity. In the 1970s, the 'edge syndrome', which had historically kept the Pacific Northwest in relative isolation, was recast as a particular West Coast identity. The Pacific Northwest was 'on the cultural edge' – a forward-looking, futuristic, inventive, technological, 'far out', unconventional place. This was represented at the 1986 World Exposition held in Vancouver, which hyped a forward looking 'world class city'. This image-making was also apparent in Portland where architects constructed what was heralded as the world's first postmodern building.

ENVIRONMENTALISM

Throughout the 1970s a seedbed of utopian and ecological thoughts grew along the west coast of North America. These thoughts were idealized in 'the liveable city' – a socially progressive and aesthetic urban ideology with a concern for quality of life and ultimately environmental sustainability. 'Liveability', which became the Pacific Northwest's catch phrase, was further hyped by the rise of environmentalism in the area. The first Livable Region Plan of the Greater Vancouver Regional District (GVRD) in 1974 advocated a variety of land use and transport measures to contain suburban sprawl, encourage the development of dense urban nodes and restrict private automobile use so as to enhance the quality of urban life in the region. Planners feared that unless residential development was contained, British Columbia's Fraser Valley would die, tarmacked over from one end to the other, choking on exhaust emissions.

During the 1980s environmentalism became a potent force in the Pacific Northwest (indeed Greenpeace was founded in Vancouver in 1971), pitting resource workers trying to earn a living in a changed resource industry against environmentalists trying to save the environment (see Chapter 4). The spotted owl controversy in the United States epitomized this jobs versus environment debate as environmentalists took out law suits to halt the logging of owl habitat in old-growth forests. The US Fish and Wildlife Service added the owl to its list of threatened species in 1990 and cited logging as the main threat to its survival. As such, logging on fed-eral lands could now be stopped legally and tensions in the areas affected ran high. In 1991, a Seattle federal judge froze timber sales to protect the owls; timber harvests plummeted and unemployed workers blamed the environmentalists, though in actual fact technological modernization and restructuring in the mills were responsible for many more job losses. In 1993 President Clinton tried to quell the furore at a 'Forest Summit' in Portland, but, given the nature of the dispute, rather predictably he was unsuccessful. Despite the arguments the logging industry is in decline anyway as the old-growth forests on which it has historically depended are systematically depleted.

The other striking feature of the forest debate in the Pacific Northwest was the way it has become globalized, involving people in far off places in the struggle to protect the forests. In British Columbia, environmentalists protesting the clear-cutting of pristine watersheds in the Clayoquot Sound, on the west coast of Vancouver Island, called on support from abroad. Activists in Europe called for an EU boycott of BC forest products. British Columbia was tagged as the 'Brazil of the North', linking it in the minds of far-off consumers with the well-known destruction of tropical rainforests. The government of British Columbia responded with its own public relations campaign, designed to allay far-off fears.

With the decline of the forest industry, resource-based communities in the Pacific Northwest are turning yet again to their environment to make a living, this time through tourism: skiing, hiking, white-water rafting, nature tours, heritage villages. In one example, the ex-resource town of Roslyn, Washington, has been used as a film set for the fictional town of Cicely, Alaska, in the CBS television series *Northern Exposure*. As a result Roslyn now attracts tourists throughout the year (Schwantes, 1996, p. 505). Similarly, Chemainus, British Columbia, another saw-mill town that saw hundreds of lay-offs in the early 1980s, has become an arts and crafts destination for tourists. The town now promotes itself as 'the little town that could'.

Vancouver, which boasts sea, mountains and some of the world's best ski resorts within a couple of hours' drive, has become 'outdoor central'.

> Every few minutes a cluster of the relatively young and toned starts up the narrow, 2.9 kilometre trail that rises 900 metres through fir and hemlock (Vancouver's Grouse Grind). Carrying water canisters, they move up the near vertical ascent ...which takes anywhere from 45 minutes to 1.5 hours. The experience is one of camaraderie, competition and physical challenge. The trail has become a symbol of Vancouver's outdoor fitness culture now that everyone from the triathlete to the computer geek heads for the mountains or out on the water during leisure time. In the last ten years, the city's Mecca-like call to hikers, climbers, cyclists and paddlers in pursuit of mountains and bodies of water has reached the mainstream. Locals have incorporated outdoor sports paraphernalia in their daily wardrobes for a look and attitude that are pure West Coast (*The Vancouver Courier*, 1997).

The city's culture is 'back to nature'. This ecotopian scene, rural though it may appear, is in fact an urban-based ideology. It is part of the rhetoric embedded in the notion of 'liveable cities' and a 'liveable region'. The passing of a no-smoking by-law, the proliferation of health food restaurants and people's awareness of environmental issues in Vancouver are all part of this cultural wave. Garreau argues that some of the roots of this new Pacific Northwest culture are to be found in the moralistic self-righteousness of those New England Puritans who settled in the Pacific Northwest. He jokes: 'It's not hard to find people in the Northwest who get as rigid with distress over the idea of a person eating an additive- and sugar-laden Twinkie as a devout ... Mormon does about someone imbibing strong drink' (1981, p. 272).

The problem with this ecotopian image is that it is attracting too many immigrants to cities on both sides of the US–Canada border. In-migrants are attracted both by the region's quality of life and relative lack of crime and by its booming service sector economy, especially the jobs in high-tech, aerospace and software industries. Only Orlando, Florida

(Disney) and Las Vegas, Nevada (gambling) can boast such booming service economies (see Chapter 11). So-called push factors are important as well, many migrants to the Pacific Northwest have fled the high taxes, crime, congestion, pollution, and inflated house prices of California, looking for a more simple, spiritual (represented in the explosion of New Religions on the West Coast) and nature-orientated life. Their relocation in rural communities particularly in Washington and Oregon has led to negative reactions from the local people. Others have fled economic recession and instability in eastern Canada and in the rest of the US. But perhaps the most visible in-migrants have been Asian, who have, likewise, sparked negative reactions from some urban communities.

THE PACIFIC RIM

The identity of the Pacific Northwest has historically been influenced by the Pacific Rim – Garreau (1981) calls it 'the northern Pacific Rim nation of Ecotopia (Fig. 18.2). Indeed, during the Second World War Pacific Northwesterners were closer to the Pacific conflict than most of the rest of Canada and the United States. The internment camps set up in both Canada and the US to intern Japanese immigrants during the war illustrate both the prejudice and fear associated with this. In more recent times though, the relationship has been less fraught and more fruitful. Looking towards the Pacific is important for the Pacific Northwest at a time when it is faced with the problem of providing employment for its own people. Canadian investment in the Pacific Rim has increased from $3.6 billion in 1984 to $12.8 billion in 1994 (Barrett, 1996, p. 1). British Columbia exports more to the Asia Pacific (38 per cent in 1990) than any of the other provinces in Canada (their average was 11 per cent in 1990). In turn, British Columbia is more dependent on Pacific Rim investment than other Canadian provinces (Cohn and Smith, 1995, p. 265–8). For example, British Columbia has been the largest lumber exporter to Japan for the last 25 years mainly for the wood-based housebuilding industry, which, in 1994, completed

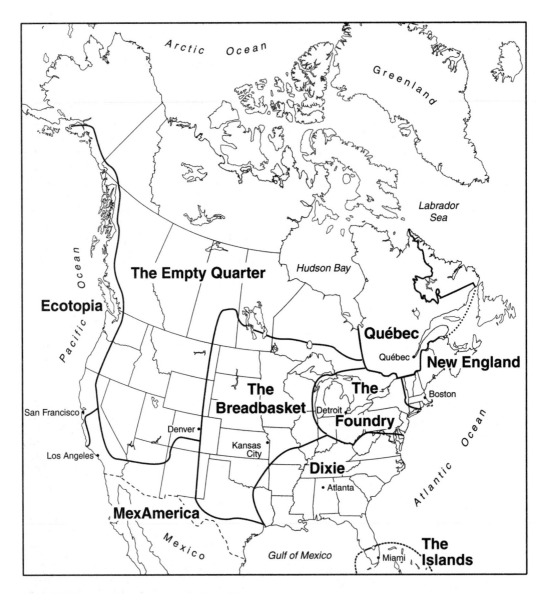

FIG. 18.2 Garreau's nine nations of North America

1.5 million units in Japan (Edgington and Hayter, 1997, p. 154). Geography aids trade as Vancouver's deep-water port is one full day closer to Asia by sea than the other West Coast seaports such as San Francisco. It also facilitates the 'Pacific Rim effect'. The Pacific time zone enables business people there to talk with Asian business people at the beginning of their workday (Bramham, 1996, p. 153).

This geographical and economic relation-ship has generated Asian in-migration into the Pacific Northwest. In the late 1980s and early 1990s, well-off Hong Kong Chinese contemplating the handover of Hong Kong from the UK to China emigrated to Vancouver in particular, engendering urban and cultural change. In Vancouver, Asian shopping malls have sprung up in the suburbs. There is an annual Chinese Dragon Boat Festival, two daily Chinese language newspapers, a Chinese TV station, and, in 1988, the Hong Kong

Bank of Canada, a subsidiary of the Hong Kong and Shanghai Bank, chose Vancouver as its Canadian headquarters and bought up the collapsed Bank of British Columbia. Vancouver has become the 'Pac Rim' capital that Los Angeles was predicted to be.

Links with the Pacific Rim have facilitated the development of a trans-Pacific residential property market. In Vancouver, Hong Kong Chinese capital has funded the redevelopment of the former Expo lands (17 per cent of downtown Vancouver) on the north shore of False Creek creating in 'Pacific Place' one of the largest Pacific Rim megaprojects thus far. Due to these dense, high-rise urban megaprojects Vancouver has been labelled VanKong or Hongcouver (Lees, 1998).

The influx of Asians and their capital into Pacific Northwest cities has not been uneventful, as with the globalization of urban communities cultural conflicts have ensued. Take Vancouver, where despite the rhetoric of multiculturalism, cultural and racial tensions simmer beneath the surface. In recent years, the most visible manifestation of this conflict has been the controversy over so-called 'monster houses'. Wealthy Asians (usually Hong Kong Chinese) have bought up old English picturesque style houses in previously Anglo, middle-class neighbourhoods, such as Kerrisdale in inner Vancouver, knocked them down, often stripping the lot of its trees and vegetation, and in their place built large houses, often occupying the whole lot. The large houses associated with Chinese immigrants had minimal facades, ostentatious Greek columns, spiral staircases, huge arched doorways, and minimal landscaping vegetation. For Chinese-Canadians this house style represented prosperity, modernity and a new life in a new country. Many Anglo-Canadians living in the same street or neighbourhood reacted with distaste at what they considered to be the inappropriateness of these new properties (Ley, 1995). The arguments between the two communities boiled over into city council meetings, which eventually resulted in city by-laws restricting the removal of vegetation, especially old, well established trees, and setting standards for house design and minimum property-line set-backs.

CASCADIA

As well as looking towards the Pacific Rim, Pacific Northwesterners are also looking inside their own region for other forms of identity. This has led to the formation of 'Cascadia' – a region held up by Bill Clinton as a model of Pacific Age co-operation (Fig. 18.1). Cascadia is a cross-border alliance between British Columbia, Washington and Oregon (sometimes Alberta and Alaska are also included) driven by the notion of a strong regional identity. It is hardly surprising that this transnational regional identity should enjoy a wide appeal on the US side of the border, where the annexation of British Columbia has been a national dream since the 1840s. But in recent years many western Canadians have also embraced the idea of Cascadia as the old east–west economic links stitching Canada together as a national economy have attenuated in the face of globalization and the North American Free Trade Agreement.

Cascadia has a common geography: a rainy climate and location looking towards the Pacific Rim, as well as a value system that promotes environmental sustainability. In terms of culture and fashion, Cascadia is the home of both gourmet coffee culture (it is the headquarters of the Starbucks coffee house chain) and grunge, the music and fashion associated with Seattle-based bands such as Nirvana. But most of all, what sets it apart is a sense that it is the envy of the world – 'Cascadia is not a state, but a state of mind' (Schell and Hamer, 1993, p. 1). Cascadia holds itself up as a twenty-first century model of a bio-region without borders whose goal is to promote a 'liveable region':

The opportunity in a larger vision of Cascadia is to redefine our economic horizon and, at the same time, be better able to manage our environmental future. … Failure to recognize our common interests reduces us to our traditional positions of western Canada as a tertiary market of Toronto and, for the northwestern states, as market outposts for Los Angeles, Chicago and New York. Commerce aside, our environment is a shared one. The vision of a larger bioregion – one forest, one waterway, one airshed – should invite us to preserve one of the last great Edens on the planet (Kelly, 1994).

This 'new regionalism' is demonstrated through an array of co-operative links between states and provinces, business councils, city and regional government bodies, task forces, policy institutes, tourist bodies, etc. Some of the main bodies include the Pacific North West Economic Region (PNWER) – a regional association of state and provincial governments to promote economic co-operation – the Pacific Corridor Enterprize Council (PACE) – a private sector regional association which aims to encourage closer business, trade and tourism links – and the Cascadia Transportation/Trade Task Force, a coalition of government and non-government organizations to develop cross-border strategies on growth management, cross-border mobility, trade and tourism (Artibise, 1995, p. 221–2).

PNWER was formally ratified in 1991. Its members include Alaska, Idaho, Montana, Oregon, Washington, Alberta, British Columbia; with Yukon joining in late 1997 (Fig. 18.1). In size the region is larger than the European Community and its combined GDP ranks it as the world's 10th largest industrial economy:

> PNWER brings together legislative, government, and private sector leaders to work toward the development of public policies that promote the economies of the PNW region and respond to the challenges of the global market place. The objective of PNWER is to build the necessary critical mass for the region to become a major player in the new global economy (www.pnwer.org/index.html).

'New regionalism' has emerged in reaction to the increased integration of national economies and to the obstacles that smaller regions face in competition for international business. Global trends in trade and the internationalization of production have in some ways forced collaboration between states and provinces 'in a way that was unthinkable a decade ago' to promote their region's products (Edgington, 1995, pp. 336–7). Cascadia's geographical location, its isolation from Ottawa and Washington DC and from North America's main commercial centres has also in part strengthened these collaborations (Edgington, 1995, p.338). Cascadia is a fast growing urban region seeking to attract global trade and investment. Indeed its population is projected to double to about 10 million by 2020. Green politicians on both sides of the border suspect that there is a hidden agenda of unrestrained growth (*The Economist*, 1994b).

Much of Cascadia's growth is occurring along the rapidly urbanizing I–5 corridor from Vancouver, BC, to Eugene, Oregon, known as 'Main Street Cascadia' (Fig. 18.1). The three core cities in the corridor (Vancouver, Seattle and Portland) are the preferred destinations of new immigrants from Asia–Pacific nations and elsewhere. Particular industries symbolize the new growth in these cities.

Vancouver has seen the growth of the film industry and, as a result, is known as 'Hollywood North' or 'Brollywood' – a pun on the city's wet climate. The American film industry has, in part, relocated from California to British Columbia because of the weak Canadian dollar, relatively weak trade union organization and lower pay rates. The *X-files* television series was filmed almost exclusively in British Columbia and Vancouver stands in for cities like New York or Los Angeles or even the wilds of Mississippi. The booming Asian film industry also makes movies in British Columbia, feeding the Asian market both in the Pacific Northwest and in Asia itself.

Seattle is dominated by two major industries: aerospace and software manufacturing. This local dominance is symbolized by Boeing, the world's largest aerospace company, and Microsoft, the world's largest personal computer software company. The aircraft industry in Seattle has been boosted by defence needs. The US government fed the industry with dollars during the Cold War years and federal investments in defence continue to aid the industry, notwithstanding its downturn in the 1970s and early 1980s. Today, Boeing provides over half the manufacturing jobs in the Seattle area and about two thirds of its sales are directed overseas. Boeing has spurred urban and industrial development in the area known as Pugetopolis, a conurbation including Seattle, Tacoma and Everett. It even yielded its own suburbs – Bellevue, known as 'Boeing's bedroom', which stands today at the centre of the area's computer software industry and is akin to California's Silicon Valley. Washington's software industry grew from 14 companies in 1975 to over 500 in 1990. Microsoft, founded by the world's currently richest man, Bill Gates, is the second largest public corporation employer in the area (Schwantes, 1996, p. 434–5).

Portland is consciously trying to shape itself as a new kind of American city (Harvey, 1996). In the late 1970s Portland drew a line around its metropolitan area and stated that on one side would be forest, farmland, and open space, and on the other, the city. The aim was to force urbanization into a compact area served by light rail, buses and cars. A silicon forest of high-tech industries grew inside the boundaries. Hewlett-Packard, Intel and Hyundai located there because they wanted an area that would attract educated workers who would be as interested in their quality of life as in the salaries they would receive.

CONCLUSION

In the 1990s the Pacific Northwest has been seen as the growth magnet on the West Coast of North America. In contrast, California, the one-time definer of the American Dream (see Chapter 15), has experienced economic recession, social and environmental problems. In particular, the media spotlighted civil riots, earthquakes and controversial court cases affected by racial overtones such as that of O.J. Simpson. California, with its massive Latino population, was looking south towards Mexico and South America when the Pacific Northwest was looking towards the Pacific Rim. The Pacific Northwest gained popularity with Asian, especially Chinese, investors when they were confronted with California's high costs, security problems and poor environmental quality. Changes to the tax laws in Oregon gave the state a competitive advantage over the nearby California (Edgington, 1995, p. 346–8). Whether this boom will continue in the face of the present (1999) economic crisis in Asia remains to be seen. Sceptical of Vancouver's recent rapid growth, journalists have labelled Vancouver as 'the city on steroids' and 'the city on prozac', implying an artificial and unsustainable growth (Canadian Broadcasting Corporation, 1996). 'The mistakes Ecotopia fears it may repeat are not those of the Foundry (the Rustbelt), but those of the boom towns of dry, sunny MexAmerica (e.g. its capital Los Angeles), (Garreau, 1981, p. 2). Whatever happens, for many people, the Pacific Northwest is a hallmark of future trends. It has led the way in restructuring its old resource-based economy into the new information age economy, associated with software and aeronautics, just as it is a leader in environmentalism and the new regional consciousness sweeping the continent.

ENDNOTE

1. The name 'Ecotopia' comes from a novel by Ernest Callenbach (1975). In this novel it is the year 1999 and northern California, Oregon and Washington have seceded from the United States, opting out of a racist, sexist, materialistic, wasteful and polluting North America. They set up their own independent nation called Ecotopia which means 'home place', sealing off their borders 'to the insidious influences of the rest of the continent'. The country is pollution- and noise-free, and an educational-social-sexual work-play ethic stresses the equal functions of men and women as tool-bearing animals capable of improving the quality of life' (Garreau, 1981, p. 250). Ironically, the contemporary Pacific Northwest is operating on very similar assumptions about enhancing the quality of life, quite different assumptions from its neighbours.

KEY READINGS

Artibise, A. 1990: Exploring the North American West: a comparative urban perspective. In Stelter, G. (ed.), *Cities and urbanization: Canadian historical perspectives.* Toronto: CCP.

Artibise, A. 1995: Achieving sustainability in Cascadia: an emerging model of urban growth management in the Vancouver–Seattle–Portland corridor. In Kresl, P. and Gappert, G. (eds), *North American cities and the global economy: challenges and opportunities,* Urban Affairs Annual Review **44**, 221–50.

Cone, J. 1996: *A common fate: endangered salmon and the people of the Pacific Northwest.* Corvallis: Oregon State University Press.

Dietrich, W. 1993: *Final forest: the battle for the last great trees of the Pacific Northwest.* Baltimore: Penguin Books.

Schell, P. and Hamer, J. 1993: *What is the future of Cascadia?* Seattle: Discovery Institute.

Schwantes, C.A. 1996: *The Pacific Northwest: an interpretive history.* Lincoln: University of Nebraska Press.

Wynne, G. and Oke, T. (eds) 1992: *Vancouver and its region.* Vancouver: University of British Columbia Press.

19

ATLANTIC CANADA: ALWAYS ON THE OUTSIDE LOOKING IN

DONALD J. SAVOIE

Eastern Canada's four Atlantic provinces of Nova Scotia, New Brunswick and Prince Edward Island (which collectively are the Maritime provinces) and Newfoundland and Labrador have had and continue to have common economic challenges, although they do not share a common history. Their economies traditionally looked to natural resource exploitation for export, starting with the fishery, also mining and timber – as both lumber and in manufactured products, initially wooden ships, later pulp and paper. However, since the end of the nineteenth century, this sector has not been able to generate much employment growth. Moreover, the region's manufacturing sector has not been able to take root and to create a sufficient amount of jobs to offset jobs lost in agriculture and other resource sectors. In short, Atlantic Canada remains a resource-dependent region and the economy is considerably weaker than that of other Canadian regions. It has long been argued that this area (Fig. 19.1) represents the gateway to North America for Europe. In addition, given its long thrust into the Atlantic, the region has insisted that it is the ideal location in which to place much of Canada's naval operations (McNaught, 1969, p. 9).

It has been customary for Canadians to treat the Maritimes and Newfoundland and Labrador as two regions rather than one given their different political histories. Until 1949, Newfoundland had few political and economic links to the Maritimes. Only upon Newfoundland joining Canada that year did it begin the process of becoming Canadian. However, the four Atlantic provinces have always had at least one thing in common: a strong reliance on government to promote economic development and even to sustain many communities. Indeed, the hand of government is visible everywhere in Atlantic Canada. The region, with a total population of 2 400 000, has four provincial governments, a strong federal presence in virtually every community and numerous municipal governments. In contrast, western Canada also has four provincial governments, but has a total population of 8 300 000. The federal government, as part of its past regional development efforts, has located several rela-

tively large operations in Atlantic Canada. For example, its superannuation branch (the office responsible for managing the pension plans of federal public servants) is located in Shediac, New Brunswick, rather than in the capital, Ottawa.

Looking back

Federal government regional development efforts in Atlantic Canada have been trying to deal with well over 100 years of economic difficulties. Indeed, Atlantic Canada, perhaps more than any other economic region, speaks to the conventional wisdom that 'in economic development success breeds success and failure breeds failure'. Unfortunately, it has been much more often than not the latter.

Regional policies

Some of the difficulties have been self inflicted. For example, the region's shipbuilding sector did not adjust in the late nineteenth century when others were quick to shift from wood to steel. Other difficulties came from the actions of outside forces. The Canadian government decided in the late 1800s that the way for Canada to deal with its strong economic neighbour to the south and to create a greater sense of nationhood was to promote a strong national economy based on east–west trade. Ottawa's nation-building policies had their start in 1879 with the introduction of the 'National Policy' through tariffs (Whalley and Trela, 1986, p. 3). In brief, the Canadian government did not hesitate to establish protective measures to promote Canadian industries and, without doubt, businesses in central Canada would benefit the most from Ottawa's nation-building policies. Ottawa's nation-building efforts would mean that the Maritime provinces would largely be left on the outside looking in at Canada's economic growth. But before too long the Maritime region would challenge the national policy, insisting that it was discriminating against local businesses. The debate gave rise to the Maritime Rights Movement in 1919.

Ottawa responded by enacting the

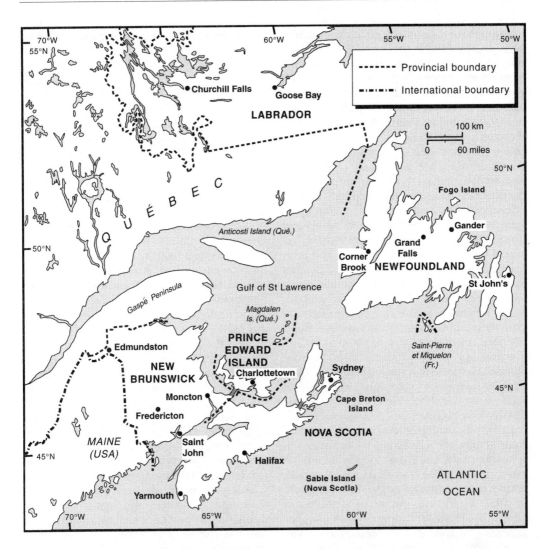

FIG. 19.1 The Atlantic Provinces of Canada

Maritime Freight Rates Act (1927) (see Forbes, 1979). The act provided for lower Maritime freight rates than elsewhere – some 50 per cent lower for selected commodities. A federal-provincial committee determined which products were eligible for the rate subsidy and the programme became costly to the federal treasury. Indeed, by the early 1990s, payments made under the Maritime Freight Rates Act (railway) and under the Atlantic Region Freight Assistance Act (railway, marine, and trucking) amounted to about Can$100 million (Dickson, 1983). Ottawa dropped the programme in its 1995 budget as part of its ambitious cost-cutting exercise, insisting that the programme had not been effective for years. The Freight Rates Act was the first of many measures the federal government introduced to assist the Maritime and, later, the Atlantic provinces on the economic front. Ottawa has, over the years, made a number of special payments and grants to the provinces to deal with specific financial problems. In the 1920s it established a royal commission to look into the continuing financial difficulties of the Maritime provinces (Royal Commission on Maritime Claims, 1926). The commission recommended a series of grants to help the Maritime provincial governments with their budget deficit and based this proposal on the need for greater equality among the provinces. Ottawa responded, but the concern there was not about the pace and location of regional economic activities. Rather, the goal was to deal with a specific public policy problem – the financial difficulties of the three Maritime provinces. However, a number of national policies and national decisions did influence the pace and location of regional economic activities. Quite apart from its national policy, the government of Canada decided to concentrate its Second World War effort in manufacturing in central Canada, with the result that central Canada's manufacturing sector was considerably strengthened in the 1940s. Financial difficulties in Atlantic Canada, coupled with strong growth in central Canada and revenues flowing from large oil and gas reserves in western Canada, led former Saskatchewan Premier Allan Blakeney to describe Canada as a 'giant mutual insurance company' (quoted in Richard Simeon, 1990, mimeo). The economic deal was straightforward – Ontario would happily support the federal government and transfer payments to have-less provinces in return for a captive market for its manufactured goods. The Canadian government established the 'fiscal equalization program' in 1957 (there are numerous studies on Ottawa's equalization programmes, see, for example, Supply and Services, 1981). The programme was designed to achieve a national standard in public services and, at the same time, to equalize provincial government revenues. Accordingly, Ottawa undertook to ensure that all provinces would have revenues sufficient to offer an acceptable level of public services without the poorer provinces having to impose an unacceptable level of taxation.

The federal government subsequently introduced expensive shared-cost programmes to promote national medical care services and social or welfare services. It also tailored some of its transfer payment programmes to individuals to favour slow-growth provinces. The federal government's unemployment insurance programme, recently renamed the employment insurance programme, remains regionally differentiated in favour of areas of high unemployment. The programme has a variable entrance requirement, which is lower, in terms of weeks employed, for areas of high unemployment and Atlantic Canada has benefited a great deal from this scheme.

Though federal transfer payments serve to ensure a national level of service and sustain demand, they do not necessarily assist in integrating slow-growth economies into the national economy and in supporting regional growth from within. To be sure, the federal government has also sought to promote economic development in Atlantic Canada. We have seen federal government efforts to rebuild depressed rural economies with an emphasis on Atlantic Canada with the Agriculture Rehabilitation and Development Act (ARDA, c. 1962) and later with the Fund for Rural Economic Development (FRED, c. 1966). In 1962 Ottawa established the Atlantic Development Board and gave it a mandate to develop measures and initiatives for promoting economic growth and development in the Atlantic region and gave it a special fund of Can$186 million to administer. Over half of the fund was spent on highway construction and water and sewerage systems throughout Atlantic Canada. Some money was also spent on electrical generating and transmission facilities and in servicing new industrial parks at various locations throughout the region. The fund, however, came under criticism for not providing assistance of any kind to private industry to locate new activities in the region (see, among others, Brewis, 1969). The federal government lost little time in rectifying this problem. A modest incentive programme for private firms to locate, modernize or expand activities in Atlantic Canada was introduced in the early 1960s. The programme was subsequently expanded many times so that by the late 1980s it was 'discovered' that by applying to different government programmes it was possible to secure over 100 per cent funding for a project if one agreed to locate in Cape Breton!

The late 1960s, however, was the high point for federal regional development programming, at least from an Atlantic perspective. Newly-elected Prime Minister Pierre E. Trudeau declared that persistent regional disparities were as threatening to Canada's national unity as English–French conflicts and the proper place of Québec in the federation. He established a new government department responsible for promoting regional economic development – the Department of Regional Economic Expansion (DREE) – and gave the department a large new budget. Trudeau appointed his close friend and trusted Québec lieutenant, Jean Marchand, as the minister responsible for DREE. Marchand established an ambitious policy agenda. In doing so, he made it very clear that the focus of their efforts would be on Atlantic Canada and eastern Québec. Marchand explained if: 'the bulk of federal government spending for regional development was not spent east of Trois Rivières [i.e. eastern Québec and Atlantic Canada], then Ottawa's regional development efforts would fail' (quoted in Savoie, 1986). However, before too long, strong political pressure was exerted on Marchand to extend the programme's regional designations further. Members of Parliament and even Cabinet ministers from the Montréal area made the point that Montréal was Québec's growth pole and that if DREE were serious about regional development, then it ought to designate Montréal under its private sector or industrial incentives programmes. Québec's economic strength, it was argued time and again, was directly linked to Montréal, and, unless new employment opportunities were created there, little hope was held for the province's peripheral areas. Montréal required special measures, it was argued, to return to a reasonable rate of growth. In the end, Marchand chose to designate Montréal as a special region for federal regional development assistance. Designating Montréal, however, proved to be a defining moment in Canadian regional development policy. If DREE could justify a presence in Montréal, why could it not also justify one in every other community, large or small, including Toronto and Vancouver?

The Trudeau government came perilously close to losing power in the 1972 election, clinging to office in a minority position,

winning only two seats more than the Progressive Conservative party. That election brought things to a head with respect to regional development policy. Marchand was transferred out of DREE and a new all-encompassing programme was introduced. The programme, essentially a series of federal-provincial agreements, was designed to enable both orders of government to introduce virtually any conceivable initiative or measure to promote economic development. The programme, by definition, applied to all 10 provinces so that both have and have-less regions would benefit from it. Once again, DREE was sending out signals that the focus of federal regional development policy was no longer eastern Québec and Atlantic Canada.

Any notion of a special geographical focus on Atlantic Canada in federal regional development efforts was completely lost in the early 1980s when DREE was renamed Department of Regional Industrial Expansion (DRIE) and a new regional industrial incentive programme was introduced. The first DRIE minister was Ed Lumley who cautioned that

> combating regional disparities is difficult even in good economic times....It is much more difficult in a period when, because of a worldwide downturn, [Canada's] traditional industries are suffering from soft markets, stiff international competition, rapid technological change and rising protectionism from the countries that make up our market (1983, pp. 1–2).

A new programme to meet these circumstances would have to be one that he could 'clearly recommend to the business community, to the Canadian public and to Members of Parliament'. DRIE, Lumley reported, had come up with such a programme. It was a 'regionally sensitized, multifaceted programme of industrial assistance in all parts of Canada....This is not a programme to be available only in certain designated regions. Whatever riding any Member of this House represents, his or her constituents will be eligible for assistance' (Lumley, 1983, pp. 1–2). The programme was also able to accommodate a variety of needs, including investment in infrastructure, for industrial diversification, for the establishment of new plants, and for the launching of new product lines. Thus by 1983 Canada's regional development policy had come full circle. The objective at first was to limit its application to eastern Québec and Atlantic Canada. Now it applies in every region and community in the country.

The federal government decided again in 1988 to overhaul its approach to regional development. The then Prime Minister Brian Mulroney had long professed to have a deep attachment to Atlantic Canada, having been educated there. Mulroney went outside government for advice on how to restructure regional development efforts. The new approach consisted of a 'stand alone' regional development agency for Atlantic Canada. The agency – the Atlantic Canada Opportunities Agency (ACOA) – would have its head office in Atlantic Canada rather than Ottawa and would report directly to its own minister. Mulroney went further on the funding front than his commissioned report on the establishment of ACOA had proposed. The report had recommended a budget of Can$200 million but Mulroney gave the agency over Can$1 billion of 'new' money and another Can$1 billion over five years of existing money by turning responsibility of some ongoing economic development programmes to the agency, together with their budgets. Upon unveiling ACOA, Mulroney declared: 'We begin with new money, a new mission and a new opportunity....the Agency will succeed where others have failed' (*Sunday Herald*, 1987). Within a few weeks, Mulroney went to western Canada to unveil a new regional development agency for that region. The Western Diversification (WD) department – remarkably similar to ACOA – would also have its head office in its region, its own ministers and a combination of new and existing money. The one difference was that WD was given a more generous budget than ACOA (Can$1.2 billion of new money versus Can$1.05 billion for ACOA) (*Globe and Mail*, 1987). Ontario and Québec would also not be outdone. The federal government established a special regional development agency for northern Ontario (FEDNOR) and for Québec (FORDQ). FORDQ programmes apply to all of Québec, including Montréal, and it too has a very generous budget compared to ACOA. The Québec agency divided the province into two regions – the central regions and the peripheral or resource regions – and designed a series of cost-shared initiatives with the province for both regions. Quite apart from the regional development agencies, the federal government also has a Department of Industry, Science and Technology which, traditionally, has concentrated its efforts and resources in central Canada, more specifically in southern Ontario. The department currently spends about Can$500 million annually on industry programmes and the bulk of its spending is in central Canada.

The Liberal government under Jean Chrétien, elected in 1993, has yet to overhaul regional development programmes. It has, however, made incentives programmes to the private sector repayable, squeezed the budgets of all the regional agencies and the Department of Industry and asked all the agencies to report through the Minister of Industry, Science and Technology.

Atlantic Canada stood on the outside looking in when the federal government introduced its 'national policy' in 1879 and again during the Second World War when the government of Canada decided to concentrate the bulk of the manufacturing sector to support the war effort in central Canada. The irony is that Atlantic Canada now has the same position relating to Ottawa's regional development policy, the very policy that was initially designed to assist the region in catching up with other regions favoured by other federal economic policies.

LOOKING AHEAD

Atlantic Canada has a number of advantages particularly with regard to quality of life. It tends to have a much slower pace of life than other regions, it has the lowest crime rate in Canada and the highest number of owner-occupied homes which are mortgage free and, also, the lowest divorce rate. Statistics

Canada reports that Atlantic Canadians tend to give more to charity than other Canadians. In addition, the region's lifestyle, interesting local culture, physical beauty, and breathtaking scenery continue to attract many tourists. They also explain why so many celebrities, such as Jack Nicholson and Paul Simon, have decided to build summer homes in the region (*Maclean's*, 1997).

However, the region also faces important challenges. Its traditional sectors are either in crisis (e.g. fishery) or in decline (e.g. forestry). Prospects for the ground fishery suggest little room for optimism, at least for the medium term. Even more worrisome is that there is also strong evidence to suggest that the region is not keeping up with the new economy. We know that research and development (R&D) is key for knowledge industries. Statistics Canada defines R&D as creative work undertaken on a systematic basis to increase the stock of scientific and technical knowledge and to use this knowledge in new applications. Atlantic Canada's track record in R&D in the late 1980s was hardly impressive. As a percentage of GDP, expenditures on R&D is 35 per cent below Canada as a whole, 78 per cent behind Ontario and 30 per cent behind Québec. The federal government accounts for 57 per cent of R&D spending in Atlantic Canada (as compared to 31 per cent nationally and 29 per cent in Ontario), while the private sector accounts for only 13 per cent (42 per cent nationally and 44 per cent in Ontario). More recent data from Statistics Canada show little progress for Atlantic Canada in R&D. In 1994, for example, 71 per cent of Canadian R&D was performed in Ontario and Québec (Statistics Canada, 1996). Moreover, there is little doubt that Atlantic Canada stands to lose more ground on the R&D front if governments are not careful as they make further cuts in their expenditure budgets, given the region's strong reliance on the public sector for its R&D investments.

Atlantic Canada still remains too dependent on the public sector. To be sure, we have seen important cuts in federal and provincial government spending as both orders of government go about repairing their balance sheets. But, notwithstanding these cuts, Atlantic Canadians are much more dependent on government transfers as a source of income than other Canadians (24.5 per cent of total income versus 18.6 per cent at the national level). In addition, public sector jobs still account for a greater share of total jobs in Atlantic Canada than is the case in other regions. Atlantic Canada is still governed by four provincial governments and progress in promoting intergovernmental co-operation in the region and in integrating the Atlantic economy has been at best tentative. Atlantic Canada has stood on the outside looking in during key developments in the Canadian economy. It again runs the risk of being on the outside looking in as Canada's new knowledge economy takes flight.

KEY READINGS

Alexander, D. 1980: New notions of happiness: nationalism, regionalism and Atlantic Canada. *Journal of Canadian Studies* **15** (2), 29–42.

Forbes, E.R. 1979: *The Maritime Rights Movement, 1919–1927*. Montréal: McGill-Queen's University Press, 149–72.

Rigaux, F. 1997: *Industrial biotechnology in the Maritime provinces*. Moncton: Canadian Institute for Research on Regional Development.

Savoie, D. 1986: *Regional economic development: Canada's search for solutions*. Toronto: Toronto University Press.

Savoie, D. 1997: *Establishing the Atlantic Canada Opportunities Agency*. Ottawa: Office of the Prime Minister.

20

THE GREAT PLAINS AND PRAIRIES

WILLIAM J. CARLYLE

PHYSICAL SETTING

The Great Plains of the United States and a northern extension into Canada called the Prairies are located in the western interior of North America (Fig. 20.1). The Great Plains–Prairies region combines three physical attributes: lack of trees, generally flat land, and a semi-arid climate. There is general agreement that the western boundary is the foothills of the Rocky Mountains. Land west of the foothills in places combines two of the physical attributes, but not all three together. The northern boundary is here taken to be the southern boundary of the Boreal Forest. This places a belt of varying width known as the parkland, which has a natural vegetation of prairie grassland intermixed with broadleaf trees, within the Canadian Prairies. The southern boundary of the Great Plains is here located along the Rio Grande River, although the three distinguishing physical features extend southwards into Mexico. The eastern boundary is the most difficult one to delineate. The 500 mm (20 in.) isohyet of annual average precipitation, the amount usually taken as separating semi-arid from humid climates, has frequently been used, as has the 98th meridian west which closely approximates to it. In his classic book, *The Great Plains*, Webb (1931) located the eastern boundary close to both these features. Others, such as Rugg and Rundquist (1981), have located the boundary considerably further east of these two lines for reasons varying from a desire to include cities which have influenced the Plains to making it easier to compile statistical data by including the states from North Dakota to Oklahoma in their entirety. The boundary chosen here is based on McKnight (1997) (Fig. 20.1).

The Great Plains–Prairies region is the largest by area of the conventional subregions of North America which are within the continuously populated ecumene – only the sparsely populated Boreal Forest and Arctic regions outside it are larger. The Plains–Prairies region spans up to 2400 km from north to south and 1000 km from east to west, making it larger than most countries of the world.

Flat to gently dipping sedimentary rocks, varying in age from the Palaeozoic to Cainozoic and increasing in elevation from east to west, form the bedrock of the region. These rocks overlie the ancient and structurally stable Precambrian or Canadian Shield. Annual average precipitation along the eastern edge is slightly more than 500 mm, i.e. subhumid, but most of the region is semi-arid, with between 250 and 500 mm a year. The frost-free season varies from more than 200 days in the south to 80 to 90 days in the north. A lack of east–west trending topographic barriers allows moist hot tropical air from the Gulf of Mexico to penetrate as far north as the Canadian Prairies during the summer, and cold dry Arctic air to move southwards as far as the Gulf Coast in the winter. Absolute temperature ranges between seasons are vast in the northern part of the region, ranging from −35°C to −40°C during most winters to +35°C to +40°C during most summers, and swings of 30°C to 40°C are not uncommon over one to two days. The natural vegetation of the region was largely prairie grassland. It varied from tall grasses, sometimes mixed with deciduous trees, on the moister eastern and northern margins through medium-length grasses to short-grasses and, in the south, shrub and grass, in the dry core of the region.

PEOPLES AND SETTLEMENT

Native North Americans were the sole occupants of the region until very recently, and the Plains Indian culture based primarily on bison (buffalo) flourished after horses and then rifles were acquired. Horse culture began with Spanish horses in the mid 1500s in the southern part of the region, and it spread northwards through trading and stealing to the Canadian Prairies by the mid to late 1700s.

European exploration of the region began in the 1500s and 1600s, largely with the intent of finding gold or jewels or a route through the region to somewhere else. Fur traders established posts in the region during the eighteenth and early nineteenth centuries, but the transformation of the land away from

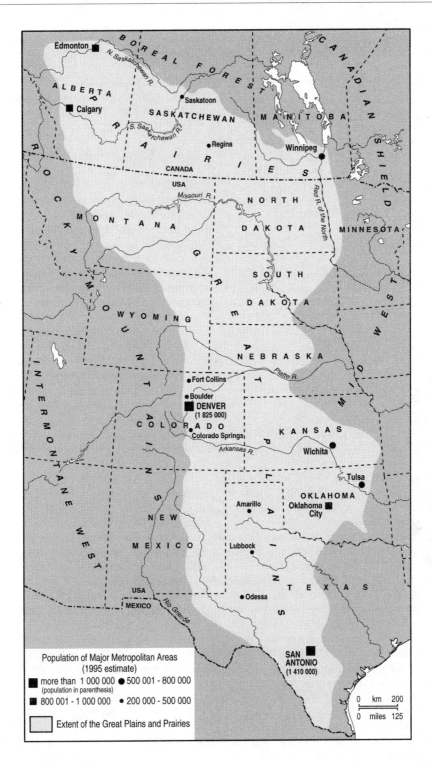

Fig. 20.1 The Great Plains and Prairies

its natural state accelerated greatly as the result of agricultural settlement from the late nineteenth century.

No region in North America, and probably in the world, was settled more quickly by farmers and ranchers than the Great Plains and Prairies. Between 1870 and 1920, millions of hectares of prairie grassland were brought under the plough, and farms, ranches, towns, and cities, interconnected by thousands of kilometres of railway and tens of thousands of kilometres of trails and roads, sprung up where a few decades before Plains Indians and bison had roamed.

In both the United States and Canada, the land was surveyed according to the Township and Range system (see Chapter 1). Virtually free homesteads of a quarter section (160 acres, approximately 65 ha) were allocated to settlers, on the condition that they actually reside on the land and make specified improvements in a given time period. This led to a dispersed settlement pattern with many isolated farmsteads. Partly because of this dispersion, many roads were needed. Publicly funded roads were developed along the section lines, and in many townships a half to all the section lines eventually had roads (36–72 miles (58–116 km) of road per township). Bantje (1992), for example, gives a figure of 160 000 km of roads in Saskatchewan for the year 1927, the vast majority of which were dirt roads constructed to serve about 130 000 farms.

Grain elevators, sizeable structures in which grain delivered by farmers is stored and then loaded on to railway cars, were established at regularly spaced points along the rail lines which were built across the Plains and Prairies during the pioneer settlement period. Most of these points quickly attracted various elementary services such as banks, lumber yards, hotels, general stores, and farm implement dealers. Thousands of such service centres sprouted up along the railways during the initial settlement period. They were closely spaced in the grain-growing districts, usually at 10–15 km intervals. Such close spacing was necessary because the farmers hauled their grain by horse and wagon over trails or primitive roads, and they needed to be near

other services offered by the railway towns. To serve the farmers, the railway network, too, was dense. In time, thousands of churches and tens of thousands of one-room elementary schools were established to serve the dispersed population.

THE REGION TODAY

Since about the Second World War, the large-scale mechanization of agriculture, more rapid transport, and the pull of urban life have severely thinned out the population of the rural areas of the region, and, with it, much of the rural infrastructure of grain elevators, towns, schools, churches, post offices, and hospitals. The population of the region has become increasingly urban, and now about three quarters of the total population lives in urban centres. Not by chance, the largest cities are located on the periphery of the region where they perform trading and service functions for this and other regions (Fig. 20.1). The largest urban centres wholly within and dependent on the resources and trade of the region itself are Regina, Saskatchewan, and Amarillo, Odessa, and Lubbock, Texas, each with metropolitan-area populations of about 200 000.

Although most of the people live in cities, especially the large ones, the economy remains based, other than on services, mainly on the products and resources of the land. Agriculture and agricultural processing are mainstays, as, too, are the fossil fuel industries. These latter include extraction and processing of coal, oil, and natural gas, together with the manufacturing of chemicals and plastics from their by-products, and a whole host of industries to supply machinery and equipment for processing and transporting the fuels. There are also many other localized manufacturing industries.

AGRICULTURE

The popular image of the Great Plains and Prairies as being an agricultural region is still true, even though most of its inhabitants live in cities. The landscapes are mostly rural and agricultural, and the agricultural output of the region is significant within the context of both North America and the world. The winter wheat belt of Kansas, Nebraska, Colorado, Oklahoma, and Texas together with the spring wheat belt of Montana, the Dakotas, and western Minnesota produce about two thirds of the total US wheat output. The Canadian Prairies spring wheat belt accounts for almost all of Canada's wheat production. Much of the wheat crop is exported, with the US followed by Canada being the two leading wheat-export nations in the world in most years.

The Prairies not only produce almost all of Canada's wheat, they also account for 75–100 per cent of its barley, oats, canola (rapeseed), and flaxseed. The Great Plains are not so dominant agriculturally within the American context mainly because a considerably smaller proportion of US agricultural land is in the Plains than that of Canada which is in the Prairies – the Prairie proportion being 82 per cent.

Even so, in addition to wheat, much of the grain sorghum and barley and a substantial amount of the alfalfa produced in the United States is grown on the Great Plains. Both the Prairies and Plains have most of the grazing land in their respective countries. About 60 per cent of the slaughtering of beef cattle in the US takes place on the Great Plains, and the proportion for the Prairies within Canada is slightly higher. The slaughtering and packing of beef cattle have become more important in the region in recent years, with a substantial shift of these industries from Ontario to Alberta within Canada and from the midwest to the middle and southern Great Plains within the US (Broadway, 1998). The pork industry, although smaller, has recently been expanding in the Prairies and Plains (see Chapter 9).

Minerals

It is one of the ironies of Canada's geography that, although the St Lawrence Lowlands of Ontario and Québec contain most of Canada's people, industry, and manufacturing, the sedimentary rocks of the region yield no coal and very little oil or gas. In contrast, in the Prairies, with a total population not much more than metropolitan Toronto, the underlying sedimentary rocks of Saskatchewan and, especially, Alberta provide most of the oil and natural gas and more than half the coal produced in Canada.

The fossil fuel industry of the United States is more dispersed (see Espenshade, 1995) but a large proportion of it is also located in the Great Plains. Strip mining of coal has been expanded in the Plains since the energy crisis of the mid 1970s, largely for use, as in the Prairies, for thermal-electrical power generation. Boom towns such as Colstrip, Montana, and Gillette, Wyoming, are a result of this recent expansion of coal mining on the Plains. Much of the oil production of the US emanates from the Great Plains. McKnight (1997) states that the Permian Basin of west Texas alone accounts for about 17 per cent of the total oil production of the country. Within the Plains, natural gas is concentrated within the extensive Panhandle field of Kansas, Oklahoma, and west Texas. Most of the helium produced in the US also comes from this field. Besides fossil fuels, potash is the most significant mineral extracted from the Great Plains–Prairies region. Two of the world's most important producing areas are in this region, one in Saskatchewan, the other in New Mexico.

This discussion of the region and its resources might give the impression that the Great Plains and Prairies is a viable and successful region. Some people do take this view. Others, however, claim that the Canadian and American experiences in the region have been failures in whole or in part.

FAILURE OR SUCCESSFUL ADAPTATION?

Failure

Ian Frazier, perhaps, makes the most harsh judgement concerning the American experience on the Plains:

This, finally, is the punch line of our two hundred years on the Great Plains: we trap the beaver, subtract the Mandan, infect the Blackfeet and the Hidasta and the Assiniboin, overdose the Arikira; call the land a desert and hurry across it to California and Oregon; suck up the buffalo, bones and all ... harvest wave after wave of immigrants' dreams and send the wised-up dreamers on their way; plow the topsoil until it blows into the ocean; ship out the wheat, ship out the cattle, dig up the earth itself and burn it in power plants and send the power down the line; dismiss the small farmers, empty the little towns, drill the oil and natural gas and pipe it away; dry up the rivers and streams, deep drill for irrigation water as the aquifer retreats (1990, pp. 209–10).

Frazier's facts are generally correct, although the negative 'spin' he puts on them is one of opinion.

While Frazier does not provide any possible solutions to perceived problems, others have done so. Deborah and Frank Popper reviewed the American experience on the Great Plains, and then made their now widely known 'Buffalo Commons' proposal. In the most distressed areas, they wrote, 'the federal government's commanding task on the Plains for the next century will be to recreate the nineteenth century, to re-establish what we call the Buffalo Commons' (1987, p. 18). This was to be achieved by large-scale purchases of farms and ranches by the federal government, who would then re-sow the land to natural prairie grasses, and re-stock them with indigenous herbivores, particularly buffalo (bison). The Poppers claimed that 'By creating the Buffalo Commons, the federal government will, however belatedly, turn the social costs of space – the curse of the shortgrass immensity – to more social benefit than the unsuccessfully privatized Plains have ever offered' (p. 18). In a succeeding article, the Poppers used population variables, poverty rate, and per capita new construction investment, indicators they had used earlier to present their Buffalo Commons proposals to audiences throughout the Plains, to identify counties in the Plains which suffered 'land use distress' (1994). They proposed that the Buffalo Commons approach be focused on these counties,

which occupy about 25 per cent of the total area of the Great Plains.

Similar views to those of the Poppers have been expressed for parts of the Canadian Prairies by Sharon Butala (1990). She suggested that the settling of farmers, as opposed to ranchers, on the shortgrass prairie of southern Saskatchewan and Alberta had been a mistake, and that it might be better to buy out the farmers, close off the land to agriculture, and develop a national park, with a rejuvenated prairie grassland restocked with antelope and bison, as a tourist attraction.

The Poppers' phrase – 'the social costs of space' in the dry areas of the Plains – has been expressed by others. Robinson (1966) claimed that the enormous social costs stemmed from the 'Too Much Mistake' – too many farms, ranches, people, towns, and railroads for the land to take. Since the 1960s, there has been a massive reduction in elements of the rural built environment and just as large a reduction in the rural population. Resources are therefore as stretched as they were, and possibly even more so (Goodman, 1996), because higher standards of living are expected now than even in the recent past. This decline of population and infrastructure in the mainly rural areas is seen by many people to be a sign of failure, failure brought on by misguided initial policies or lack of subsequent policies to deal with change.

Successful adaptation

Other people have examined the same facts as those used by the 'failure' group, but have drawn very different conclusions. This second group claims that the Canadian and American experiences on the Prairies and Plains have been ones of successful, although often painful, adaptations to changing conditions. For example, the claim that there was 'too much' of most settlement features on the Plains and Prairies and that the thinning out of these features signifies regrettable loss or failure has been countered. As Stabler *et al.* (1992), Stabler and Olfert (1996), Rees (1988), and Hudson (1979) point out, the settlement infrastructure developed in the region during the initial settlement period suited that period, and was developed without the knowledge that forces were soon to come into play which would render much of that infrastructure inappropriate.

For Hudson (1996), the rapid adoption of increasingly larger and more powerful farm machinery on the Plains, bringing with it much-enlarged but far fewer farms, is one aspect of successful adaptation in the region. He claims that on the Plains

> What invariably succeeded ... was the adoption of the most capital-intensive approach to exploiting the available resources at the fastest pace. This, invari-ably, has meant depopulation wherever there has not been the means to transform the environment itself, as through irriga-tion (1996, p. 8).

He goes on to argue that in the future even fewer people will be needed to cultivate the land in dry-farming areas. To those who view this as failure, he poses the questions: 'Why is a certain level of population density or level of social services indicative of settlement suc-cess or failure in the Plains? More specifically, why equate depopulation with failure?' (p. 6). His view is that loss of farms and population is not a sign of trouble, but one of successful adjustment to changing circumstances. Claw-son (1981) expresses similar views.

Other elements which were parts of Robinson's (1966) 'Too-Much Syndrome', have been very much reduced since the 1950s and 1960s, and can also be viewed as success-ful adaptations instead of being signs of decline or failure. As pointed out by Hudson (1996), larger grain cars and modern grain-handling technology allow a single grain ele-vator today to handle a quantity of grain which had required a dozen elevators as recently as the 1960s. Thousands of grain ele-vators have, for this reason, been closed across the Prairies and Plains. In the Canadian Prairies alone, the number of primary grain elevators has declined from 5204 in 1962–3 to 1016 in 1999[1]. The closings pose no great hard-ship for grain farmers because they now use larger grain trucks and can travel 100 km an hour on much-improved roads.

Moreover, as pointed out by Carlyle (1991) and Hudson (1996), the decline of thousands of small towns across the region, even to the point of disappearance of some, cannot be attributed to removal of grain elevators or, in places, the rail lines themselves. The towns were already in decline, largely because improved roads and faster rates of travel have, since the 1940s, increasingly allowed rural folk wider horizons. One result is that they bypass small local service centres for larger ones farther away. The decline of many, and in places most, of the towns can be viewed, as by Stabler and Olfert (1996), as a necessary adjustment.

The intensification of agriculture which has taken place in recent years on the Ogal-lala aquifer, which occupies parts of eight states and stretches from Texas to South Dakota, and in other irrigated areas can also be viewed as a sign of success, not resource exploitation or mismanagement (see Chapter 9). White (1994) notes that only about 5 per cent of the Ogallala's pre-irrigation volume of 3.25 billion drainable acre-feet (0.4 billion ha/m) of water had been tapped by 1980. He does acknowledge that the availability of surface water in part of the Ogallala has declined, but ascribes this to better on-farm conservation measures as well as to ground-water depletion. White (1994) also points out that groundwater availability provides for growth of towns of more than 500 people in west Kansas – 'oases of growth' – and these towns depend on apparently deserted lands of irrigated crops and dryland farms as their hinterlands. In reference to the Poppers' pro-posals, he claims that 'Restoration of a buffalo commons would bring certain death to these oases, not to mention the forced eviction of tens of thousands of people there and else-where on the High Plains. Abandonment is a regional development policy that looks good only if you are a buffalo' (White, 1994, p. 42).

Also regarding the Poppers' proposals, De Bres *et al.* (1993) point out that the variables the Poppers used to identify counties in 'land use distress' were not land use variables. Partly because of these and other criticisms, the Poppers have recently (1994) modified their original views. In the revised version, they downplay the need for large-scale fed-eral intervention. They instead rely on the actions of many Plains dwellers, private agencies, and various levels of government to create piece by piece the Buffalo Commons, and they give numerous examples of work completed or in progress to achieve this end.

These initiatives can be viewed as part of the 'success' story, yet other examples of adaptation to changing conditions on the Plains. It is scarcely a Buffalo Commons, however. Estimates of the number of buffalo on the Great Plains and Prairies are in the order of 150000 to 200000. The number of buffalo slaughtered in a year in the United States is probably about the same as the number of beef cattle slaughtered in a day according to *The Economist* (1994a). It is nevertheless important that new forms of stock raising are being experimented with in the region, including buffalo, elk, deer, and antelope for meat. Other uses include recreational hunting, and buffalo rearing for cultural or spiritual reasons in the case of Native peoples.

Regarding the Canadian Prairies, Paul (1992) points out that close to half of the dry shortgrass region, referred to by Butala (1990) above, has already been withdrawn from cultivation or was never farmed, in yet another instance of successful adaptation to the region. Most of the land is leased to ranchers for grazing beef cattle and sheep; there are very few buffalo.

The extraction of fossil fuels and other minerals from the region can also be viewed in a positive light. Strip-mining of coal does produce a scarred landscape, although, also, much-needed electrical power, but something close to the original landscape can be and is being restored by in-filling and re-seeding of grasses. Although oil, natural gas, coal, and potash are shipped out of the region, they do provide employment and income for people within it.

CONCLUSION

The Great Plains and Prairies region has undergone a rapid transition over the past 100 to 125 years. It has gone from being a very sparsely-populated region inhabited by Native peoples, with their horse and bison culture, through a phase of rapid agricultural settlement to an again sparsely-populated rural region for the most part, with the bulk of the population living in urban centres on the periphery of the region. Some people consider many of the changes which have occurred to be a sign of failure, while others view them as signs of a series of continuing and, on the whole, successful adaptations to a demanding environment.

ENDNOTE

1. About 1000 were 'lost' due to a reclassification in the 1980s; the real decline was just over 3000 elevators between 1962–3 and 1999.

KEY READINGS

Butler, W.F. 1872: *The great lone land: a narrative of travel and adventure in the north-west of America*. London: Sampson Low.

Haines, F. 1970: *The buffalo*. New York: Thomas Y. Crowell.

Kraenzel, C.F. 1955: *The Great Plains in transition*. Norman: University of Oklahoma Press.

Rees, R. 1988: *New and naked land*. Saskatoon: Western Producer Prairie Books.

Stegner, W. 1990: *Wolf willow: a history, a story, and a memory of the last Plains frontier*. New York: Penguin Books.

Webb, W.P. 1931: *The Great Plains*. Boston: Ginn.

21

THE AMERICAN SOUTH

JAMES O. WHEELER

A dark shadow lies upon this part of the map of the United States (Winston S. Churchill, *History of the English-speaking peoples*, 1958, Vol. 4, p. 204)

The American South, for two centuries at the periphery of the US economy, has in the last quarter of the twentieth century undergone massive transformations in urbanization and industrialization, in cultural and societal viewpoints, in agriculture and in politics (Bartley, 1995; Prunty, 1977). In many ways, the South has become the most changed region in all of the US and Canada. Migration to the region has been massive. Florida, which has been a well-known migrant destination for retirees, now attracts people of all ages. Atlanta, the magnet of the Southeast, grew from migration by over 100 000 between 1995 and 1996 (see Chapter 14). Dallas–Fort Worth and Houston, the two mid-continent metropolises, are engaged in a giant's battle for dominance in the western part of the South. Most of Florida is classified as metropolitan.

At the same time that this marvel of change has been occurring in metropolitan economies, many rural, peripheral areas remain poor and participate only little in the new South prosperity. These are those non-metropolitan counties that continue to lose population through out-migration, where the age composition of the existing population is elderly, where there is often a high percentage of African Americans, and where education and job skills are low. Such poor rural areas extend across the entire region, highlighting a wide gap between the two Souths.

The South, as defined in this chapter, is consistent with the region as delineated by the US Bureau of the Census, extending from Texas and Oklahoma in the West to Maryland and Delaware to the North and East. The South is typically considered more than a collection of contiguous states; for all its diversity it shares something of a common culture and economic history. This strong regional sense of identity has been based on traditional patterns of living, though 'closer contact with the rest of the nation is rapidly blurring some of the more pronounced idiosyncrasies which once characterised the region' (Hart, 1976, p. 17). Whereas Atlanta, for example, is basically a northern city with the teeming flows of in-migrants from the North, the smaller towns and countryside in the South retain certain traditionalist features, themselves being rapidly eroded: a predominantly Native-born Protestant population, a conservative Democratic party, a distinctive heavy-fat diet (fried chicken and pork rather than beef), class distinctions – tied to church affiliation and family – and a high percentage of poor African Americans (Hart, 1976).

PHYSIOGRAPHIC PROVINCES OF THE AMERICAN SOUTH

Except for a thin slice of the Great Basin physiographic province in far western Texas, the Central Lowlands in western Texas and most of Oklahoma, and the Ozark Plateau and the Ouachita Mountains in northwestern Arkansas, three major physiographic subdivisions of the North American continent characterize the US South: (1) the Coastal Plain, (2) the Appalachian Highlands, and (3) the Interior Low Plateau (see Chapter 2). By far the most extensive is the Coastal Plain, including every state in the region except West Virginia. The Coastal Plain is underlain by raised sedimentary rocks formed beneath the sea, and this flat region slopes gently toward the Atlantic Ocean and the Gulf of Mexico. The second major physiographic province is the Appalachian Highlands, consisting of four important divisions. The Piedmont Province is a rolling plain underlain by crystalline (metamorphic) rocks, extending nearly 1800 km from central Alabama through Maryland and into New York. The Piedmont Province joins the Ridge and Valley Province to the west in Alabama and northwest Georgia and the Blue Ridge Mountains in north Georgia, northwestern South Carolina, western North Carolina and Virginia. Lying to the west of the Ridge and Valley Province is the Appalachian Plateau, essentially horizontal layers of eroded sedimentary rock. The third major physiographic province in the South is the Interior Low

Plateau of Kentucky and Tennessee, including the northernmost tip of Alabama. Except for the Bluegrass and Nashville basins – eroded structural domes – the landscape is hilly and composed of sedimentary strata that has not undergone glaciation as has the area to the north in Illinois, Indiana, and Ohio.

SOUTHERN SETTLEMENT

The first European settlement on the North American Atlantic mainland (thus discounting earlier Viking penetration into Newfoundland) was a Spanish colony of 1565 which occurred at St Augustine, Florida. Despite this, it was the northern Atlantic coastal colonies which grew more rapidly in population. The large protected harbours at what would become Boston, New York, Philadelphia and Baltimore also had the advantage of being somewhat closer to Great Britain than the southern Atlantic ports of Charleston, Savannah, and the Gulf Coast port of New Orleans. Most European immigrants streamed to the North Atlantic settlements, which became increasingly familiar destinations to fresh immigrants seeking out New World habitation. By contrast, the South lacked people to spread settlement. As a result, today, the most densely-settled area of North America is known as Megalopolis, the continuously urbanized area between south of Washington, DC, and north of Boston (Gottmann, 1961) (see Chapter 7).

In the South settlement proceeded from coastal regions toward the interior (Fig. 21.1). By 1790, the date of the first US population census, most of Maryland, Virginia, North Carolina, and South Carolina were described as 'settled', though the density of population was extremely low. Except for the fringe around St Augustine in the northeast, Florida was only inhabited by Indians, European settlers viewing it as essentially a swampy, excessively humid, and agriculturally unproductive area. Florida was later settled from north to south. Only the eastern strip of Georgia had attracted settlement by 1790, associated with migration up the Savannah River, which separates Georgia and South Carolina. The Bluegrass region of north central Kentucky had been occupied by 1790 as settlers moved down the Ohio River and used Louisville as a 'jumping off' point for pursuing agriculture on the fertile limestone soils of the Bluegrass. Likewise, the Nashville Basin of Tennessee had attracted some early pioneers, as had a fringe area around New Orleans, at the mouth of the Mississippi River. Except for Native American presence, all of Texas, Oklahoma, Arkansas, Mississippi, and Alabama remained unsettled by 1790, as did most of Louisiana, Tennessee, Kentucky, and West Virginia. Settlement spread very slowly from east to west in Georgia, into the Appalachian region, and into Texas, as well as into northwestern Louisiana. Oklahoma was Indian territory and was not opened to other settlement until 1889.

The South had not only a low density of population compared with the more urbanized and industrialized Northeast and later the Midwest, but also had a distinctive agricultural system. This was the plantation economy with its associated practice and ideology of slavery. The New England states of Maine, Vermont, New Hampshire, and Massachusetts, which initially had some slaves, had none by 1810, though New York, Pennsylvania, New Jersey, Connecticut and Rhode Island continued to have slaves and free blacks (less than 10 per cent of the population) as late as 1820 (Brown, 1948, pp. 131–2). In the early 1800s, the greatest concentrations of African American slaves were (1) in eastern Virginia and North Carolina, (2) coastal South Carolina and Georgia and (3) the lower Mississippi River of Louisiana.

By 1860, just prior to the Civil War that would end slavery, the South had slaves in all states (except Oklahoma's Indian Territory). There were slaves also in Missouri (Fig. 21.2). The greatest concentrations were in eastern Virginia and North Carolina, the Coastal Plain and Piedmont of South Carolina, the Piedmont of Georgia, the 'Black Belt' (named for the rich, black soils) of Alabama (Webster and Samson, 1992), the Mississippi River basin of Louisiana and Mississippi, as well as the Nashville Basin of Tennessee and Bluegrass region of Kentucky.

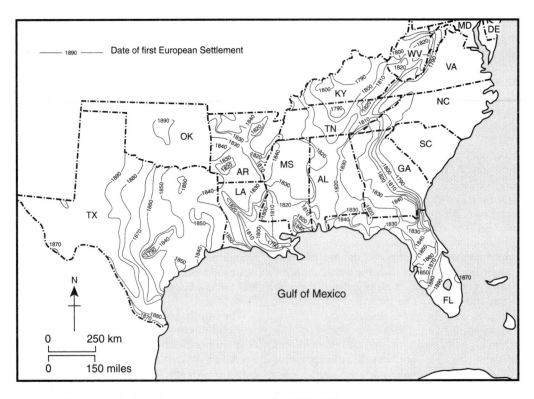

FIG. 21.1 The spread of settlement across the South, 1790–1890
Source: Marschner (1959)

Florida, south-central Georgia, northwestern Arkansas, western Texas, and the mountainous Appalachian region had few slaves. The distributions of slaves largely corresponded to plantation agriculture in 1860.

The plantation economy, which occupied large tracts of agricultural land, existed alongside small farmers who typically could not afford to own slaves. In Georgia, for example, a ragged line running from Elberton through Athens and Atlanta had plantations with scattered yeomen farmers lying to the south and basically yeomen farmers and almost no plantations north of this line (Fisher, 1973, p. 83). The small owner-occupied farms were, in fact, more numerous than the large plantations. Even today, the upper Georgia Piedmont and North Georgia in general has few African Americans compared with the rest of the state.

The southern plantation represented not only a particular agricultural system, it also manifested a distinctive way of life that impacted economic development. Whereas the popular image of the pre-American Civil War southern plantation is simply a large landholding, plantation agriculture involves a more complex set of factors. In addition to a large landholding, the plantation required a large labour force (slaves prior to the Civil War) to take advantage of economies of scale (unit savings resulting from large output). For similar reasons, plantations specialized in a particular cash crop, principally cotton. The embryonic locations of southern plantations concentrated on tobacco production in coastal Virginia and Maryland, rice growing in coastal South Carolina and sugar-cane cultivation in the Mississippi Valley of Louisiana (Aiken, 1998, p. 5). The plantation complex also needed careful management organized from top to bottom to carry out its competitive operations. Each of these characteristics of the plantation implied a high level of capitalization. Last, plantations exhibited a unique settlement arrangement, based on a

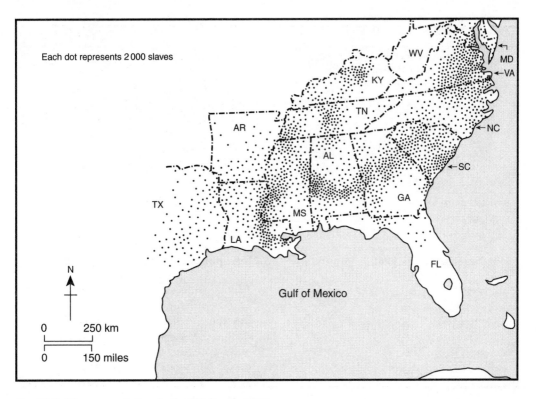

Each dot represents 2 000 slaves

Gulf of Mexico

0   250 km

0   150 miles

FIG. 21.2 Slave population in the US South, 1860
Source: Hilliard (1984)

nucleation focused on the plantation house, with slave quarters and equipment and mule and horse sheds located nearby (Aiken, 1998, pp. 5–7).

This particular plantation settlement pattern, with economic activity focused on the plantation house nucleation, discouraged the development of towns and villages. As Doyle noted:

> The old South did not have a large urban population because it did not need a large one. Plantation agriculture, and the cotton crop in particular, required only minimal urban development. Before the Civil War, the urban South consisted of only a small number of cities with a low share of the population; this weakly integrated scattering of cities often had stronger links to cities outside the region than to one another (1990, p. 3).

The retarding effect of plantations on urban development, first under slavery and then the tenancy system, was one handicap to the South's development. Another was the fact that, because of the more advanced industrialization in the North, the South lagged far behind and could not readily compete with the already-established Northeast core region. It was not until 1880 that more than 10 per cent of the South's population was classified as urban, in contrast to over 50 per cent for the Northeast in the same year. In 1860, on the eve of the Civil War, New Orleans was the largest city in the South, and Covington, just south of Cincinnati, ranked tenth with 14 000 people (Table 21.1). In 1900, among the 10 largest cities, only Atlanta and Richmond were not on navigable water bodies.

AGRICULTURE

The Civil War did little to alter the South's reliance on cotton production, though a new

TABLE 21.1 The South's 10 largest cities and metropolitan areas, 1860–1996

Cities 1860	Population	Cities 1890	Population
1. New Orleans, LA	168 000	1. New Orleans, LA	287 000
2. Louisville, KY	68 000	2. Louisville, KY	205 000
3. Charleston, SC	40 000	3. Memphis, TN	102 000
4. Richmond, VA	38 000	4. Atlanta, GA	90 000
5. Mobile, AL	29 000	5. Richmond, VA	85 000
6. Memphis, TN	23 000	6. Nashville, TN	80 000
7. Savannah, GA	22 000	7. Charleston, SC	56 000
8. Petersburg, VA	18 000	8. Savannah, GA	54 000
9. Nashville, TN	17 000	9. Norfolk, VA	47 000
10. Covington, KY	14 000	10. Houston, TX	45 000

Metropolitan areas 1960	Population	Metropolitan areas 1996	Population
1. Houston, TX	1 243 000	1. Dallas-Fort Worth, TX	4 575 000
2. Dallas, TX	1 084 000	2. Houston, TX	4 253 000
3. Atlanta, GA	1 017 000	3. Atlanta, GA	3 541 000
4. Miami, FL	935 000	4. Miami–Ft Lauderdale, FL	3 514 000
5. New Orleans, LA	868 000	5. Tampa–St Petersburg, FL	2 200 000
6. Tampa–St Petersburg, FL	772 000	6. Norfolk–Virginia Beach, VA	1 540 000
7. Louisville, KY	725 000	7. San Antonio, TX	1 490 000
8. San Antonio, TX	687 000	8. Orlando, FL	1 417 000
9. Birmingham, AL	635 000	9. Charlotte, NC	1 320 000
10. Memphis, TN	627 000	10. New Orleans, LA	1 313 000

Source: US Bureau of the Census

system of labour relations had to be put in place following the elimination of slavery. That system was tenancy. Tenant farmers owned no land and had varied working relations with the owner, depending on whether the farmer had his own farm implements, merely rented the land, or paid the owner 'cash' (usually in bales of cotton grown). Tenancy continued the reliance on labour for agricultural production. Many former slaves stayed on as tenant farmers. Plantations continued to operate alongside small farms but within the tenant labour system, which reduced the control of the land owner.

During the early twentieth century the plantation system of the previous century changed. 'One route led to mechanization and modernization, the other to decline and the demise of plantation agriculture' (Aiken, 1998, p. 63). The decline of plantation agri-

culture resulted from a decline in landowner management and the tenant system, soil erosion, and the cotton boll weevil. Great numbers of African Americans migrated from the rural South to factory jobs in industrial cities in the North, such as Detroit, Chicago, St Louis, Cleveland, Baltimore, Philadelphia, and New York.

Labour-saving machinery, particularly the tractor, was slow to take hold in the cotton South, where mule- and horse-power had so long been dominant. Cotton picking remained a hand task; it was not until 1956, at a time when little cotton was grown in the old cotton regions of the Southeast, that 'the mechanical cotton harvester had become an important part of cotton production' (Aiken, 1998, p 106). Cotton production in the South underwent tremendous reduction starting in the early twentieth century, declining by nearly 50 per cent between 1924 and 1944 (Prunty, 1951). Cotton production shifted to eastern Texas and then to west Texas and California. Alabama, Georgia, North Carolina, and South Carolina accounted for less than 5 per cent of US cotton production in the early 1980s, though there has been a resurgence in the late 1990s, largely as the result of rising cotton prices and the eradication of the boll weevil (Lord, 1996; Fournier and Risse, 1996).

Contemporary southern agriculture specializes in particular crops and products for which the South has a comparative advantage over many other areas of the United States (Hart, 1976). Winsberg succinctly summarized contemporary agriculture in the region:

> Today southern agriculture is practised on a far larger scale than in the past, and much of it has become highly mechanized. In areas of the region's agriculture that have experienced little mechanization, Hispanics ... now meet much of the labour demand. Contract labour is replacing hired because farmers find it cheaper. Despite modernization of much of the region's agriculture, there has been little change in its geographical distribution. As in the past, many southern counties continue to depend heavily upon the sale of one agricultural commodity (1997, p. 193).

Whereas in 1950 agricultural income was 18 per cent of total income in the South, by 1994 the figure had declined to 1.2 per cent primarily because of the enormous rise in the information-based service sector. Yet, today, the South accounts for 19 per cent of US agricultural production, up from 17 per cent in 1950.

The leading farm product in sales in the South is poultry, i.e., broilers produced for meat, comprising 27 per cent of all agricultural sales in the Southeast (Winsberg, 1997, p. 198). Broiler production is highly localized in selected areas of the South, where major broiler processing companies are located and where contracts with local farmers (who own land no longer profitable for crop agriculture) ensures a steady flow of birds through the processing plants (Hart and Mayda, 1998). Especially notable concentrations of broiler production occur on small owner-occupied farms in areas of Arkansas, the leading broiler producer state, and north Georgia, the second leading state, as well as the Delaware Peninsula, south-central Mississippi, northern Alabama, central North Carolina, and an area of northern Virginia.

According to the 1992 Census of Agriculture the South produces 90 per cent of all US broilers. Almost all US tobacco is produced in the South, particularly in North Carolina and Kentucky. Sugar cane, rice, and peanuts are also distinctive southern crops, all accounting for over 70 per cent of US production, with sugar cane and rice showing dramatic increases after the Second World War (Table 21.2). Each of these crops is localized, with sugar cane grown principally in Florida and Louisiana, rice in Louisiana and north coastal Texas, tobacco in eastern North Carolina, the Kentucky Bluegrass, and in south Georgia and peanuts in southwestern Georgia and southeastern Alabama. In contrast, cattle raising, the second leading agricultural product in sales in the South, is widely scattered throughout the region. The

TABLE 21.2 The South's share of US agricultural production, 1950 and 1992

Commodity	1950 (%)	1992 (%)
Tobacco*	88	95
Sugar cane*	43	79
Peanuts*	77	78
Rice*	49	72
Poultry**	14	54
Nursery products**	14	25
Vegetables**	19	23
Cattle and calves*	17	17
Soya beans*	10	17
Corn*	15	6
Total farm sales**	17	19

Notes: * Measured by quantity produced
** Measured by sales
The data exclude the states of Delaware, Maryland, Oklahoma and Texas
Source: Winsberg (1997, p. 197, Table 2)

resurgence of cotton, reaching in the late 1990s levels that have exceeded production of the early twentieth century, has taken place largely on the Coastal Plain, rather than on the Piedmont.

INDUSTRIALIZATION

Given the traditional emphasis on agriculture in the South as a way of life, it is not surprising that manufacturing lagged far behind the industrial port cities of the Northeast and the Great Lakes cities of the Midwest. With the demise of cotton growing in the early twentieth century in the Southeast, manufacturing became attracted to rural areas that had an abundance of 'left-over' farm workers who were eager to take industrial employment. The type of manufacturers attracted to the rural South sought out low-cost, low-skilled, and abundant labour (Johnson, 1997, p. 170). Many plants or mills were owned by companies headquartered in the North, and southern manufacturing was often characterized as a branch-plant economy. The manufacturing sectors that were dominant in the South were textiles, wood and paper products, furniture, tobacco, food processing, and inexpensive garments. These industries paid relatively low wages, required few skills, often employed a high percentage of females, and relied on local raw materials. The textile industry, which developed rapidly in the Southeast in the early twentieth century, continues to dominate southern manufacturing, having growing value added production levels in the face of declining employment totals. The textile industry increasingly has substituted capital equipment for labour. Except for the historically important but now defunct iron and steel industry in Birmingham, Alabama, a high percentage of southern manufacturing occurred in rural areas, not in the cities (Lonsdale and Browning, 1971).

Today, many manufacturing jobs have been lost in rural areas of the South, particularly in the textile, apparel, and wood-processing industries. At the same time, the suburban parts of large metropolitan areas have grown in manufacturing importance, especially as capital investment in machinery and equipment has replaced labour, as already noted in the textile industry. Whereas most of the traditional industrial cities of the North have lost manufacturing jobs since the early 1980s, southern metropolitan areas – which were never really major industrial centres – have not been impacted as adversely as have their northern counterparts. Atlanta and, to a lesser extent, Houston and Dallas–Fort Worth have actually gained modestly in manufacturing employment over the last two decades. Foreign automobile makers have helped fuel recent industrial growth in the South. The assembly plants have located in rural settings with interstate highway access for movement of auto parts to the plants and the assembled automobile to markets (Chapter 10). Toyota has a plant in Georgetown, Kentucky, near Lexington; Nissan assembles vehicles in Smyrna, Tennessee, near Nashville; Mercedes-Benz makes cars in Vance, Alabama, near Birmingham; and BMW is located in Greer, South Carolina, near Greenville–Spartanburg. These foreign automobile assembly manufacturers have obviously avoided the traditional Detroit region in their locational decisions, have established parts plants in the region with easy interstate highway access, and have relied on just-in-time transport (whereby precisely timed parts deliveries are substituted for large warehouse inventories) (Rubenstein, 1992).

RACE AND ETHNICITY

The South has the highest concentration of African American population of any US region. The Hispanic population, however, has been increasing much faster than the rest of the population and, for the US as a whole, is expected to exceed African Americans by the year 2005. Cubans in Miami have grown from 5 per cent of the population in 1960, just after Castro gained power, to some 60 per cent today. Texan cities, especially Houston and San Antonio, are well known for their large Hispanic (largely Mexican) populations. In the 1990s, a rapidly growing Hispanic population has settled in non-metropolitan areas throughout the South

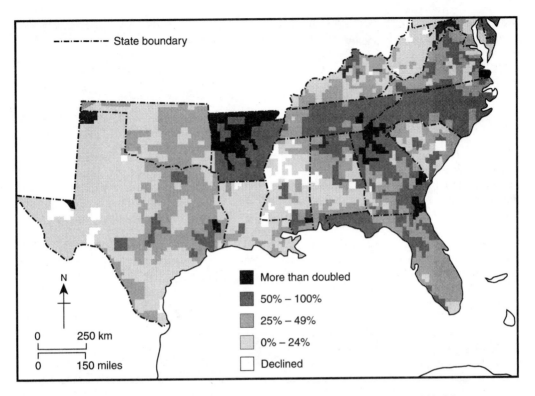

Fig. 21.3 Percentage growth of Hispanic population in the South by county, 1990–96

where they work in food processing (chiefly poultry), landscaping, and other labour-intensive jobs (Fig. 21.3). Extensive areas of rural Florida, Georgia, North Carolina, Tennessee, Arkansas, and Texas have experienced this growth during the 1990s, most notably the chicken-processing areas of Georgia and Arkansas. It is anticipated that the Hispanic population will continue to growth rapidly in the South, and gain significantly in large metropolitan areas outside Texas and Florida as well as in rural areas throughout the region.

THE NEW SERVICE ECONOMY AND SOUTHERN METROPOLISES

Almost all metropolitan areas in the South with a population greater than one million have grown at a rate faster than the US metropolitan growth average during the 1990s (Table 21.3; Fig. 21.4). In percentage increase,

the fastest-growing centre was Austin, Texas, associated with the high-tech 'Silicon Prairie' and the University of Texas at Austin. Atlanta, the service and distribution centre of the Southeast (Allen, 1996), was the fastest growing large (3.5 million in 1996) metropolitan area in the South, ahead of Houston, Dallas–Fort Worth, and Miami–Fort Lauderdale (see Chapter 14). Raleigh–Durham, in the high-tech Research Triangle, was the third-fastest growing metropolitan area, followed by Orlando and its associated theme parks of Disney World and the Epcot Center.

The recent expansion of southern metropolitan centres has been primarily the result of in-migration and only secondarily through natural increase (more births than deaths). As Pandit pointed out, it was

During the 1970s [that] the South experienced a remarkable reversal of its historical migration patterns. Traditionally a region with net out-migration to the rest of the country, the South began receiving

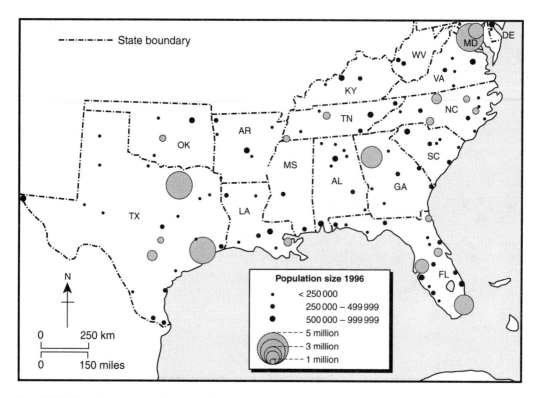

Fig. 21.4 Population size of metropolitan areas in the US South, 1996

large volumes of migrants from the Midwest and Northeast and showing a net gain through migration .... National economic restructuring in the 1970s and 1980s prompted a regional shift of industry and employment from the North to the South, [which] coincided with the coming of age of the Baby Boom generation .... The growth of elderly migration and the unprecedented reverse migration of blacks to the South gave further impetus to the Southward movement (1997, p. 238).

Migration to large metropolitan areas of the South was not only the consequence of manufacturing decline in the older industrial areas of the Midwest and Northeast, such as Chicago, Cleveland, Buffalo, Baltimore, and Philadelphia. The new, booming, information-based service economy of the South was a major attraction (Table 21.1). Many of the migrants were relatively young, well educated, and upwardly mobile. Companies per-

ceived the South to have a positive business climate. The restrictive racial barriers of the past were eased and many educated, middle class African Americans, whose families one to three generations before had left the agrarian South for northern industrial cities, now returned to southern metropolitan areas which offered good salaries in the advanced service economy. Major corporations grew prosperous in the southern metropolitan environments and some firms relocated to the South (Table 21.4). The South in 1997 constituted 25 per cent of *Fortune* magazine's 500 largest US corporations and accounted for 21.6 per cent of their revenues. Many other corporations which are headquartered elsewhere found it necessary to open regional offices in the South, principally in Atlanta and Dallas–Fort Worth. The South, at the backwater of North American settlement since the very beginning, suddenly found itself at the attractive forefront of and, along with the US West, a part of the prospering Sunbelt Society.

TABLE 21.3 Change in size in southern metropolitan areas (greater than one million population), 1990–96

Rank	Metropolitan area	% change	Absolute change
1	Austin, TX	23.1	195 103
2	Atlanta, GA	19.7	581 730
3	Raleigh–Durham, NC	19.4	167 768
4	Orlando, FL	15.7	192 477
5	Houston, TX	14.0	522 399
6	Charlotte, NC	13.7	158 928
7	Nashville, TN	13.4	132 152
8	Dallas–Fort Worth, TX	13.3	537 279
9	San Antonio, TX	12.5	165 362
10	Jacksonville, FL	11.2	101 906
11	Miami–Fort Lauderdale, FL	10.1	321 678
12	Greensboro–Winston Salem, High Point, NC	8.7	90 934
13	Memphis, TN	7.0	70 855
14	Norfolk–Virginia Beach, VA	6.6	95 542
15	Tampa–St Petersburg, FL	6.3	131 272
16	New Orleans, LA	6.9	27 628
	US metropolitan areas	6.9	13 615 123

Source: US Bureau of the Census

TABLE 21.4 The number and total revenue of *Fortune 500*, the largest US corporations in major southern metropolitan areas, 1997

Metropolitan areas	Number of corporations	Total revenues ($billion)
Dallas–Fort Worth, TX	15	163.9
Washington, DC	8	155.5
Atlanta, GA	12	153.4
Houston, TX	15	134.2
Charlotte, NC	4	56.6
Richmond, VA	9	54.1
Wilmington, DE	3	48.8
San Antonio, TX	3	40.9
Nashville, TN	2	22.5
Jacksonville, FL	2	17.3
Miami, FL	4	17.1
Birmingham, AL	4	17.0
Total US South	125	1194.4

Source: Fortune (1998)

CONCLUSION

The US South has probably undergone the most profound changes of any US or Canadian region in the past 25 years. For two centuries, a socio-economically lagging area with an agrarian orientation, urbanization came late to the South. The labour-intensive nature of agriculture, at first based on slavery and later the tenant system, and the single cash-crop farming practices retarded the rise of industrial centres such as those which occurred in the North. The industry that did prosper in the South typically located in rural areas, paid low wages, and was disproportionately involved in making textiles and inexpensive apparel, in food processing and tobacco and in forest products.

In recent decades, the US South, along with the West, has become a favoured migration destination, both for whites and African Americans. The shift to the knowledge-rich information service economy that grew so rapidly in the non-industrial South has made metropolitan areas such as Dallas–Fort Worth, Washington, Atlanta, and Houston favoured locations for large corporations and the business and professional services they require. Southern financial institutions have also prospered. The South provided two out of four presidents of the US in the late twentieth century in Jimmy Carter of Georgia and Bill Clinton of Arkansas. The Sunbelt Society of the South has flourished in metropolitan areas, especially in the suburbs, placing the region now as part of the US economic core. At the same time, some central city areas and rural locales remain among the poorest segments to be found anywhere in American society.

KEY READINGS

Aiken, C.S. 1998: *The cotton plantation South since the Civil War*. Baltimore: Johns Hopkins University Press.

Bartley, N.V. 1995: *The new South: 1945–1980*. Baton Rouge: Louisiana University Press.

Doyle, D.H. 1990: *New men, new cities, new South: Atlanta, Nashville, Charleston, Mobile*. Chapel Hill: University of North Carolina Press.

Johnson, M.L. 1997: To restructure or not to restructure: contemplations on postwar industrial geography in the US South. *Southeastern Geographer* **37**, 162–92.

Pandit, K. 1997: The southern migration turnaround and current patterns, *Southeastern Geographer* **37**, 238–50.

Winsberg, M.D. 1997: The great southern agricultural transformation and its social consequences. *Southeastern Geographer* **37**, 193–213.

22

THE CANADIAN NORTH

WILLIAM C. WONDERS

The Canadian North is by far the largest region in North America, and the least populated. Elsewhere man's impact has altered and even destroyed the natural environment, but in the North that impact has been limited until very recently and then mainly confined to some southern margins.

For Canadians and foreigners alike it is the North that symbolizes Canada – the 'true North' of the national anthem. 'The North' still conjures up in their minds colourful and heroic images of a vast, snowbound landscape sparsely inhabited by a sprinkling of aboriginal peoples and a few hardy white incomers. Yet over the past half-century this 'old North' has been altered dramatically. More than twenty-five years ago I titled a book of readings *Canada's changing North* (Wonders, 1971). Today it would more accurately be titled 'Canada's changed North'. Economically, socially, and even politically in places, today's North is a very different place from what it was only a generation ago.

The Canadian frontier of the late nineteenth century was the West. In the present century it is synonymous with the North. Its role has been the traditional one for frontier areas – a provider of natural resources for the primary benefit of the nation's older, more densely settled regions, the role of 'hinterland' to 'heartland' in the terminology of current geographers (McCann, 1982) and historians (Careless, 1979).

Major technological improvements in engineering, transport and communications have made it possible for resource developments to be extended into the North to an unprecedented degree. At the same time there have emerged in the past quarter of a century two new constraints which now affect such developments – the environment protection movement and the growing influential role of the aboriginal peoples in major areas of the North. In the twenty-first century, northern resources will become increasingly valuable and technologically accessible, but development will occur only if it reflects a new attitude towards this vast frontier region.

DEFINING THE NORTH

'One of the most basic and troublesome problems in dealing with the Canadian North is the difficulty of defining it satisfactorily' (Wonders, 1984a, p. 226). The most successful of the many approaches is the one devised by Hamelin (1972; 1979). He used ten criteria, six environmental and four cultural, to measure the *nordicity* (degree of northernness) of localities. With a possible 100 points for each criterion, the North Pole is assigned a theoretical 1000 points. To be included in the North 200 points are necessary. Thus defined, its southern boundary lies north of the transcontinental Canadian National Railway and beyond the zone of contiguous agricultural settlement ('Base Canada'). Three zones are identified within the North: 'Extreme North' (over 800 points), 'Far North' (500–800 points), and 'Middle North' (200–500 points) (Fig. 22.1).

It still is possible for individual exceptional outliers to occur within the general regional boundaries, e.g. Fort McMurray, Alberta (192 points and hence part of Base Canada) situated some 200 km north of Base Canada within the Middle North. Such exceptions usually reflect localized cultural changes, in this case associated with oil sands development. A similar change occurred at Fort George–Chisasibi on the east coast of James Bay with a drop from 338 points to 263 as a result of hydroelectric development (Hamelin, 1979, p. 36). Over time there has been an irregular and erratic northward expansion of Base Canada as a result of resource developments.

Today, approximately 70 per cent of Canada's total area is included within the North. The largest amount (3.9 million km^2) is included in the 'territorial' North beyond 60°N which is still largely dependent upon the federal government in Ottawa, though increasing political direction now comes from the two territorial capitals in Yellowknife (Northwest Territories) and in Whitehorse (Yukon) with their elected legislatures. A third territorial legislature has been located in Iqaluit (Frobisher Bay) with the separation of Nunavut Territory from the

FIG. 22.1 The Canadian North

existing Northwest Territories (NWT) in April 1999.

Almost as much provincial area in total is also northern, ranging from 77 per cent of Newfoundland–Labrador to 31 per cent of British Columbia (Table 22.1). This political fragmentation of the Canadian North greatly complicates any overall policy and planning for the region as a whole. In the 1970s, Rohmer (1970, p. 13) promoted the 'Mid-Canada Development Corridor' concept as 'an approach to the orderly planning and development of the mid-north, for people and for industry and for the services necessary to support them'. Despite considerable enthusiasm at the time, this national overall strategy withered and separate political

approaches continue to prevail, with the provincial Norths remaining 'lands of divided dreams' in the words of Coates and Morrison (1992, p. 130).

THE NORTHERN ENVIRONMENT

Canada's three largest physiographic regions stretch into the North (see Chapter 2). The ice-scoured, lake-strewn Canadian Shield arcs around Hudson Bay in the eastern and central sectors. It includes the great upland plateau of Labrador Ungava, the low rocky plains west of Hudson Bay, and the sediment-veneered flatlands of northern Ontario. To the west of the Shield, a narrowing northward

TABLE 22.1 The 'North' in Canada's provinces

Province	'Northern area' (km^2)*	% of area within the 'North'*	Northern population 1991#
Québec	1 079 000	70	511 058
Manitoba	488 000	75	75 591
Ontario	428 000	40	438 744
Saskatchewan	326 000	50	26 735
British Columbia	294 000	31	231 143
Alberta	264 000	40	98 452
Newfoundland–Labrador	31 300	77	55 387
Total	2 910 300		1 437 110

Note: Northwest Territories (with Nunavut) and Yukon are entirely within the 'North'. The former's population is 57 649 (area 3 379 684 km^2) and the latter's 27 797 (area 482 515 km^2). None of Nova Scotia, New Brunswick or Prince Edward Island is in the 'North'.
Sources: *Hamelin (1979) # Census of Canada 1991

projection of the Interior Plains extends that level-bedded landscape to the Beaufort Sea. In the far west the Cordilleran region consists of a complex of plateaus and mountain landscapes with the highest of the latter ice-covered (Fig. 22.2).

North of the continental mainland a unique maze of islands makes up the Arctic Archipelago (Zazlow, 1981). The largest of these, Baffin Island (507 450 km^2), is twice the size of Britain. Cape Columbia on Ellesmere Island, at 83°07' N, is within 800 km of the North Pole. Broadly speaking, archipelago bedrock ranges from Precambrian in the southeast through progressively younger formations to the northwest, and its terrain from glacier-covered mountains in the eastern islands to plateau and low-lying plains in the west. Heavy sea ice limits marine transport for large ships to only two or three months in most areas of the Archipelago and denies it entirely in the extreme northwest.

Length and severity of winter cold and low total precipitation are the dominant climatic conditions of Canada's northern climates (French and Slaymaker, 1993) (see also Chapter 3 and Fig. 3.2a). Internally the North is subdivided into Arctic and Subarctic (or Boreal) regions, with the lower extreme winter temperatures of the latter region being balanced against its much warmer summer

temperatures than in the Arctic. The boundary between the two regions approximates the treeline, the divide between tundra and forested landscapes, which on the ground is more a transitional zone than a well-defined line (Timoney *et al.*, 1992). The chilling impact of Hudson Bay and the offshore Labrador Current in the east are reflected in the southward deflection of the treeline. Permafrost underlies most of the Canadian North, contributing to poor drainage and complicating construction.

NORTHERN PEOPLES

In the enormous area of the Canadian North there are remarkably few people. In 1986 Bone (1992, p. 75) placed the figure at about 1.5 million, though two thirds of these were included in a southernmost transition zone within the Middle North which Hamelin excluded. So far as it is possible to estimate (boundaries of some census divisions were changed in 1991), the same area still had a population of only 1.5 million in 1991 (Table 22.1). About five sixths of this population is found within provincial boundaries in the Middle North and most of this is in the southernmost margins of the region. Population in the Yukon and in the

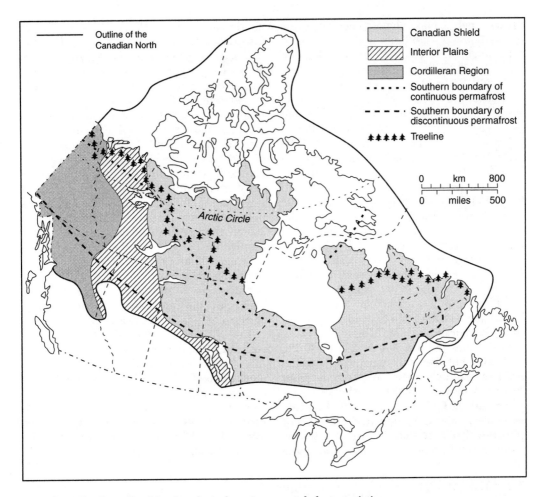

Fig. 22.2 The Canadian North: selected environmental characteristics

NWT increased significantly (18.3 per cent and 10.4 per cent respectively) over the decade 1981–91, well above the national average (7.9 per cent), but these populations are still so small, 27 797 and 57 649 respectively, relative to their areas that the population density is extremely low – 0.1 and 0.02 per km^2 respectively (Statistics Canada, 1992a).

In Canada overall, aboriginal peoples make up only 4 per cent (one million) of the total population (Table 22.2). In the Canadian North however, the relative numbers of these peoples increase significantly, from 10–20 per cent of the total population of northern Alberta to 60–80 per cent of the population of Ungava Québec (*Nouveau-Québec*). In the

Yukon aboriginal peoples make up 23 per cent of the total population, but in the NWT they constitute 61 per cent. Their predominance in the latter area is reflected in representation in the territorial legislature, a unique situation in the Canadian political scene, with important repercussions in policy decisions affecting this one third of Canada (Wonders, 1987).

Historically, the treeline has separated the major ethnological groups in the North, with the Inuit occupying the Arctic and the Indians and Métis (mixed-bloods) the forested Subarctic. This separation still prevails except for a very few localities, e.g. the Mackenzie River delta, NWT, Churchill, Manitoba, and Kuujjuarapik (Great Whale

River), Québec. Nevertheless, there are areas of overlapping traditional and current land use which have created complications in recent land claims negotiations, particularly in the NWT (Wonders, 1984b, 1990).

Of the 49 255 Inuit living in Canada in 1991, 43 per cent were in the NWT, 17 per cent in Québec, 13 per cent in Newfoundland–Labrador, and 11 per cent in Ontario. The Indian population of the Subarctic includes a large number of 'First Nations' as former tribes now prefer to be known (Fig. 22.3). In the NWT this diversity of ethnicity has resulted in the designation of six aboriginal languages as 'official languages' so far as territorial government is concerned, in addition to the two national official languages, English and French. The largest of the Indian nations within the NWT are the Dogrib (2690) and the Slave (2655), and within the Yukon the Tutchone (1420) and the Tlingit (925) (Statistics Canada, 1992b).

Birth rates in the North are declining somewhat but they still remain high. In the NWT for example, they have declined from 29.0 per thousand in 1983 to 23.9 in 1995, compared with the Canadian average for the same years of 14.7 and 12.9 respectively (Bureau of Statistics, 1997). The northern population is also a youthful one, particularly among the

aboriginal peoples. While 21 per cent of Canada's total population is under 15 years of age, amongst all Canadian aboriginals it is 36 per cent, amongst Yukon aboriginals it is 34 per cent, and amongst NWT aboriginals 40 per cent (Statistics Canada, 1992b). Such figures cause great pressure on educational facilities and on the need for employment opportunities. Despite increased educational opportunities however, large numbers of aboriginal youths drop out of schooling at an early age, handicapping their employment possibilities. In 1991 only 105 aboriginals held university degrees in the Yukon and 205 in the NWT (Statistics Canada, 1995). Though the percentages of employed aboriginals aged 15 and over has increased in both territories, these remain well below those for non-aboriginals, and their average income is well below the Canadian and territorial averages overall (Tables 22.3a and 22.3b).

In 1973 the Supreme Court of Canada recognized land rights based on 'aboriginal title'. The following year the federal government established an Office of Native Claims to negotiate claims settlements of two types: comprehensive claims where there had been no earlier treaties, and specific claims arising from alleged non-fulfilment of previous Indian treaties. Funding for researching,

TABLE 22.2 Aboriginal peoples in Canada, 1991

	Canada	NWT	YT	Nfld	PEI	NS	NB
Total	1 002 675	35 390	6 390	13 110	1 880	21 885	12 815
Indian	783 980	11 100	5 870	5 845	1 665	19 950	11 835
Métis	212 650	4 310	565	1 605	185	1 590	975
Inuit	49 255	21 355	170	6 460	75	770	450

	Qué	Ont	Man	Sask	Alta	BC
Total	137 615	243 550	116 200	96 580	148 220	169 035
Indian	112 590	220 135	76 370	69 385	99 650	149 570
Métis	19 480	26 905	45 575	32 840	56 310	22 295
Inuit	8 480	5 250	900	540	2 825	1 990

Note: NWT = Northwest Territories; YT = Yukon Territory; Nfld = Newfoundland and Labrador; PEI = Prince Edward Island; NS = Nova Scotia; NB = New Brunswick; Qué = Québec; Ont = Ontario; Man = Manitoba; Sask = Saskatchewan; Alta = Alberta; BC = British Columbia. Due to multiple counting, the sum of the individual categories is greater than the total population reporting on Aboriginal ancestry.
Source: Census of Canada 1991

FIG. 22.3 Aboriginal peoples of the Canadian North

developing and negotiating native land claims is provided by contributions and loans from the government.

Ten comprehensive claims agreements have been settled since 1973, involving aboriginals in extensive areas of the North (Fig. 22.4).[1] The largest areas remaining for agreements to be reached include the southwestern sectors of the NWT, much of northern British Columbia, and Labrador (Usher *et al.*, 1992). Comprehensive claims agreements typically provide cash financial benefits, land ownership of certain areas and a share of resource royalties, guaranteed hunting and trapping rights, and a greater role in the management of wildlife, land and the environment, and often in areas of social ser-

vices. Aboriginal corporations have been created to administer their resources.

NORTHERN RESOURCES

Historically, the Canadian North has been a major source of furs for the world's fashion industry, and until recently these have been the primary source of income for the aboriginal population. Fluctuating world demand and cyclical animal numbers always made for considerable unpredictability of livelihood, but this has been aggravated in recent years by the growth of the 'animal rights' movement and a fur-ban lobby. It has had a severe impact on those aboriginals who still

TABLE 22.3A Employment distribution of population 15 years and over, Yukon and Northwest Territories, 1981 and 1991

	1981		1991	
	Aboriginal	**Non-aboriginal**	**Aboriginal**	**Non-aboriginal**
Yukon				
Population 15+	2 615	14 379	4 330	16 525
Employed	1 165	10 820	2 300	12 740
% Employed	39.6	80.6	53.1	77.1
Northwest Territories				
Population 15+	15 380	14 285	21 390	17 165
Employed	6 090	11 520	9 740	14 740
% Employed	39.6	80.6	45.5	85.9

Source: Indian and Northern Affairs Canada 1996c

TABLE 22.3B Average income of population 15 years and over, Canada, Yukon and Northwest Territories (NWT), 1990

	Canada	**Yukon**	**Yukon aboriginals**	**NWT**	**NWT aboriginals**
	$	**$**	**$**	**$**	**$**
Males	30 205	31 236	19 281	31 231	18 469
Females	17 577	21 765	16 519	20 816	13 585
All	24 001	26 803	17 872	26 467	16 151

Source: Census of Canada 1991

prefer (or need) to continue as hunters and trappers. An extreme example of this was experienced by the Inuit whose sealskin industry collapsed in the 1980s forcing many onto welfare, when the European market evaporated due to the boycott of Newfoundland sealskins. The total value of sealskins in the NWT dropped from $890 278 in 1980–81 to $54 471 in 1984–5 (Anon., 1986).

Despite the fact that most aboriginal peoples in the North now live in numerous small, permanent communities rather than out on the land, many continue to hunt and trap, though often on a part-time basis. The reasons are varied. Some of the better trappers are still able to make an acceptable income; many others seek 'country food' as a means of offsetting the high costs of store-purchased food in the North, while still others view the 'subsistence economy as the foundation of Native culture' (de Lancey, 1985, p. 5). In the western Arctic it was found that when the cash revenues and imputed incomes from country food harvests were compared between wage-earning and full-time hunting Inuit, the 'combined income (cash + imputed food value) of the two groups was very similar' (Smith and Wright, 1989, p.93). Similar results were found among Indians in the Yukon (Wein and Freeman, 1995) and in northern Saskatchewan (Tobias and Kay, 1994). Yet, significantly, a study of fur trapping in the NWT concluded that 85 per cent of the participants consisted

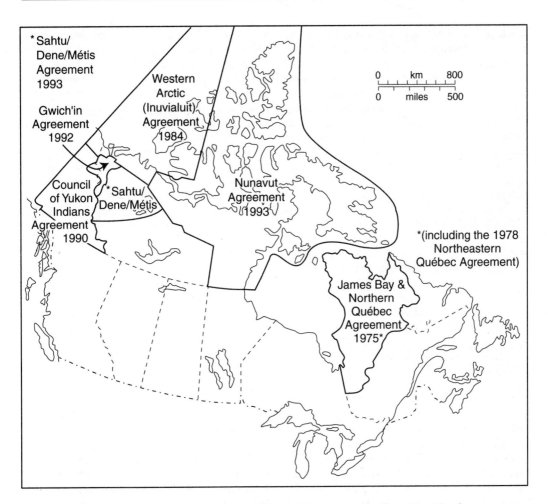

Fig. 22.4 Recent aboriginal comprehensive claims settlements in the Canadian North

of those 'whose participation is best explained by the lack of alternative employment opportunities' (Stabler *et al.*, 1990, p. 1).

Although commercial fur trapping is no longer the primary economic activity of many of the aboriginal peoples of the NWT (and to varying degree, elsewhere in the North), both the Canadian government and the aboriginal peoples have vigorously opposed the European Union's plans to ban Canadian furs.

In the present century other northern resources have come to the fore as more accessible sources have been utilized (Wonders, 1981). Although over 40 per cent of Canada's northern Boreal Forest is classed as 'forest and barren' (Gray, 1995, p. 2) and of no potential value for commercial forestry, the remaining 290 million hectares provide a major national asset. The lumber segment of the wood industry began in the marginal south-central parts of the region, primarily in Ontario and Québec, but its focus now has shifted northwestwards to British Columbia and even into southeastern Yukon. Pulp wood and pulp and paper now form the basis of many communities scattered across the southern edges of the North from the Gulf of St Lawrence to north-central British Columbia. Unlike many of the older lumbering centres, many of these communities are well planned and intended as permanent centres, making use of local hydroelectric power.

Minerals have provided the other major economic resource of the Canadian North (Nassichuk, 1987). The threefold structural base of the region is reflected in its varying mineral potential – metallics generally in the Canadian Shield and Cordilleran regions, fuels in the sedimentary Interior Plains and western Arctic. Mining communities form another settlement type right across the North from Labrador to the Yukon. To justify high development costs in the more remote sectors, the mineral deposits must be of unusually high value, e.g. gold in the territories and the current hectic activity for diamonds in the Lac de Gras area 320 km northeast of Yellowknife, or in enormous abundance which can be extracted on a massive scale, e.g. the Labrador iron ores or the lead-zinc deposit at Faro, Yukon.

In times past northern mines often were temporary features in the landscape, providing minimum facilities, using white incomers' labour with high turnover rates, and abandoning the sites when the ore was depleted. Today's mining communities seek a more permanent, married workforce by providing planned, modern amenities, or for the most remote sites, flying workers in and out on a rotation basis.

Northern mines are very sensitive to world market demand. Because of the slump in world mineral prices in the early 1980s, the entire town of Schefferville in Ungava Québec was shut down (Bradbury and St-Martin, 1983) and the Yukon lost an estimated 25 per cent of its population because of mine closings (*Christian Science Monitor*, 8 April 1983). Ultimately all ore bodies are exhausted. When this happens the community dependent upon it often disappears, e.g. Uranium City in Saskatchewan. The fortunate ones may find an alternative economic base, e.g. Elliott Lake, Ontario, as a retirement centre, or because of a location convenient to serve new mines in the general area, e.g. Timmins, Ontario, and Yellowknife, NWT.

Despite the finite life of any mine, governments encourage such developments as a means of providing income and offsetting the high costs of providing services to northern populations. The first diamond mine in

the NWT, for example, is expected to contribute $2.5 billion to the territorial economy over its projected 25-year life, with an average of 800 people, two thirds northern residents, employed during the operating phase and supporting an additional 640 jobs in the economy (Indian and Northern Affairs Canada, 1996a).

Northern oil and gas resources are impressively large. The Norman Wells oilfield in the Mackenzie Valley has been in production since 1921; in 1995 it produced 1.7 million m^3 of oil and 130 m^3 of natural gas. In the 1960s and 1970s an active search for new sources of oil and gas extended into the Mackenzie Delta–Beaufort Sea area and into the Sverdrup Basin in the northwestern Arctic Islands. Total discovered resources to December 1995 amounted to some 345 million m^3 of oil and 811 billion m^3 of natural gas (Indian and Northern Affairs Canada 1996b, p. 4), but there has been relatively little production due to the high costs involved.

Oil sands near Fort McMurray in the Athabasca Valley of northern Alberta contain 1.7 trillion barrels of oil, a third of the world's oil, of which 300 billion barrels are recoverable with today's technology. Two megaplants have been in operation since 1967 and 1977 respectively, producing over 20 per cent of Canada's oil (Petroleum Communication Foundation, 1996). In 1997 two additional billion-dollar plants were announced.

CHANGING PERSPECTIVES

In the 1970s new factors came to play a major role in northern development. Aboriginal rights in law obliged governments and private companies to take into consideration Native views on proposed developments, to provide increased employment in any such developments, and in areas of the new agreements at least, considerable joint management (Campbell, 1996). In September 1997 the Nunavut Inuit took the federal Department of Fisheries and Oceans to court, for example, because of its failure to consult Inuit authorities before setting turbot quotas in Davis Strait.

Widespread popular support for environmental protection added a strong ally to the aboriginal community in most cases. The federal government has applied environmental assessments of projects where it has been involved (often extending within provincial boundaries) since 1974 and issued guidelines in 1984; provincial governments also now require environmental impact studies of developments.

The 'new era' was first reflected in connection with the proposed Mackenzie Valley Pipeline to transport Arctic oil south, a proposal that was delayed and that still has not been implemented (Berger, 1977). Before the federal government approved later construction of the Norman Wells Pipeline which since 1985 carries oil south to Edmonton, the project was subject to strict environmental guidelines (Bone, 1992, pp. 145–55).

In earlier years some well-publicized mega-projects of provincial governments proceeded with minimal attention to these new factors (Gill and Cooke, 1974). Water power is a third major resource of the North which has assumed great value in recent decades, both to meet increasing electricity demand from Base Canada and from adjacent US markets. Such massive northern hydroelectric developments as the Peace River project in British Columbia (Reinelt, 1971), the Churchill-Nelson project in Manitoba (Waldram, 1984; Krotz, 1991), and the Churchill Falls project in Newfoundland–Labrador (Smith, 1975) were put in place with little or no attention to the environmental impact or to the disruption of traditional aboriginal land use in extensive areas of the North. Quinn (1991, p.132) commented that such developments 'have operated to the advantage of hydroelectric, forestry and energy corporations and at the same time to the serious disadvantage of Native people. The latter have only recently learned to organize resistance to a dominant culture which continues to treat their homelands as resource hinterlands.'

The monumental James Bay Power Project in northwestern Québec, like the Mackenzie Valley Pipeline in the west, marked the start of a new epoch for the North. The Québec government, in 1971, announced its intention to develop the hydroelectric potential of the rivers draining into James Bay and to develop all the resources of this area, approximately the size of Germany. Legal opposition by the aboriginal peoples forced the government to reach agreement with both the Cree and the Inuit of the area in 1975 (Anon., 1976), and with the Naskapi to the northeast in 1978. The massive river diversions and dam constructions of the La Grande River complex, the world's largest hydroelectric development with a capacity of over 10 000 megawatts, radically affected the Cree (Salisbury, 1986; Vincent and Bowers, 1988; Gorrie, 1990; Beyea, 1990; Peters, 1992).

In 1990 the Québec government announced its intention of proceeding with the second stage of the project which would have dammed the Great Whale River. The Cree launched legal actions to block it and mounted a skilful international public relations campaign, backed by environmental groups and Native-rights advocates, which resulted in New York State cancelling its contract for power from Hydro Québec. In 1994 the Québec government postponed indefinitely its plans for the Great Whale River development. (An unrelated mega-project proposal on a comparable scale had suggested converting James Bay itself into a fresh water reservoir, to pump water to continental markets (Rouse et al., 1992).)

To an increasing extent, aboriginal values are influencing 'northern development' (Keith, 1995–96; McDonald et al., 1997). In northern Labrador the 1600 Innu (Montagnais–Naskapi) have been successfully blocking development of the rich nickel deposits discovered at Voisey's Bay until their environmental concerns and claims are met. In 1994 the 8700 Inuit of Ungava Québec ('Nunavik') took over all or part of the justice, social service and education systems north of the 55th parallel. Most dramatic of all, in 1990 the Inuit of the central and eastern Arctic persuaded the federal and territorial governments and the majority of NWT residents to agree to the separation of almost 2 million km^2 of the NWT, to form a new territory, 'Nunavut' ('Our Land'), where aboriginal values will prevail in government and development (Pelly, 1993; Dickerson and McCullough, 1993).

Major challenges facing the Canadian North include the need to provide more employment opportunities for northerners, broaden the economic base and extend the life of resource-based communities (Robinson, 1962; Porteus, 1984; Duerden, 1992; Randall and Ironside, 1996). Still others include the increasing pollution of the Arctic from distant industrial regions (Twitchell, 1991), and the future of northern Québec if the province should separate politically from Canada – the aboriginal peoples of the region already have declared their intention of remaining in Canada in such a situation! The greatest challenge of all may well prove to be the impact of global warming. A six-year government/private sector study of the Mackenzie Basin suggests that the region already is warming and that this could increase by 4°C to 5°C by the middle of the twenty-first century, with generally adverse effects (thawing permafrost, increased forest fires, reduced forest growth) offsetting any potential benefits from a longer growing season or less severe sea ice conditions (Cohen, 1997).

ENDNOTE

On 11 May 1999 it was announced that the Labrador Innuit Association had reached agreement in principle with the federal government on their comprehensive land claim.

KEY READINGS

Bone, R.M. 1992: *The geography of the Canadian North: issues and challenges.* Toronto: Oxford University Press.

Coates, K. and Powell, J. 1989: *The modern North: people, politics and the rejection of colonialism.* Toronto: James Lorimer & Co.

French, H.M., and Slaymaker, O. 1993: *Canada's cold environments.* Montréal and Kingston: McGill-Queen's University Press.

Hamelin, L-E. 1979: *Canadian nordicity: it's your North too.* Barr, W. (trans.). Montréal: Harvest House.

Wonders, W.C. (ed.) 1988: *Knowing the North: reflections on tradition, technology and science.* Edmonton: University of Alberta, Boreal Institute for Northern Studies.

Zazlow, M. (ed.) 1981: *A century of Canada's Arctic islands, 1880–1980.* Ottawa: Royal Society of Canada.

23

THE GREAT LAKES AND THE ST LAWRENCE LOWLAND

MAURICE YEATES

The Great Lakes–St Lawrence Lowland is a physical region of immense importance. Homeland of many indigenous peoples, since 1600 AD it has been a prime avenue for European penetration into the North American continent. The region contains a fifth of the world's supply of freshwater in lakes and rivers, and provides major supplies of renewable and non-renewable natural resources. It includes much of the traditional manufacturing heartland of the continent, and is currently responding to decades of despoliation and contamination arising from economic activities and human settlement. Today, it combines parts of two countries, one of which (Canada) operates with two official languages, incorporating a French language concentration in Québec which has embedded within it a core population that wishes to form a separate state. The Great Lakes–St Lawrence Lowland extends over a territory (1.4 million km^2; half land, half water) the size of Western Europe, and in 1996 it was inhabited by an estimated 43.6 million people who generate a GRP (gross regional product) of about $800 billion. If the region were a separate country it would have the 10th largest economy in the world.

THE PHYSICAL REGION

The Great Lakes Basin is the upper watershed of the St Lawrence River, which flows in a northeasterly direction to its gulf and the North Atlantic Ocean (Fig. 23.1). These are simple observations to make, but they have enormous implications because the other major river networks in North America flow in a south–north direction (the Mackenzie River system to the Arctic Ocean), and north–south direction (the Missouri–Mississippi river system to the Gulf of Mexico). Part of the headwaters of the Great Lakes–St Lawrence and Mississippi systems are remarkably proximate west of Chicago – so close, in fact, that a shallow ditch at Chicago connecting Lake Michigan with the headwaters of the Mississippi was in use by 1846. In 1900, the newly-constructed Chicago Sanitary District reversed the flow of the Chicago River from

Lake Michigan to the Mississippi system to carry effluents away from the lake and shore.

The Mackenzie system lies within Canada, the Missouri–Mississippi system within the United States, but the Great Lakes–St Lawrence system is divided by part of the Canada–US boundary. Thus, a vital element of the story of the Great Lakes–St Lawrence system lies in the way in which the US, and the territories that became Canada in 1867, first went their separate and competitive ways in their exploitation of the resources of the region, but from around 1950 the two countries began to co-operate with its use and care. For example, the Erie Canal, connecting the Mohawk River to Lake Erie at Buffalo, opened in 1829 – providing an all US water route to the developing Midwest for east coast ports, boosting the fortunes of Syracuse, Rochester, and Buffalo, and capturing much of the actual and potential hinterland of Montréal for New York. More recently, the St Lawrence Seaway, which opened in 1959, has been a co-operative venture built with funds provided by the Canadian and US governments to expand and deepen the existing system of locks and canals to allow larger and longer sea-going and lake vessels to penetrate into the heart of the continent. The seaway is operated by a joint commission.

In Fig. 23.1 the Great Lakes Basin–St Lawrence Lowland is defined by its watershed (CF/IRPP, 1990, p. 2). To the west and south of the Great Lakes Basin this watershed lies remarkably close to the lakes themselves, to the north it is more extensive. The region embraces a variety of surface geologies, physiographic forms, and sub-climates, which together affect settlement and land use.

The Shield, which has the appearance of a dissected plateau varying in height from about 200 to 1000 m in its most mountainous sections, is a complex mixture of hard igneous and metamorphic rocks which has been much eroded by continental glaciation. Located primarily in the Canadian portion of the basin, it surrounds much of Lake Superior, and a prong extends into the Adirondack Mountains in the United States. This prong, marked by the 'Thousand

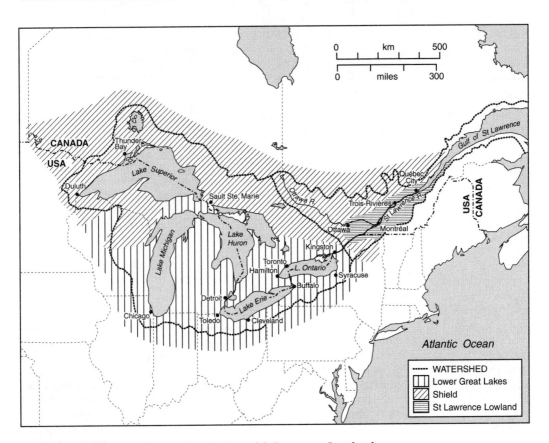

Fig. 23.1 Watersheds, Great Lakes Basin and St Lawrence Lowland

Islands' portion of the river just east of Kingston, effectively separates the basin from the St Lawrence River lowland. Although the shield has little good topsoil, and generally a climate with long and severe winters, making it inimical for settlement, it is the location of: pockets of mining, such as nickel and copper in the Sudbury area and iron ore in the Mesabi range west of Duluth; forestry and pulp/paper manufacturing, which has suffered from severe resource depletion leading to economic hardships in many single-industry towns (for example, Deep River and Sturgeon Falls in Ontario); and recreational activities, which include ski resorts and summer recreation areas in the Laurentians, Gatineau Hills, and Adirondacks, and second homes in the Muskoka district and the Thousand Island area.

The Lower Great Lakes sub-region has been affected by several physical processes, most recently (about 15 000 years ago) by retreating glacial action. In some areas, such as southwestern Ontario, southern Michigan, and upstate New York, glacial deposits have produced quite fertile soils. In other areas such as northern Michigan and central Ontario, sand and gravel, often deposited in the form of drumlins and eskers, makes for poor farmland. A humid continental climate with a six-month growing season prevails over much of the sub-region – conducive for wheat and general farming in southwestern Ontario and upstate New York, and corn/wheat general farming in southern Michigan and lands adjacent to the southern shores of Lakes Erie and Michigan. Particular micro-climates in the Niagara Peninsula and Finger Lakes area are sufficient for the production of some tender fruits, such as peaches and grapes.

The **St Lawrence Lowland** sub-region extends from the Thousand Islands east to Québec City, and includes the Ottawa River valley. The river is tidal as far as Trois Rivières, and navigable without locks as far as Montréal. Between Lake Ontario and Montréal there is a vertical drop of 72 m over a distance of 300 km – which has required many dams and locks, and much dredging, to maintain a depth of 10.7 m in the shipping channel (Government of Canada, 1991, 19: 5). In general, the sub-region from Montréal to Québec City consists of broad areas of alluvial and old marine clay interspersed with pockets of sandy soil. The area south of the St Lawrence River (*cantons de l'est*), in particular, is excellent for agriculture. The growing season, however, is marginally shorter than that in southwestern Ontario, and farmers have generally concentrated on dairying, with higher yield vegetable crops produced close to Montréal. The southern margin of the sub-region is demarcated by an upland extension of the Appalachian physiographic region which is generally inimical to settlement and agriculture, but has a few declining mining towns (for example, Asbestos).

POPULATION AND URBAN GROWTH

Though the population of the Great Lakes–St Lawrence region is large and increasing, it is not increasing as fast as the population of North America as a whole. Between 1970 and 1996 the population of the region increased from 40.6 million to 43.6 million, but its share of the continental population decreased quite steadily from 17.9 per cent in 1970 to 14.8 per cent in 1996. This decline in share of the continental population appears to have also been accompanied by a relative decrease in wealth in the region. As far as can be estimated from state and provincial figures, disposable per capita incomes within the region were generally 1.2 per cent above the continental average in 1980–81 but 2.1 per cent below the average in 1990–91. Thus, if population change and disposable income are inter-

preted jointly as indicators of economic health, the region has become economically less 'healthy' than the continent as a whole during the 1970–96 period. Is this comparative decline pervasive throughout the region in the 1990s?

Given that 90 per cent of the population of the region resides in metropolitan areas with more than 100 000 people (Table 23.1 and Fig. 23.2), variations in population change in these urban areas within the major physical sub-regions cast some light on the generality of the decline. Urban areas in the shield exhibit no real increase in population in the 1990s. The St Lawrence Lowland urban areas show some population increase (4.4 per cent), but this is less than the continental average (1990–96: 7 per cent) – the minor aberration is Ottawa–Hull, the capital of the Canadian Federation. The bulk of the population is located in the Lower Great Lakes sub-region where the growth of population in the 1990s is, on average, less (3.9 per cent) than in the St Lawrence Lowland sub-region. There is, however, a noticeable difference in growth rates between the Ontario and US parts of the Lower Great Lakes sub-region – the Ontario part experienced a growth rate of almost 8 per cent in the early 1990s, whereas the United States part has a population increase of about 3 per cent. Why are population growth rates within the region comparatively low? And, why is the growth rate in the Ontario part of the Lower Great Lakes so much greater than elsewhere within the region?

THE REGION: THE IMPACT OF RECENT SOCIAL, ECONOMIC AND POLITICAL TRENDS

The answers to the above questions are rooted in a number of trends that have had a particular impact on the region since 1970.

Socio-economic changes in employment

Four trends in general have had a significant impact on employment. First, the increase in

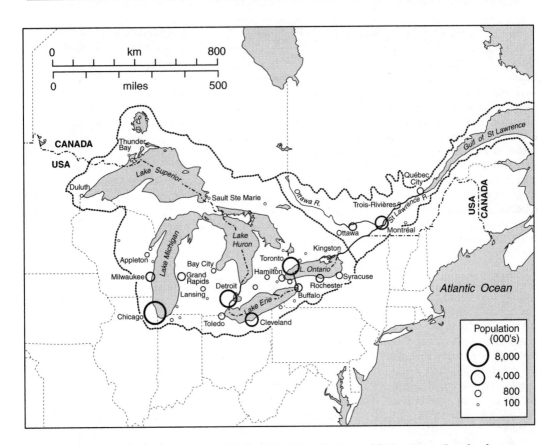

FIG. 23.2 Metropolitan areas (over 100 000): Great Lakes Basin and St Lawrence Lowland

participation of women in the paid labour force has meant that while the population increased by only 4.7% between 1970–71 and 1990–91, and male employment changed little, total employment in the region increased by about 28 per cent (Table 23.2). The employment base has thus widened considerably, increasing competition for jobs. Second, resource depletion and resource substitution has generated a decline in employment in mining and forestry based industries which is directly related to the lack of population growth in the Shield sub-region (Mackenzie and Norcliffe, 1997). Jobs have often had to be maintained through the relocation of governmental activities to this sub-region, such as Revenue Canada to Sudbury, and the Ontario Lottery Corporation to Sault Ste Marie. The shift in employment emphasis to office and clerical jobs has increased employment opportuni-

ties for women, but has adversely affected less well educated men who can no longer find reasonably well paid jobs in resource-based industries.

Third, the lowering of international tariffs on a variety of manufactured products and raw materials under the General Agreement on Tariffs and Trade (GATT) since 1963 has made it increasingly possible for lower wage areas (within North America or elsewhere) to focus on the production of low value added products, such as textiles, and export them to higher wage areas – thus directly competing with a similar industry if it existed in the higher wage country. For example, prior to 1970, much of the textile industry in Canada was located in the St Lawrence Lowland sub-region, and since 1970 it has been decimated by competition from lower cost parts of the world (Yeates, 1998a).

Fourth, the US–Canada Automotive

TABLE 23.1 Great Lakes–St Lawrence region: metropolitan area populations, 1990/91–96

	1990 US' ('000)	1991 Canada ('000)	1996 US + Canada ('000)	Change 1990/91–96 (%)
Shield				
Thunder Bay CMA		125	126	0.5
Sault Ste Marie (Canada + US)		100	100	0.0
Sudbury CMA		158	160	1.8
Chicoutimi–Jonquière CMA		161	160	−0.3
Duluth–Superior MSA	240		240	0.0
Sub total	240	544	786	1.0
Lower Great Lakes				
Toronto CMA		3 899	4264	9.4
Hamilton CMA		600	624	4.1
London CMA		382	399	4.5
Kitchener CMA		356	383	7.4
St. Catharines–Niagara CMA		365	372	2.2
Windsor CMA		262	279	6.3
Oshawa CMA		240	269	11.9
Kingston CA		136	143	5.1
Barrie CA		97	119	22.2
Guelph CA		98	105	7.9
Brantford CA		97	100	3.2
Peterborough CA		98	100	2.2
Ontario part		6 630	7 157	7.9
Chicago–Gary–Kenosha CMSA	8 240		8 600	4.4
Detroit–Ann Arbor–Flint CMSA	5 187		5 284	1.9
Cleveland–Akron CMSA	2 860		2 913	1.9
Milwaukee–Racine CMSA	1 607		1 643	2.2
Buffalo–Niagara Falls CMSA	1 189		1 175	−1.2
Rochester MSA	1 062		1 088	2.4
Grand Rapids MSA	938		1 015	8.2
Syracuse MSA	742		746	0.5
Toledo MSA	614		611	−0.4
Lansing–East Lansing MSA	433		448	3.5
Saginaw–Bay City–Midland MSA	399		403	0.9
Appleton–Oshkosh–Neenah MSA	315		341	8.1
Utica–Rome MSA	317		302	−4.5
Erie MSA	276		281	1.8
South Bend MSA	247		258	4.3
Green Bay MSA	195		213	9.5
Elkhart–Goshen MSA	156		169	8.3
Benton Harbor MSA	161		161	0.0
Jackson MSA	148		154	4.1
Jamestown MSA	142		141	−0.1
Wausau MSA	115		122	5.5
Sheboygan MSA	104		110	5.5
United States part	25447		26 178	2.9
St Lawrence Lowland				
Montréal CMA		3 209	3 327	3.7
Ottawa–Hull CMA		942	1 010	7.3
Québec–Levis CMA		646	671	4.1
Sherbrooke CMA		141	147	4.7
Trois Rivières CMA		136	140	2.7
Sub total		5 074	5 295	4.4
Metropolitan total	37 935		39 416	3.9

Source: US and Canadian censuses

Trades Agreement ('Auto Pact', 1965) designed to foster integration of the North American automobile industry, led, by 1990, to a high level of integration of the economies of the Canadian and the US parts of the Lower Great Lakes sub-region. In 1970, this industry was concentrated in Detroit, Cleveland, Toledo, Oshawa, and Windsor. Furthermore, the Auto Pact accentuated the dependence of the sub-region on the automobile industry (GTA, 1996; Hayward and Erickson, 1995). Thus, when new production and process technologies (pioneered primarily in Japan) led after 1970, to increased price and quality competition in the automobile and related industries, such as iron and steel (Gary, Indiana; Hamilton, Ontario; and Sault Ste Marie, Ontario), the sub-region was particularly hard hit. In response to this competition, industrial restructuring has led to the establishment of more efficient production facilities (often in greenfield locations), intensive use of highway dependent just-in-time delivery systems, automobile assembly related job loss and wage/benefit reductions, and component outsourcing to lower cost producers (Warf and Holly, 1997) (see Chapter 10). In consequence, manufacturing employment within the region decreased slightly over the 1970–90 period (–2.3%) – hence the low rate of population growth throughout much of the Lower Great Lakes sub-region and, again, a loss of job opportunities for less well-educated males (Table 23.2).

The favouring of Canada by the Auto Pact has, on balance, tipped the location of production facilities, particularly of parts, to Ontario, and hence created comparatively more jobs in that part of the Lower Great Lakes. The reasons for this are quite mixed, ranging from health care costs covered by a province-wide insurance scheme (the largest single cost item for US-based auto firms is company contributions to employee health insurance), to a downward shifting Canada–US exchange rate since 1977 which has had the effect of replacing a bi-national tariff barrier with an exchange rate barrier. One of the reasons for the formation of the US–Canada Free Trade Agreement in 1989 – metamorphosed into the North America Free

TABLE 23.2 Great Lakes–St Lawrence region: labour force by major economic group, 1970/71 and 1990/91

Economic group	1970/71		1990/91		Change 1970–90 (%)
	Number (millions)	Share (%)	Number (millions)	Share (%)	
Primary	1.3	7.9	1.1	5.2	–15.3
Manufacturing	4.2	25.4	4.1	19.5	–2.3
Consumer services	4.2	25.5	5.7	26.9	35.7
Producer services	1.8	10.9	3.5	16.6	94.4
Government and public	3.4	20.6	4.8	22.7	41.2
Infrastructure and interaction	1.6	9.7	1.9	9.1	18.8
Total	16.5	100	21.1	100	27.9

Note: this is in essence a simulation based on a blending of Statistics Canada experienced labour force data by province and Bureau of the Census civilian labour force data by state for the respective census periods, adjusted on a pro rata basis to labour force totals derived from data pertaining to MSAs, CMAs and CAs. 'Consumer services' include employment in retail and wholesale trade, accommodations, restaurants and personal services. 'Producer services' include employment in finance, insurance and real estate (FIRE) and business services. 'Government and public' includes federal, state and local government employment, jobs in education, and health related employment. 'Infrastructure' is construction and 'interaction' is communications.

Trade Agreement (NAFTA) in 1994 to include Mexico – was to provide a means for addressing the perception in the US that the Auto Pact favoured Canada (Cole, 1991; Holmes, 1992).

Thus, the shift to post-industrialism since 1970 has changed radically the employment structure of the North American economy (Semple, 1997). This is evidenced not only by the static nature of manufacturing employment, but also by the growth of employment in consumer services, producer services, and governmental and other public services within the region (Table 23.2; Chapter 11). The increased importance of capital mobility and hence financial/equity/commodity trading in the world since 1970 has led to an almost doubling of employment in the producer service industries within the region. Producer service employment tends to concentrate in metropolises, such as Chicago and Toronto, that link extensive domestic regions to the global financial community (Knox, 1997).

Political trends

The separatist movement in Québec, which has the objective of establishing an independent francophone nation in North America, has gained in strength considerably since 1970. The political uncertainty that this movement has generated has had a negative effect on the vitality of the economy within the St Lawrence Lowland sub-region, particularly Montréal which has been traditionally the motor for economic development within the Province of Québec (Waddell and Gunn, 1998). This negative effect has been accentuated by the general economic shift to post-industrialism and the increasing importance of producer services, for whereas prior to 1970 Montréal and Toronto were both locations of national financial and corporate headquarter activities, since that time much of what may have been Montréal's share has been assumed by Toronto, and, secondarily, Calgary and Vancouver (Coffey, 1994) (see Chapter 16).

The effect of political uncertainty is difficult to measure because it involves not only decisions made, but also those not made. But,

after two decades an impact has to be recognized, embedded though it may be within other larger scale trends such as those related to the shift of labour intensive industries to low wage areas elsewhere. In 1971, the Montréal CMA had a population of 2.7 million and the Toronto CMA 2.6 million. Despite similar changes in CMA definition following suburban growth, the Toronto CMA in 1996 has one million more people than the Montréal CMA. Montréal, and the rest of Québec, has plainly experienced a slower growth rate than it may have had in the absence of political uncertainty. Furthermore, the Province of Québec has exercised its prerogative to implement more severe limits on immigration than those exercised in the rest of the country – which again has dampened population growth in its largest metropolis.

Demographic trends

Net internal migration and net immigration have become more important components of population growth since 1960 as rates of net natural population increase decreased. The strong pattern of net internal migration in the United States since 1970 is from the 'Snowbelt' (of which the US part of the Lower Great Lakes sub-region is a definite member) to the 'Sunbelt'. This net internal flow involves firms (and hence jobs), as well as increasing numbers of retirees (and hence all kinds of retirement income and health care services). In Canada the general long-standing net internal flow has been from east to west, with, since 1970, British Columbia, and in most years Ontario and Alberta, incurring positive in-migration (Yeates, 1998b, Fig. 4.6). Thus, the only part of the Great Lakes–St Lawrence Lowland consistently benefiting since 1970 from internal migration is the Ontario part of the Lower Great Lakes sub-region.

During the early 1990s, net immigration accounts for about 23 per cent of net population increase in Canada; and legal plus illegal immigration must have accounted for at least 30 per cent of the 1990–96 net increase in the United States. Most immigrants to Canada settle initially in Toronto, Vancouver, and Montréal. In 1996, 41 per cent of the population of the Toronto CMA and 17.4 per cent of the Montréal CMA were born outside of Canada. Most post–1970 immigrants to the US have not generally settled in metropolises in the US part of the Lower Great Lakes sub-region (Carlson, 1994), with the exception of Chicago (Greene, 1997). Thus, net internal migration within Canada contributes to the higher rate of population increase within the Ontario part of the Lower Great Lakes sub-region, and net immigration significantly reinforces population increase in the metropolitan areas of Toronto and Chicago. In fact, net immigration has probably been the strongest growth dynamic in these metropolises since 1990.

THE ENVIRONMENT OF THE GREAT LAKES BASIN AND ST LAWRENCE LOWLAND

The idea of sustainable development recognizes the mutually supportive nature of the environment and development, and emphasizes that economic growth should not compromise the needs of future generations (CF/IRPP, 1990). The Great Lakes Basin–St Lawrence Lowland region is a natural ecosystem that has accumulated many environmental stresses during the past 200 years. These stresses are, in some cases, so severe that local, provincial/state, national, and transnational programmes have been devised to address the negative environmental impacts from this legacy of urban and industrial development.

Industrial development in the St Lawrence Lowland sub-region began during the early nineteenth century, and with the construction of locks on the St Lawrence River above Montréal it extended into the Great Lakes Basin. The Great Lakes themselves became progressively interconnected for ships and barges with the opening of the Welland Canal in 1829 (bypassing Niagara Falls), and the Sault Canal (1855) connected the Lower Great Lakes with the iron and copper deposits which are found in the Upper Peninsula of Michigan. The basin was also connected by the Erie Canal (1829), and soon by the railroad, with the Hudson River and

New York City. The integrated waterway provided the means for bringing bulky raw materials together for manufacture at lake-side locations; and the continent-wide expansion of the railroad system became based in large part on east–west routes located either side of the Lower Great Lakes.

By 1920–21, when the population within the region was about 22 million, heavily concentrated in the largest cities, environmental concerns were coming to the fore. The Great Lakes and St Lawrence River had been used as both a municipal and industrial sewer, as well as a source of drinking and industrial water, for about a century. Concerns emanated primarily from issues related to public health – informed citizens around 1900 noted that 'Whereas in the major European cities at this time the annual death rate from typhoid was around 5 per 100 000 population, in the principal US cities in the Great Lakes Basin it was four times higher or more' (CF/IRPP, 1990, p.53). As typhoid fever was known to be transmitted by contaminated water, municipalities gradually installed water filtration and chlorination 'purification' systems to provide safer water, but did not provide (or require of industry) much by way of waste treatment with respect to outflows which were causing the contamination in the first place.

After 1920, a growing petrochemical industry (Chicago, Cleveland, Detroit, Buffalo, Toronto, Montréal), and increasing domestic use of cleansers, added many more contaminants to the environment. The result has been extensive pollution in the Great Lakes and St Lawrence River. This was illustrated by the near-death of Lake Erie by the 1960s – which at that time was choking through massive growth of algae stimulated by phosphorous and nitrogen nutrients being discharged into the lake. Concern over water quality, and realization that bi-national action was called for, eventually led to the US–Canada Great Lakes Water Quality Agreement (1972) and The Great Lakes Water Agreement (1978) – the objectives of which are 'to restore and maintain the chemical, physical, and biological integrity of the waters of the Great Lakes Basin ecosystem' (IJC, 1988, Article II).

This agreement, administered by an International Joint Commission (IJC), is a landmark because it provides one of the few examples of a programme jointly conceived and funded by adjacent nation-states to address sustainable development issues relating to a large transnational ecosystem. The programme has proven quite successful – the Great Lakes are far healthier in the mid 1990s than they were in the 1970s. The St Lawrence River is less healthy because there has not been the same level of concerted action in that sub-region – headway has been made in reducing pollution from pulp and paper and refinery plants, but waste from metallurgical and chemical industries, and many municipalities, remains to be properly regulated (Government of Canada, 1991, Chapter 18).

Industrial decline also creates a whole series of issues that impact negatively on the environment. Wastes from the chemical industry were often buried in what were thought to be safe sites. The problem with many chemical effluents is that they cannot be disposed of so easily, and they frequently mix through seepage in water and air, creating a chemical soup that may have unknown future consequences (CF/IRPP, 1990, pp. 165–9). Although negative impacts on plant and animal life can be regarded as early warnings to humans, such as seepages from the Love Canal which generated mutations and death in Lake Ontario fish, it is direct widespread linkages to human health that generate the greatest concern and the most immediate responses (Levine, 1982). The Superfund cleanup programme in the United States, and provincially-based remediation initiatives in Canada, are designed to decontaminate the worst of these hazardous sites.

CONCLUSION

There are, therefore, significant economic and administrative forces fostering cohesion, and political forces for division, within the Great Lakes Basin–St Lawrence Lowland physical region. The forces for cohesion emanate from a need for both Canada and the United States to maximize: (1) their use of the waters in the region for transport, and for human and industrial use – the St Lawrence Seaway project and Great Lakes Water Agreement are excellent and successful examples of bi-national capital investment and joint administration; and, (2) the productive capacities of an integrated industry that is concentrated within the Lower Great Lakes sub-region – the US–Canada Automotive Trades Agreement has been highly effective in this regard, and it appears that NAFTA is reinforcing this economic interdependence.

Forces for division are derived from: (1) the partition of the region into two political units – which means that regional co-operation requires considerable and difficult negotiation, and, as in the case of environmental concerns, will occur only when matters are reaching a crisis; and (2) the well-established separatist movement in Québec which may yet force the formation of a third country operating within the region that would control the St Lawrence River passage, and Canadian land/air routes, between the Atlantic Ocean, the Maritime provinces, and the Great Lakes. The Great Lakes Basin–St Lawrence Lowland is, therefore, a region of immense economic, cultural, environmental, and political consequence in North America.

KEY READINGS

CF/IRRP 1990: *Great Lakes, great legacy.* Washington, DC: The Conservation Foundation, and Montréal: Institute for Research on Public Policy.

Gentilcore, R.L. *et al.* (eds) 1993: *Historical atlas of Canada, Vol. II: the land transformed, 1800–1891.* Toronto: University of Toronto Press.

Kerr, D. and Holdsworth, D. (eds.) 1990: *Historical atlas of Canada, Vol. III: addressing the twentieth century.* Toronto: University of Toronto Press.

McCann, L. and Gunn, A. (eds) 1998: *Heartland and hinterland: a regional geography of Canada.* Toronto: Prentice Hall Canada, Chapters 3, 4, 5 and 7.

Sinclair, R. 1994: Industrial restructuring and

urban development: an examination in metropolitan Detroit. In Braun, G.O. (ed.), *Managing and marketing of urban development and urban life*. Berlin: Reimer, 205–19.

Warf, B. and Holly, B. 1997: The rise and fall and rise of Cleveland. *Annals of the American Academy of Political and Social Science* **551**, 208–21.

Section G

The continent and the world

24

NORTH AMERICA AND THE WIDER WORLD

JOHN A. AGNEW

'Again and again in history, ideas have cast off their swaddling clothes and struck out against the social systems that bore them' (Max Horkheimer, 1947, p. 178)

The twentieth century has been the North American century. The United States and its neighbour and largest trading/investment partner, Canada, have achieved average income levels and a general quality of life since the 1930s unrivalled elsewhere for such large numbers of people. Both countries were members of the alliances that won the two world wars. After the Second World War, the US, with Canada as a key ally, formed NATO to contain the former Soviet Union and its allies. This required significant military and political commitments beyond the boundaries of North America, changing previous intermittent involvement in world affairs into a permanent and decisive presence. There are important differences between them, the most notable being those of Canada's greater reliance on resource industries and foreign (American) markets and a significantly higher provision of services by government. In a global context, however, Canada lies in the shadow of the US and although it has had distinctive foreign policies, for example, a greater commitment to international institutions such as the United Nations, it has been largely reactive to and dependent on American positions. Inevitably, therefore, a survey of North America in the wider world becomes an examination of the rise to global political and economic prominence of the US and the impact of contemporary challenges to this on the US and Canada. The story, however, is not so much that of the simple rise of a new imperial or hegemonic state in succession to previous ones but rather the creation of a global economy under American auspices, reflecting important features of the dominant ideology arising from the founding of the country, and the feedback of this system on North America.

The chapter is broken into four sections. The first provides a brief description of the dominant ideology that emerged from the founding of the United States and how this evolved from a 'marketplace imperialism' to 'globalization'. The second section considers how the Cold War years provided the context for the growth of an increasingly global economy under American auspices. A third part surveys the impact of this global economy on the contemporary US. A fourth, and final, section suggests that the American political system now has difficulty in responding to the tension between a globalizing world economy, on the one hand, and the problems of a territorial economy and its population, on the other.

FROM MARKETPLACE IMPERIALISM TO GLOBALIZATION

It is a commonplace now to see the genius of the American Constitution of 1787, as expressed most eloquently and persuasively in the writings of James Madison, as tying freedom to empire. Madison maintained that in place of the British colonial system the best solution for the American rebels would be the creation of a powerful central government which would provide the best security for the survival of republican government. This government would oversee geographical expansion into the continent and this would guarantee an outlet for a growing population that would otherwise invade the rights and property of other citizens. In this way, republican government was tied to an ever-expanding system. Madison had brilliantly reversed the traditional thinking about the relationship between size and freedom. Small was no longer beautiful. Of course, Thomas Jefferson and others less tied to the fortunes of land acquisition and growing markets initially opposed the logic of expansionism. But eventually they, too, came around. Indeed, when he became President, Jefferson justified the acquisition of Louisiana and the prospective addition of Canada and Cuba by claiming the extension of an 'empire for liberty' (Williams, 1969).

Although couched in the language of political rights and citizenship, the association of freedom with geographical expansion reflected two important economic principles. The first was that expansion of the marketplace is necessary for political and social well-being. The second was that economic liberty is by definition the foundation for freedom *per se*. So, the political system was designed to combine these two principles: on the one hand the central government guaranteeing the capacity for expansion into the continental interior and into foreign markets and on the other both the existence of lower-tier legislatures (the states) and the division of powers between the branches of central government, limiting the power of government to regulate and limit economic liberty. The founders of the United States could find ready justification for their institutional creation in the timely publication of Adam Smith's *Inquiry into the nature and causes of the wealth of nations* in 1776. Smith stood in relation to the founding as Keynes was to do to the political economy of the New Deal in the 1930s: a systematizer of an emerging 'common sense' for the times.

As the dominant social group at the time of independence and for many years thereafter, American farmers saw themselves as intimately involved in marketplace relations. They never saw themselves in terms of self-sufficiency. Their commercial outlook had its origins in the spatial division of labour organized under British mercantilism, in which they came to serve distant markets rather than engage in subsistence agriculture. Much of the basis for American independence lay in the struggle to expand the boundaries for individual economic liberty within a system that was more orientated to a sense of an organic whole: the British Empire. Madisonian 'marketplace imperialism', therefore, was not a pure intellectual production, but arose out of a context in which it served the identity and interests of a dominant social group.

The continental expansion of the United States in the nineteenth century secured a contiguous land mass and resource base unmatched by other empires of the time except for European Russia's involvement with Asia. A national policy of conquest, settlement and exploitation was initially geared almost entirely to agricultural development, but by the Civil War years (1860–65), the creation of an integrated national manufacturing economy was under way. Indeed, the

Civil War itself can be seen, at least in part, as a struggle over the economic trajectory of the US; each side representing not only opposing positions on slavery but also distinctive views on the development of manufacturing industry. The victory of the North ensured the effective transformation of the American economy from agrarian to industrial. The South and the West became resource peripheries for the burgeoning Manufacturing Belt of the Northeast, feeding foodstuffs and industrial raw materials to the factories of the victorious North. Alan Trachtenberg captured the creation of a truly national and incipient international market this entailed:

> Following the lead of the railroads, commercial and industrial businesses conceived of themselves as having the entire national space at their disposal: from raw materials for processing to goods for marketing. The process of making themselves national entailed a changed relation of corporations to agriculture, an assimilation of agricultural enterprise within productive and marketing structures. Agricultural products entered the commodities market and became part of an international system of buying, selling, and shipping (1982, pp. 20–21).

The entire edifice was underwritten by a rapid expansion of consumption based on the growth of a 'sales culture' in which purchasing goods was established as a means to achieve satisfaction and happiness. The United States pioneered both in advertising and salesmanship as ways of incorporating the population into mass markets for the goods pouring out of its factories. The ethos of mass production for mass consumption was an American invention. The limits set to consumption by the size of the national market, gigantic as it might seem, could, then, only be overcome by expansion over the horizon.

The creation of an integrated national economy, however, did not mean that there were not important differences between the major sections or regions of such a vast country over the direction that American expansion beyond continental borders should take. Indeed, American political disputes over trade, investment, banking and military policies have always taken a sectional cast given the different needs and expectations of the populations associated with the major sections of the country (Trubowitz, 1998). For example, the dominance of manufacturing in the Northeast from the Civil War to the 1950s necessarily encouraged a more positive attitude to tariffs in that region on the import of manufactured goods than did the resource-based economy of the South where such tariffs were seen as raising the costs of manufactured goods in the region without commensurate compensation to the needs of the regional economy. What is not at issue, however, is that such disputes have always taken place within a dominant discourse that has privileged the presumed benefits of continuing economic growth and the need to expand economically beyond national shores to realise that objective. In other words, inter-regional disputes over foreign policy have been principally over means more than ends.

Certainly, by the 1890s the United States had, in the eyes of influential commentators and political leaders from all over the country, fulfilled its 'continental destiny'. The time was propitious, they believed, to launch the US as a truly world power. One source of this tendency was a concern for internal social order. Not only did the late nineteenth century witness the growth of domestic labour and socialist movements that challenged the pre-eminence of business within American society, it also saw a major period of depression and stagnation, the so-called Long Depression from the 1870s to 1896, in which profit rates declined and unemployment increased. This combination was seen as a volatile cocktail, ready to explode at any moment. Commercial expansion abroad was viewed as a way of both building markets and resolving the profits squeeze. Unemployment would decline and the appeal of subversive politics would decrease. Another source was more immediately ideological. US history had been one of expansion: why should the continent set limits to the 'march of freedom'? To Frederick Jackson Turner, the historian who had claimed the internal 'frontier' as the source of America's difference with other societies, the US could only be

'itself' (for which one reserved the term 'America', even though it applied to the entire continent, not just the part occupied by the US) if it continued to expand. An invigorated American foreign policy and investment beyond continental shores were the necessary corollaries:

> For nearly three hundred years the dominant fact in American life has been expansion. With the settlement of the Pacific Coast and the occupation of the free lands (*sic*), this movement has come to a check. That these energies of expansion will no longer operate would be a rash prediction; and the demands for a vigorous foreign policy, for an interoceanic canal, for a revival of our power upon the seas, and for the extension of American influence to outlying islands and adjoining countries, are indications that the movement will continue (Turner, 1896, p. 289).

The outburst of European colonialism in the late nineteenth century was also of importance in stimulating American designs for expansion beyond continental limits. Home markets were no longer enough for large segments of American manufacturing industry, particularly the emerging monopolies such as, for example, Standard Oil and the Singer Sewing Machine Co. Without following the Europeans, the fear was that American firms would be cut out of overseas markets that exercised an increasing spell over the American national imagination, such as China and South East Asia. The difference between the Americans and the Europeans, however, was that for the Americans business expansion did not necessarily entail territorial expansion. Guaranteed access was what they craved. Indeed, colonialism in the European tradition was generally seen as neither necessary nor desirable. Not only was it expensive for governments, in many cases it also involved making cultural compromises and deferring to local despots of one sort or another; costs Americans were not anxious to bear.

The American approach tended to favour direct investment rather than portfolio investment and conventional trade. Advantages hitherto specific to the United States in terms of economic concentration and mass markets – the cost-effectiveness of large plants, economies of process, product and market integration – were exportable by large firms as they invested in overseas subsidiaries. For much of the nineteenth century capital exports and trade were what drove the world economy. By 1910, however, a largely new type of expatriate investment was increasingly dominant: the setting up of foreign branches by firms operating from a home base. US firms were overwhelmingly the most important agents of this new trend. They were laying the groundwork for the globalization of production that has slowly emerged, with the 1930s and 1940s as the only period of retraction, since then (Agnew, 1987).

But American expansionism after 1896 was never simply economic. It was always political and cultural. There was a 'mission' to spread American values and the American ethos as well as to rescue American business from its economic impasse. These were invariably related to one another by American politicians and commentators as parts of a virtuous circle. Spreading American values led to the consumption of American products, American mass culture broke down barriers of class and ethnicity and undermining these barriers encouraged the further consumption of products made by American businesses. American foreign policy largely followed this course thereafter, with different emphases reflecting the balance of power between different domestic interests and general global conditions: making the world safe for expanding markets and growing investment beyond the borders of the United States.

THE COLD WAR AND THE GLOBAL ECONOMY

American involvement in the First World War in alliance with Britain and France did not lead, as President Woodrow Wilson had hoped, to a commanding American presence in the world, with the United States as the sponsor of a great new scheme of collective security, the League of Nations. Rather, the US withdrew into its shell, the dominant political forces of the time, suspicious of foreign entanglements and central government activism ('Jeffersonians despite Jefferson', they might be called), viewing markets as natural phenomena not in need of political stimulation and protection. The onset of the Great Depression at first reinforced this mentality, encouraging a flurry of protectionist legislation and a search for economic salvation within the territorial economy of the country. Though the election of Franklin Delano Roosevelt as President in 1932 did lead to a reassertion of central government authority over the national economy, only the arrival of the Second World War truly rescued the American economy from its slump. The wartime economy encouraged an unprecedented federal direction of the economy, laid the basis for a military–industrial sector that the US (unlike most other industrial countries) had never really had and established as conventional wisdom the view that the Great Depression and the conditions leading to the war itself had been exacerbated, if not caused, by the competitive national protectionism of the 1930s.

The outcome of the Second World War had two consequences that deepened attachment in the United States to the idea of an America committed to a world in which international boundaries would have lessened significance. The first was the geographical penetration of the Soviet Union, representing a system of political–economic organization widely seen as antithetical to that of the US, into Germany and other parts of Western Europe. The second was the almost complete collapse of the industrialist–capitalist world except for the US, which meant that for the security of the American economy and society in the face of Soviet competition, the US government had to help rebuild the war-shattered economies of both erstwhile allies and enemies alike (Agnew and Corbridge, 1995).

Irrespective of the new ideological conflict with the Soviet Union, however, the United States political economy had three features that underpinned the internationalism of US government policy. One was that in almost every industry control over domestic markets was exercised by relatively few, but very large, firms. To expand production and profitability they could not simply depend upon such strategies as advertising and product differentiation ('n' types of washing powder or toothpaste) but needed to expand access to markets elsewhere, particularly where potential demand would be high, but competition low, i.e. in recently-devastated industrial countries. A second was a largely autonomous financial sector which, during the Great Depression, had been made independent of industrial corporations, that saw tremendous benefits in capturing the role of international mediator that British financial interests had been forced to relinquish over the previous 20 years. A third, and final, feature was a commitment to the 'Fordist' model of combining high mass production and mass consumption through paying relatively high wages to assembly-line workers and co-ordinating labour–management relations through a system of government-mandated regulations. This initially encouraged the 'export' of Fordism as a (new) part of the American ethos, if nothing other than to make sure that competitors for American firms did not face a more tractable labour market. Later, particularly after the late 1960s and as production costs increased, it encouraged the movement of labour-intensive manufacturing production away from the US.

Given American economic dominance within its sphere of influence after 1945 it is not surprising that it should have had a major impact on the workings of the world economy. But the American impact was not in the form of a mere recapitulation of the pre-war world economy based on rivalry between competing territorial-imperial blocs. Rather, it was something new. Abandoning territorial imperialism, 'Western capitalism [under American auspices] resolved the old problem of overproduction [which produced periodic massive downturns in economic growth], thus removing what Lenin believed was the major incentive for imperialism and war, (Calleo, 1987, p.147). The driving force behind this was the emergence of high mass consumption across the industrial world,

from North America to Western Europe and Japan. Major industries increasingly traded across these economies, setting up subsidiaries and generally reducing the significance of borders for production. By means of the Bretton Woods system of semi-fixed exchange rates, the US dollar provided the main currency for the world financial system from 1944 to 1973. Meanwhile, even as they expanded economically, Germany and Japan were tied to the US both by the fear of the Soviet Union and also by the military 'shield' that the US was seen as providing them.

The Cold War ended in 1989–91 with the collapse of the Soviet Union, as its model of development failed to deliver both military goods and a higher standard of living at the same time. But the virtuous circle (from an establishment American point of view) of American expansion feeding a globalizing world economy feeding the American economy had already come to a close with the American economic crisis of 1969–72. US producers had faced more effective foreign competition within its domestic market after the lowering of tariffs on many imports after 1963, the European Community provided the economies of scale for European producers heretofore enjoyed only by American ones, US military spending was a major drain on both the central government budget (partly because of the war in Vietnam) and Japanese and other foreign producers were becoming more technologically innovative than their American counterparts. The Bretton Woods system of exchange rates was abrogated unilaterally by the US in 1972 as a response to this new reality. It was replaced by a system of floating exchange rates so that the dollar's value against the currencies of other countries could be used more easily as an instrument for US economic-policy making; for example to lower the price of exports by lowering the value of the dollar against the currencies of destination markets. The net effect, however, has been to give more power to markets in major financial centres such as New York, London and Tokyo to set exchange rates and in other respects govern the world economy. The American response to crisis, therefore, was to remove one of the major instruments of government control over the world economy and create conditions for the further financial and productive globalization of that world economy.

THE EMPIRE STRIKES BACK

The expansionism inherent in the American experience has recently come into question in the United States. Elsewhere (including Canada), the benevolence of American omnipotence has long been openly problematic. What has focused minds has been the seemingly negative impact that the globalization brought by American global hegemony is now having at home. On one side the dramatically increased polarization of incomes and wealth between the rich and everyone else have been attributed by many commentators and some politicians to the effects of globalization, particularly the loss of the relatively high-paying assembly line jobs that characterized Fordist America (Agnew, 1994; O'Loughlin, 1997). It seems somewhat more than a coincidence of timing that US median male earnings peaked in 1973 and have stagnated ever since with a massive earnings gap opening up during the same period between the richest 1 per cent of the population (who garner their incomes largely from stocks and real estate) and everyone else (whose incomes come from work). Though unemployment in the US has been consistently lower than in Europe over most of this period, this difference is largely accounted for by the growth of part-time employment and the proliferation of low-paid consumer service jobs in the US relative to the old continent. From a different angle of vision, the point has been made that there is increasing pessimism on the part of Americans about achieving one of the fundamental 'promises' of the American historical experience: the high likelihood that your children will have a higher standard of living than you do. The generations coming to maturity after the Second World War and employed in the burgeoning Fordist industries of the epoch enjoyed continuing real growth in their incomes. From 1973 to 1989, however, this pattern was replaced by net declines in real incomes for production workers (Fig.

24.1). Now, this is not to suggest that this promise is necessarily a good thing in the context of growing concern about the sustainability of high levels of economic growth under conditions of environmental degradation, but only that the promise of increasing incomes is very much a part of the American ethos of expansionism. Its potential loss is of major cultural and political significance (see, for example, Newman, 1993; Dudley, 1994). It is this 'compact' between populace and expanding economy that has bonded so many Americans to the American model of economic development down the years.

There is a very marked regional aspect to the pattern of income decline and polarization. The Midwest region, particularly around the Great Lakes, has experienced the highest levels of job loss in the middle-income category, with much lower-income service sector jobs as the only available substitutes (see Chapter 23). Elsewhere, particularly in New England and on the West Coast, though business cycle effects produced downturns in the early 1990s, the general trend has been less negative. Overall, average income differences between US regions, after converging in the years since the 1930s, have started to diverge once more, suggesting the increased prospect of inter-regional political conflict over the course of the US economy as a whole. It is not surprising that the most fervent proponents of free trade tend to be from California and New England, with those counselling greater degrees of protectionism and opposing the development of more free trade accords with other countries tending to come from the Midwest and those parts of the South with the greatest to lose from a further globalization of labour markets by mobile capital (Phillips, 1991) (see Chapter 21).

A plausible account of the phenomena of income stagnation and fading promise would stress first of all the cutting of the Fordist knot that tied together production and consumption. The globalization of labour markets has meant that businesses without local markets are relatively free to move at will to wherever they can obtain the best 'deal'. Under such conditions, expansionism no longer guarantees a return for

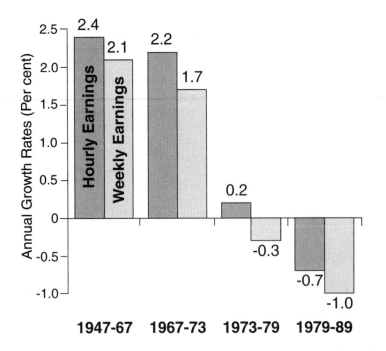

Fig. 24.1 Earnings growth for production workers in the US, 1947–89. Data from Mishel and Frankel (1991, p. 74)

most of those people left at home. Only those who earn their livings from investments are beneficiaries. At the same time this also discourages businesses from investing in capital, especially in productivity-enhancing equipment (Wolman and Colamosca, 1997, pp. 76–7). This accounts for the decline in capital/output ratios in the United States (a measure of the relative importance of labour and technology in production) since the early 1980s (Fig. 24.2). The capital/labour ratio measures the average amount of capital that is used to produce a given output. The rate of growth in the amount of capital available to each worker in the US has been falling. This suggests that the American economy is producing its goods and services in a more labour-intensive way than it did in the past. In other words, work has been substituted for capital; toil has increased as investment in capital equipment has declined. The long-term dilemma is that without sufficient capital investment median incomes will not rise and, without rising incomes, labour will not be able to purchase the products of globalized production. At first sight this might

sound surprising. For example, has there not been a vast investment in computer technology that has presumably improved productivity and reduced the burden on labour? In fact, considerable evidence suggests that computer technology has not produced the productivity gains widely predicted. Information overload, rapid obsolescence, the lack of impact of information processing in many industries and technical-interface conflicts have conspired to reduce the overall effect of the new technology in the US economy (see, for example, Blinder and Quandt, 1997.

Globalizing labour markets can be seen at work in two separate ways. One is in the increasing choice of locations for businesses with a wide range of labour-market requirements at different phases of production or attracted to diverse markets so as to spread investment risks. As countries 'liberalize' their economies, they reduce regulations and constraints on business practices. They thus impose fewer constraints (environmental regulations, labour-force restrictions, product liability claims, etc.) on mobile businesses. The other is in the recruitment and

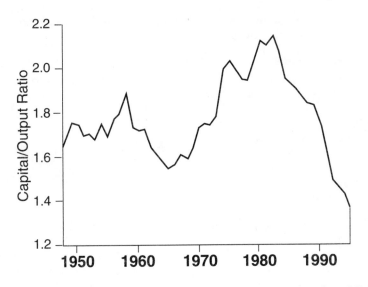

Fig. 24.2 The course of the capital/output ratio in the US, 1948–95. Data from Mishel and Frankel (1991, p. 69)

migration of both skilled and unskilled workers. In the first case movement happens both within transnational corporations and in sectors such as health care where contracts reflect, respectively, the needs of businesses and global labour shortages. The second case is the movement of large numbers of unskilled workers from poorer countries to fill jobs unattractive to locals in industrial and other wealthy economies. These trends not only have economic consequences, in terms of potentially reducing the bargaining power of resident workers, but also cultural and social impacts as societies such as the United States become increasingly multicultural with marked ethnic divisions of labour. Since 1990, however, there is evidence that international migration of both skilled and unskilled labour has declined whereas the movement of foreign direct investment has tended to increase. It is the first feature of globalizing labour markets, therefore, that is now more important relative to the US than the second one.

The increasing breakdown of the geographical matching of production and consumption has been exacerbated by increasing pressures on governments to facilitate 'market access' and to make 'their' businesses 'lean and mean' for the rigours of global

competition. There are fewer and fewer 'policy buffers' between countries and the global business cycle (Drache, 1996, p. 50). Even large and previously 'sheltered' economies such as that of the United States find themselves subject to economic shocks that are increasingly beyond the powers of central government to manage. At the same time, in order to conform to the discipline of financial markets, countries must restrict their welfare-state expenditures. Yet, as Rodrik (1997) has persuasively suggested, traditionally more open economies have had bigger relative welfare expenditures than more closed ones. The US with its minimalist welfare state (by industrial country standards) is faced, therefore, by a prospective social problem as welfare cutbacks parallel increased openness of the national economy.

Not surprisingly, the sense of increased economic fragility has given rise to a questioning of the conventional wisdom or American 'common sense' upon which the American ethos has long been based. One dimension of this in particular has come into question. This is the faith in the benefits of free trade. Increasingly heard are criticisms of a free-trade regime in which wages are set at the lowest cost location without attention to the collective consequences of expanding production without commensurate increases in the earnings capacity for global consumption. Now, this is a global as much as it is an American problem. It is just that Americans are no longer sheltered from its effects. Even though the United States is less open to and less dependent on foreign trade than any other major industrial country (Tables 24.1 and 24.2) its relative position has changed significantly in the years since 1960.

Of course, it is misleading to portray the United States and its population as simply one more set of 'victims' of globalization. In the first place, many Americans and some American localities have benefited mightily from the growth of trade and foreign investment. This is particularly the case for those workers and places with successful export firms and those benefiting from flows of foreign direct investment. Critically, for example, the growth of trade with Asia and the Pacific Rim as the Newly Developing Countries of those regions have inserted themselves in the circuits of globalized production has particularly benefited the western states of the US. Second, and well worth reiterating, the contemporary world economy is largely a product of US design and ideology. From its origins as a new state political–economic expansionism has lain at the heart of the American experience. In a position to realize its old ambitions after the Second World War, US government and business became enthusiastic supporters of a shrinking world in which capital would know no boundaries. Americans now also live with the consequences: a world much more open to trade, investment and cultural influences but also one of increased economic competition and insecurity.

MADISONIAN ENTROPY IN A GLOBALIZING WORLD

The American system of government, as set up under the influence of the ideas and compromises of the founders, has always been in tension with the centralized conduct of international relations. Not only has such a vast country in population and area had distinct regional economies with different identities and interests that had somehow to be managed collectively, the episodic yet undeniable trend towards an increasingly powerful central government has had to cope with a constitutional framework designed to frustrate the achievement of concentrated public power. The federal division of powers between the states and the central government and the separation of powers at the centre between the legislative, executive and judicial branches pose a challenge to the creation of any coherent response to the dilemmas facing the United States in a globalizing world. A governmental system set up to facil-

TABLE 24.1 Trade openness of major industrial economies (exports + imports as a percentage of GDP)

Country	1960	1972	1985	1990
Canada	33.0	44.4	52.3	60.4
United States	8.5	11.9	17.8	22.0
Japan	14.7	21.2	28.7	36.5
West Germany	28.1	43.1	66.2	76.3
France	22.6	36.3	45.1	52.6
Italy	22.5	42.1	43.6	51.0
United Kingdom	42.9	53.0	56.3	62.6
Spain	14.7	29.9	36.1	45.1
Portugal	41.5	60.9	78.7	112.1

Source: OECD National Accounts, Vol. 2, detailed table, Paris, 1979 and 1992

TABLE 24.2 Trade dependence of the major industrial economies (exports as a percentage of GDP)

Country	1960	1972	1985	1990
Canada	17.2	22.0	28.4	29.2
United States	5.2	5.8	7.1	10.5
Japan	10.7	10.6	14.6	18.1
West Germany	19.0	20.9	32.4	39.7
France	14.5	16.7	23.9	25.2
Italy	13.0	17.7	22.8	23.8
United Kingdom	20.9	21.8	29.1	29.4
Spain	10.2	14.6	20.1	19.6
Portugal	17.3	27.2	37.3	47.3

Source: OECD National Accounts, Vol. I, main aggregates 1960–88, Paris, 1990

itate the expansion of private economic interests, yet restrict the exercise of collective political power, has built-in disadvantages for dealing with adjustments to a new world economy in which it is no longer the singular centre.

Philip Cerny has provocatively referred to this as the problem of 'Madisonian entropy' in the face of external challenges to American economic well-being. By this he means that the 'US system of government is characterized by a great deal of energy which is absorbed or dissipated through the internal workings of the structure, and which is unavailable for the policy tasks which modern states must perform' (1989, p. 54). Entropy in physics refers to the measure of energy within a system unavailable for work. Madisonian signifies the dominant influence of James Madison on the final form taken by US institutions at the time of ratification of the US Constitution.

In Cerny's view the US geopolitical position during the Cold War was based on three institutional developments within the US government that went around the blockages built into the system. But, in the absence of these developments, the institutional fragmentation characteristic of US government will return to frustrate the capacity to rescue the country from internal conflicts through external expansion. This has been the leitmotif of American foreign policy from the late eighteenth century to the present but achieving realization only after the Second World War. The first was the development of the 'imperial presidency' in which the President emerged as a monarch-like figure, short-circuiting the other branches of government because of the imminent threat of nuclear war and the need to plan rapid responses to foreign crises. The second was the long-term maintenance of a cross-party consensus about most major foreign-policy issues with congressional deference to presidential authority over foreign affairs. Third, and finally, was the superior capacity of the federal executive branch to manage trade and monetary issues without interference from the legislative branch or the states.

Each of these has been undermined at the same time as globalization has increased. The Vietnam War began the erosion of both the imperial presidency and bipartisan consensus between the Democratic and Republican parties over the conduct of foreign affairs. The Watergate scandal in the 1970s and the Contra scandal in the 1980s further undermined presidential authority to carry out foreign policy initiatives unimpeded by Congress. Weakened parties in which politicians increasingly represent geographical and sectoral constituencies have retreated from the consensus that marked US foreign policy in the 1950s and 1960s. Though the US dollar has retained considerable importance in the post-Bretton Woods system of floating exchange rates, its liberation by the US government in 1972 removed the central plank from the international stage upon which the domestic financial power of the federal executive branch rested. Markets have replaced central-governmental power in this critical area. Increasingly, the various states are also striking out on their own to attract external investment, pursue industrial and trade policies and place limits (for example, with respect to immigration controls) in areas traditionally reserved to the federal government. The state of California, for example, which on its own constitutes the world's seventh largest economy and has 10 trade representative offices around the world from Mexico City to Tokyo, has tried to create its own immigration policy; it has its own welfare and health care systems for the poor, and passes laws restricting investments by the state in other countries and protesting US policies on human rights abroad (e.g. Lasher, 1998). With states asserting their powers and given the balance of regional interests in Congress, the achievement of a national industrial policy or coherent national-level response to the globalization of labour markets seems next to impossible to imagine. As Cerny concludes:

> The United States may not be a 'stalemate society', to borrow Stanley Hoffman's well-known description of France under the Third Republic. Indeed, American society is, as always, energetic and creative on many levels. What absorbs and dissipates that energy, however, is the complexity, duplication and cross-purposes which were not only built into the Madisonian system, but which have also been increasing faster than the problem-solving capacity of the system, especially where international interdependence is concerned. The US suffers from a kind of dynamic immobilism, creative stagnation or stalemated superstructure, which I have called Madisonian entropy (1989, p. 59).

CONCLUSION

The US commitment to external expansion to manage conflicts at home and serve the com-

mercial ethos of dominant social groups has come back to haunt its creators. The US now appears as an 'ordinary' country subject to external pressures more than simply the source of pressure on others. The US relationship to the external world, therefore, has changed profoundly since the days of Madison and Jefferson. Moving from a peripheral to a central position within the world geopolitical order, the US brought to that position its own ethos. Its absolute economic predominance and a number of key institutional changes produced in the critical years during and after the Second World War allowed the successful projection of the American ethos beyond its shores. Expansion seemed to produce not only economic but also political returns for the American system. The benefits proved more ephemeral for the majority of the American population than ever was contemplated. Globalization of the world economy under American auspices has shifted control over parts of the American economy to ever more distant seats of power. The increasing fragmentation and entropy of the American governmental system suggest that regaining control will be an almost impossible task. Many Americans consequently find themselves with the increased sense of insecurity and loss of control that others have long experienced. This is doubly felt in the American case, however. From its origin as a country the promise of America has been that through expansion almost everyone would benefit. That they no longer do calls into question the very system that produced that promise in the first place.

KEY READINGS

Agnew, J.A. and Corbridge, S. 1995: *Mastering space: hegemony, territory and international political economy.* London: Routledge.
Engelhardt, T. 1995: *The end of victory culture: Cold War America and the disillusioning of a generation.* New York: Basic Books.
Europa 1998: *The USA and Canada 1998.* London: Europa Publications.
Lazare, D. 1996: *Frozen republic: how the constitution is paralyzing democracy.* New York: Harcourt Brace.

Nikovitch, F. 1994: *Modernity and power: a history of the domino theory in the twentieth century*. Chicago: University of Chicago Press.

Perlmutter, A. 1997: *Making the world safe for democracy: a century of Wilsonianism and its totalitarian challengers*. Chapel Hill: University of North Carolina Press.

Trubowitz, P. 1998: *Defining the national interest: conflict and change in American foreign policy*. Chicago: University of Chicago Press.

Williams, W.A. 1969: *The roots of the modern American empire: a study of the growth and shaping of social consciousness in a marketplace society*. New York: Random House.

REFERENCES

Agnew, J.A. 1987: *The United States in the world economy: a regional geography*. Cambridge: Cambridge University Press.

Agnew, J.A. 1994: Global hegemony versus national economy: the United States in the new world order. In Demko, G.J. and Wood, W.B. (eds), *Reordering the world: geopolitical perspectives on the twenty-first century*. Boulder: Westview Press, 269–79.

Agnew, J.A. and Corbridge, S. 1995: *Mastering space: hegemony, territory, and international political economy*. London: Routledge.

Aiken, C.S. 1998: *The cotton plantation South since the Civil War*. Baltimore: Johns Hopkins University Press.

Aleinikoff, A.T. 1998: A multicultural nationalism? *The American Prospect* 36, 80–6.

Allen, F. 1996: *Atlanta rising: the invention of an international city*. Atlanta: Longstreet Press.

Angell, H.M. 1996: Demographic shifts. *The Globe and Mail*, 5 June, A16.

Anon. 1976: *This land is our life*. Yellowknife: n.p.

Artibise, A. 1995: Achieving sustainability in Cascadia: an emerging model of urban growth management in the Vancouver–Seattle–Portland Corridor. In Kresl, P. and Gappert, G. (eds), *North American cities and the global economy: challenges and opportunities*. Urban Affairs Annual Review **44**, 221–50.

Bantje, R. 1992: Improved earth: travel on the Canadian Prairies, 1920–1950. *Journal of Transport History* **13**(2), 115–40.

Barr, B.M. and Lehr, J.C. 1982: The western interior: the transformation of a hinterland region. In McCann, L.D. (ed.), *A geography of Canada: hinterland and heartland*. Scarborough, Ontario: Prentice Hall Canada.

Barrett, C. 1996: *Special briefing – Canada and the future of the Asia Pacific economy: unending boom?* Ottawa: The Conference Board of Canada.

Bartley, N.V. 1995: *The new South: 1945–1980*. Baton Rouge: Louisiana University Press.

Bathelt, H. 1991: Employment changes and input–output linkages in key technology industries: a comparative analysis. *Regional Studies* **25**, 31–43.

Baumol, W.J. 1967: Macroeconomics of unbalanced growth: the anatomy of urban crisis. *The American Economic Review* **57**(3), 415–26.

Bayor, R.H. 1996: *Race and the shaping of twentieth century Atlanta*. Chapel Hill: University of North Carolina Press.

Beauregard, R.A. 1993: *Voices of decline*. Cambridge, MA: Blackwell.

Bell, D. 1973: *The coming of the post-industrial society*. New York: Basic Books.

Berger, T.R. 1977: *Northern frontier, northern homeland: the report of the Mackenzie Valley Pipeline Inquiry, Vols I and II*. Ottawa: Department of Supply and Services.

Berry, B.J.L. (ed.) 1976: *Urbanization and counter-urbanization*. Beverly Hills: Sage Publications.

Berry, M.J., Jacoby, W.R., Niblett, E.R. and

Stacey, R.A. 1971: A review of geophysical studies in the Canadian Cordillera. *Canadian Journal of Earth Sciences* **8**, 788–801.

Best, M. 1990: *The new competition: institutions of industrial restructuring.* Cambridge, MA: Harvard University Press.

Beyea, J. 1990: Long-term threats to Canada's James Bay from hydroelectric development. *Information North* **16**(3), 1–7.

Beyers, W.B. 1991: Trends in the producer services in the USA: the last decade. In Daniels, P.W. (ed.), *Services and metropolitan development: international perspectives.* London: Routledge, 146–72.

Beyers, W.B. 1992: Producer services and metropolitan growth and development. In Mills, E.S. and McDonald, J.F. (eds), *Sources of metropolitan growth.* New Brunswick, NJ: Center for Urban Policy Research, 125–46.

Beyers, W.B. 1998: Trends in producer service employment in the United States: the 1985–1995 experience. A paper presented at the 1998 North American Regional Science Meetings, Santa Fe, New Mexico.

Beyers, W.B. and Alvine, M.J. 1985: Export services in postindustrial society. *Papers of the Regional Science Association* **57**, 33–45.

Beyers, W.B. and Lindahl, D.P. 1996a: Explaining the demand for producer services: is cost-driven externalization the major factor? *Papers in Regional Science* **75**(3), 351–74.

Beyers, W.B. and Lindahl, D.P. 1996b: Lone eagles and high fliers in rural producer services. *Rural Development Perspectives* **12**(3), 2–10.

Beyers, W.B. and Lindahl, D.P. 1998: *Services and the new economic landscape.* Vienna: European Regional Science Association.

Birdsall, S.S. and Florin, J.W. 1992: *Regional landscapes of the United States and Canada.* New York: Wiley.

Blakely, E. and Snyder, M.G. 1997: *Fortress America.* Washington, DC: Brookings Institution Press.

Blinder, A.S and Quandt, R.E. 1997: The computer and the economy. *The Atlantic Monthly,* December, 26–32.

Bliss, M. 1994: *Right honourable men: the descent of Canadian politics from Macdonald to Mulroney.* Toronto: Harper Collins.

Bluestein, H.B. 1993: *Synoptic-dynamic meteorology in midlatitudes, Vol. II: observations and theory of weather systems.* New York: Oxford University Press.

Bogue, D. 1953: *Population growth in Standard Metropolitan Areas, 1900–1950.* Washington, DC: Housing and Home Finance Agency.

Bohannan, P. 1967: *Introduction.* In Bohannan, P. and Ploy, F. (eds), *Beyond the frontier.* Garden City, NY: Natural History Press.

Bone, R.M. 1992: *The geography of the Canadian North: issues and challenges.* Toronto: Oxford University Press.

Bonnanno, A., Busch, L., Friedland, W.H., Gouveia, L. and Mingulone, E. (eds) 1994: *From Columbus to Conagra: the globalization of agriculture and food.* Lawrence: University of Kansas Press.

Borchert, J. 1991: Futures of American cities. In J.F. Hart (ed.), *Our changing cities.* Baltimore: Johns Hopkins University Press, 218–50.

Bourhis, R.Y. (ed.) 1984: *Conflict and language planning in Québec.* Clevedon, Philadelphia: Multilingual Matters.

Bourne, L.S. 1989: Are new urban forms emerging? Empirical tests for Canadian urban areas. *The Canadian Geographer/Le Géographe Canadien* **33**, 312–28.

Bourne, L.S. 1993: Close together and worlds apart. An analysis of changes in the ecology of income in Canadian cities. *Urban Studies* **30**, 1293–1317.

Bourne, L.S. and Flowers, M. 1996: *The Canadian urban system revisited: a statistical analysis.* Research Paper 192. University of Toronto: Centre for Urban and Community Studies.

Bourne, L.S. and Olvet, A. 1995: *New urban and regional geographies in Canada.* Major Report 33. University of Toronto: Centre for Urban and Community Studies.

Bradbury, J.H. and St-Martin, I. 1983: Winding-down in a Québec mining town: a case study of Schefferville. *The Canadian Geographer/Le Géographe Canadien* **27**(2), 128–44.

Bramham, D. 1996: The capital of capital. In Pierce, J.K. *et al., Cascadia: a tale of two cities, Seattle and Vancouver, BC.* New York: Harry N. Abrams, 152–7.

Brewis, T.N. 1969: *Regional development in Canada in historical perspective.* Toronto: Macmillan.

Britton, J.N.H. and Gilmour, J.M. 1978: *The weakest link: a technological perspective on Canadian industrial underdevelopment*. Background Study 43. Ottawa: Science Council of Canada.

Broadway, M.J. 1998: Where's the beef? The integration of the Canadian and American beefpacking industries: *Prairie Forum*, **23** (1), 19–29.

Broadway, M.J. and Ward, T. 1990: Recent changes in the structure and location of the US meatpacking industry. *Geography* **75**, 76–9.

Brown, R.H. 1948: *Historical geography of the United States*. 2nd edn, New York: Harcourt, Brace, & World, Inc.

Browne, W.P. 1988: *Private interests, public policy and American agriculture*. Lawrence: University of Kansas Press.

Browning, H.C. and Singlemann, J. 1975: *The emergence of a service society*. Springfield: National Technical Information Service.

Bryant, C.R. and Johnston, T.R.R. 1992: *Agriculture in the city's countryside*. London: Belhaven.

Bunting. T. and Filion, P. (eds) 2000: *Canadian cities in transition*. Toronto: Oxford University Press.

Bureau of Statistics 1997: *Statistics quarterly*, December 1996. Yellowknife: Government of Northwest Territories.

Butala, S. 1990: Field of broken dreams. *West* **2**(6), 30–9.

California Department of Food and Agriculture 1997: *California agricultural resource directory*. Sacramento: Department of Food and Agriculture.

Callenbach, E. 1975: *Ecotopia: the notebooks and reports of William Weston*. Berkeley: Banyan Tree Books.

Calleo, D.P. 1987: *Beyond American hegemony: the future of the western alliance*. New York: Basic Books.

Campbell, T. 1996: Co-management of aboriginal resources. *Information North* **22**(1), 1–6.

Canada's Future Forest Alliance 1993: *Brazil of the North*. New Denver.

Canadian Broadcasting Corporation 1996: *Special feature: 'Paradise Lost'*. CBC Evening News, 6 November.

Card, D. and Freeman, C.B. 1993: *Small differences that matter: labor markets and income*

maintenance in Canada and the United States. Chicago: University of Chicago Press.

Careless, J.M.S. 1979: Metropolis and region: the interplay between city and region in Canadian history before 1914. *Urban History Review* **78**(3), 99–118.

Carlson, A.W. 1994: America's new immigration: characteristics, destinations and impact. *Social Science Journal* **31**, 213–36.

Carlyle, W.J. 1991: Rural change in the Prairies. In Robinson, G.M. (ed.), *A social geography of Canada*. Toronto and Oxford: Dundurn Press, 330–58.

Carney, G.O. (ed.) 1994: *The sounds of people and places. A geography of American folk and popular music*. 3rd edn, Lanham, MD: Rowman & Littlefield.

Carney, G.O. (ed.) 1995: *Fast food, stock cars and rock-n-roll: place and space in American pop culture*. Lanham, MD: Rowman & Littlefield.

Cary, A. 1898: How to apply forestry to spruce lands. *Paper Trade Journal* **27**, 11 February, 157–62.

Castells, M. 1996: *The information age: economy, society, and culture, Vol. I: the rise of the network society*. Cambridge, MA: Blackwell Publishers.

Cerny, P.G. 1989: Political entropy and American decline. *Millennium* **18**, 47–63.

Cernetig, M. 1997: Salmon fishery caught in current of change. *The Globe and Mail*, 16 August, A1.

CF/IRRP 1990: *Great Lakes, great legacy*. Washington, DC: The Conservation Foundation, and Montréal: Institute for Research on Public Policy.

Chinitz, B. 1991: A framework for speculating about future growth patterns in the US. *Urban Studies* **28**(6), 939–59.

Chiotti, Q. 1992: Sectoral adjustments in agriculture: dairy and beef livestock industries in Canada. In Bowler, I., Bryant, C. and Nellis, M.D. (eds), *Contemporary rural systems in transition, Vol. I: agriculture and environment*. Wallingford: CAB International, 43–57.

Churchill, W.S. 1958: *A history of the English-speaking peoples, Vol. IV: the great democracies*. London: Cassell.

Citizenship and Immigration Canada 1998: A profile of immigrants in Canada (/profile/9608imme.html); facts and figures

1996: immigration overview (/facts96/ 1h.html). Information found at: *http://cicnet. ci.gc.ca/ english/ref*

Clark, J. 1996: Public opinion rules and sadly it is often ill-informed. *The Globe and Mail*, 11 March, A13.

Clawson, M. 1981: Natural resources of the Great Plains in historical perspective. In Lawson, M.P. and Baker, M.E. (eds), *The Great Plains: perspectives and prospects*. Lincoln: University of Nebraska Press, 3–10.

Clay, G. 1973: *Close-up: how to read the American city*. Chicago: University of Chicago Press.

Coates, K. and Powell, J. 1989: *The modern North*. Toronto: James Lorimer.

Coates, K., and Morrison, W. 1992: *The forgotten North: its nature and prospects*. Toronto: James Lorimer.

Coffey, W. 1994: *The evolution of Canada's metropolitan economies*. Montréal: The Institute for Research on Public Policy.

Coffey, W. and Polèse, M. 1991: Culture barriers to the location of producer services: the Montréal–Toronto rivalry and the limits to polarization. *Canadian Journal of Regional Science* 14, 433–46.

Coffey, W. and Shearmur, R.G. 1997: The growth and location of high order services in the Canadian urban system, 1971–1991. *The Professional Geographer* 49(4), 404–17.

Cohen, S.J. (ed.) 1997: *Mackenzie Basin impact study (MBIS): final report*. Environment Canada, Ottawa: Ministry of Supply and Services.

Cohn, T. and Smith, P. 1995: Developing global cities in the Pacific Northwest: the cases of Vancouver and Seattle. In Kresl, P. and Gappert, G. (eds), *North American cities and the global economy: challenges and opportunities*. Urban Affairs Annual Review 44, 251–85.

Cole, S. 1991: Indicators of regional integration and the Canada–US Free Trade Agreement. *Canadian Journal of Regional Science* 13, 171–8.

Committee on the Role of Alternative Farming Methods in Modern Production Agriculture 1989: *Alternative agriculture*. Washington: National Resource Council.

Conzen, M.P. (ed.) 1990: *The making of the*

American landscape. Boston, MA: Unwin Hyman.

Cosgrove, D. 1984: *Social formation and symbolic landscape*. Ottawa: Barnes & Nobles Books.

Courchene, T. and Telmer, C. 1998: *From heartland to North American region state*. Toronto: University of Toronto Press.

Crawford J. (ed.) 1992: *Language loyalties: a source book on the official English controversy*. Chicago: University of Chicago Press.

Cronon, W. 1983: *Changes in the land: Indians, colonists, and the ecology of New England*. New York: Hill & Wang.

Cronon, W. 1991: *Nature's metropolis*. New York: W.W. Norton.

Cronon, W. 1995: The trouble with wilderness; or, getting back to the wrong nature. In Cronon, W. (ed.), *Uncommon ground: toward reinventing nature*. New York: W.W. Norton, 69–90.

Crosby, A. 1972: *The Columbian exchange: biological and cultural consequences of 1492*. Westport: Greenwood.

Crosby, A. 1986: *Ecological imperialism: the biological expansion of Europe 900–1900*. Cambridge: Cambridge University Press.

Damas, D. (ed.) 1984: *Handbook of North American Indians, Vol. V: Arctic*. Washington, DC: Smithsonian Institution.

Davies, W.K.D. and Murdie, R.A. 1993: Measuring the social ecology of cities. In Bourne, L.S. and Ley, D.F. (eds),*The changing social geography of Canadian cities*. Montréal and Kingston: McGill-Queen's University Press.

Davis, M. 1990: *City of quartz: excavating the future in Los Angeles*. New York: Verso.

de Bres, K.J., Kromm, D.E. and White, S.E. 1993: The buffalo commons debate. *Focus* 43(4), 16, 20, 320.

deLancey, D. 1985: Trapping and the aboriginal economy. *Information North* Winter, 5–12.

Demeritt, D. 1998: Science, social constructivism, and nature. In Castree, N. and Willems-Braun, B. (eds), *Remaking reality: nature at the millennium*. New York: Routledge, 177–97.

de Vries, J. 1994: Canada's official language communities: an overview of the current demolinguistic situation. *International Journal of the Sociology of Language* **105/106**, 37–68.

Diamond, J. 1997: *Guns, germs and steel.* London: Vintage.

Dickerson, M.O. and McCullough, K.M. 1993: Nunavut ('Our Land'). *Information North* **19**(2), 1–7.

Dickson, C.S. 1983: Notes for an address: skippers' perspective of Atlantic transportation subsidies. Transportation Seminar Series, University of New Brunswick. Fredericton: Atlantic Provinces Transportation Commission, mimeo.

Dietrich, W. 1992: *The final forest: the battle for the last great trees of the Pacific Northwest.* New York: Penguin.

Dinnerstein, L., Nichols, R.L. and Reimers, D.M. 1996: *Natives and strangers: a multicultural history of Americans.* New York: Oxford University Press.

Dockery, D.W., Pope, C.A., Xu, X. *et al.* 1993: An association between air pollution and mortality in six US cities. *New England Journal of Medicine* **329**, 1753–9.

Dowd, M. 1996: Everyone's too jumped up on caffeine to be laid-back in Seattle. *The Vancouver Sun*, 10 August, A19.

Downs, A. 1994: *New visions for metropolitan America.* Washington, DC: The Brookings Institution.

Doyle, D.H. 1990: *New men, new cities, new South: Atlanta, Nashville, Charleston, Mobile.* Chapel Hill: University of North Carolina Press.

Drache, D. 1996: From Keynes to K-mart: competitiveness in a corporate age. In Boyer, R. and Drache, D. (eds), *States against markets: the limits of globalization.* London: Routledge, 31–61.

Dudley, K.M. 1994: *The end of the line: lost jobs, new lives in postindustrial America.* Chicago: University of Chicago Press.

Duerden, F. 1992: A critical look at sustainable development in the Canadian North. *Arctic* **45**(3), 219–25.

Duff, P.McL.D. 1993: *Holmes' principles of physical geology.* 4th edn, London: Chapman & Hall.

Dunn, C. 1995: *Canadian political debates: opposing news on issues that divide Canada.* Toronto: McClelland & Stewart, Inc.

Easterlin, R.A. 1980: Immigration: economic and social characteristics. In Thernstrom, S. (ed.), *Harvard encyclopaedia of American ethnic groups.* Cambridge, MA: Harvard University Press, 476–86.

Eckstrom, F.H. 1926: History of the Chadwick Survey from Fort Pownall in the District of Maine to the Province of Québec in Canada in 1764. *Sprague's Journal of Maine History* **14**, 62–89.

Eckstrom, F.H. 1941: *Indian place names of the Penobscot Valley and the Maine Coast.* Orono: University of Maine Press.

Economist, The 1994a: Buffalo-ranching: back to the frontier. 30 April, 30–1.

Economist, The 1994b: Welcome to Cascadia. 21 May, 52.

Edgington, D. 1995: Trade, investment and the new regionalism: Cascadia and its economic links with Japan. *Canadian Journal of Regional Science* **18** (3), 333–56.

Edgington, D. and Hayter, R. 1997: International trade, production chains and corporate strategies: Japan's timber trade with British Columbia. *Regional Studies* **31**(2), 151–66.

Eichenlaub, V.L. 1979: *Weather and climate of the Great Lakes region.* Notre Dame: University Press.

Ennals, P. and Holdsworth, D. 1988: The cultural landscape of the Maritime provinces. In Day, D. (ed.), *Geographical perspectives on the Maritime provinces.* Halifax, Nova Scotia: Saint Mary's University.

Espenshade, E.B. (ed.) 1995: *Goode's world atlas.* 19th edn, Chicago: Rand McNally.

Fainstein, S., Gordon, I. and Harloe, M. 1992. *Divided cities: New York and London in the contemporary world.* Cambridge, MA: Blackwell.

Farley, R., Schuman, H., Bianchi, S., Colasanto, D. and Hatchett, S. 1978: Chocolate city, vanilla suburbs: will the trend toward racially separate communities continue? *Social Science Research* **7**, 319–44.

Ferguson, N. 1998: *Virtual history: alternatives and counterfactuals.* London: Macmillan.

Fisher, J.S. 1973: The Piedmont: old and new. In Heyl, R.J. (ed.), *The South: a vade mecum.* Atlanta: Association of American Geographers, 75–85.

Fong, E. 1996: A comparative perspective on racial residential segregation: American and Canadian experiences. *The Sociological Quarterly* **37**, 199–226.

Forbes, E.R. 1979: *The Maritime Rights Movement, 1919–1927*. Montréal: McGill-Queen's University Press, 149–72.

Fortune 1998: Fortune 500 largest US Corporations. April 27, F1–20.

Fournier, E.J. and Risse, L.M. 1996: Cotton returns to the South: evidence from Georgia. *The Southeastern Geographer* **36**, 207–14.

Frazier, I. 1990: *Great Plains*. New York: Penguin Books.

French, H.M. and Slaymaker, O. 1993: *Canada's cold environments*. Montréal and Kingston: McGill-Queen's University Press.

Frey, W. 1995: The new geography of population shifts. In Farley, R. (ed.), *State of the union: America in the 1990s, Vol. XI*. New York: Russell Sage Foundation, 271–334.

Frey, W. and Liaw, K-L 1998: Immigrant concentration and domestic migrant dispersal: is movement to nonmetropolitan areas white flight? *Professional Geographer* **50**(2), 215–32.

Frey, W. and O'Hare, W. 1993: Vivan los suburbios. *American Demographics*, April, 30–7.

Frey, W. and Speare, A. Jr. 1992: The revival of metropolitan growth in the US. *Population and Development Review* **18**(1), 129–46.

Frisken, F. (ed.) 1994: *The Canadian metropolis: a public policy perspective*. 2 vols. Toronto: Canadian Urban Institute, and Berkeley: University of California.

Gad, G. 1995: The major cities of Ontario in the context of the Canadian heartland. Paper presented at International Seminar on Québec–Ontario, Mexico City.

Garreau, J. 1981: *The nine nations of North America*, Boston: Houghton Mifflin Co.

Garreau, J. 1991: *Edge city: life on the new frontier*. New York: Doubleday.

Gentilcore, R.L., Measner, D., Walder, R.H., Matthews, G.J. and Moldofsky, B.1993: *Historical atlas of Canada, Vol. II: the land transformed, 1800–1891*. Toronto: University of Toronto Press.

Geological Survey of Canada 1969: Tectonic map of Canada: Map 1251A, 1:50 000. Ottawa: Geological Survey of Canada.

Getis, A. and Getis, J. (eds) 1995: *United States and Canada: the land and the people*. Dubuque, IA: Wm. C. Brown Publishers.

Gibbins, R. 1994: *Conflict and unity: an introduction to Canadian political life*. Scarborough, Ontario: Nelson Canada Ltd.

Gill, D. and Cooke, A.D. 1974: Controversies over hydroelectric developments in sub-Arctic Canada. *Polar Record* **17**(107), 109–27.

Gillis, R.P. and Roach, T.R. 1986: *Lost initiatives: Canada's forest industries, forest policy, and forest conservation*. Westport: Greenwood Press.

Globe and Mail 1987: Can$2 billion for western Canada. 5 August, A8.

Goddard, E.N., Billings, M.P., Levorsen, A.I. et al. 1965: *Geologic map of North America*. Washington, DC: United States Geological Survey.

Goldberg, M. and Mercer, J. 1986: *The myth of the North American city*. Vancouver: University of British Columbia Press.

Goodacre, A.K., Grieve, R.A.F. and Halfpenny, J.F. 1987: Bouguer gravity anomaly map of Canada. *Canadian Geophysical Atlas*, Map 3.

Goodman, D., Sorj, B. and Wilkinson, J. 1987: *From farming to biotechnology*. New York: Basil Blackwell.

Goodman, L.R. 1996: *The economic health of North Dakota*. Grand Forks: LRG Properties Ltd with the University of North Dakota Alumni Association.

Gorrie, P. 1990: The James Bay power project. *Canadian Geographic* **110**(1), 20–31.

Gottmann, J. 1961: *Megalopolis: the urbanized northeastern seaboard of the United States*. New York: Twentieth Century Fund.

Gottmann, J. 1979: Office work and the evolution of cities. *Ekistics* **274**, 4–7.

Gottschalk, P. and Smeeding, T. 1997: Cross-national comparisons of earnings and income inequality. *Journal of Economic Literature* **35**, 633–87.

Government of Canada 1991: *The state of Canada's environment*. Ottawa: Ministry of the Environment.

Grass, E. and Hayter, R. 1989: Employment change during recession: the experience of forest product manufacturing plants in British Columbia, 1981–1985. *Canadian Geographer/Le Géographe Canadien* **3**, 240–52.

Gray, S.L. 1995: *A descriptive forest inventory of*

Canada's forest regions. Information Report PI-X-122. Chalk River, Ontario: Petawawa National Forestry Institute.

Great Northern Paper Company n.d.: *Map of the West Branch region: the working forest of Great Northern Paper*. Millinocket: Bowater-Great Northern Paper.

Greater Vancouver Regional District 1974: *The Livable Region Plan*. Vancouver: Greater Vancouver Regional District.

Greeley, W.B. 1925: The relation of geography to timber supply. *Economic Geography* **1**, 1–14.

Greene, R. 1997: Chicago's new immigrants, indigenous poor, and edge cities. *Annals of the American Academy of Political and Social Science* **551**, 178–90.

GTA 1996: *Report of the Greater Toronto Area Task Force* [The Golden Commission]. Toronto: Queen's Printer for Ontario.

Gwynn, R. 1995: *Nationalism without walls: the unbearable lightness of being Canadian*. Toronto: McClelland & Stewart, Inc.

Hallberg, M.C., Spitze, R.G.F. and Ray, D.E. (eds) 1994: *Food, agriculture and rural policy into the twenty-first century*. Boulder: Westview Press.

Hamelin, L-E. 1972: L'écoumène du Nord Canadien. In Wonders, W.C. (ed.), *The North: studies in Canadian geography*. Toronto: University of Toronto Press, 25–40.

Hamelin, L-E. 1979: *Canadian nordicity: it's your North too*. Barr, W. (trans.). Montréal: Harvest House.

Hamers, J.F. and Hummel, K.M. 1994: The francophones of Québec: language policies and language use. *International Journal of the Sociology of Language* **105/106**, 127–52.

Hanratty, M. 1992: Why Canada has less poverty. *Social Policy* **23**, 32–7.

Haraway, D. 1992: The promises of monsters: a regenerative politics for inappropriate/d others. In Grossberg, L., Nelson, C. and Treichler, P.A. (eds), *Cultural studies*. New York: Routledge, 295–337.

Hardy, R.L. 1971: Multiquadric equations of topography and other irregular surfaces. *Journal of Geophysical Research* **76**, 1905–15.

Harmon, R.L. 1996: *Reinventing the business: preparing today's enterprise for tomorrow's technology*. New York: The Free Press.

Harrington, J.W., MacPherson, A.D. and Lombard, J. 1991: Interregional trade in producer services: review and synthesis. *Growth and Change* **22**(4), 75–94.

Harris, R.C. and Matthews, G.J. 1987: *Historical atlas of Canada, Vol. I: from the beginning to 1800*. Toronto: University of Toronto Press.

Harris, R.C. and Warkentin, J. 1974: *Canada before confederation*. New York: Oxford University Press.

Harrison, B. 1984: Regional restructuring and 'good business climates': the economic transformation of New England since World War II. In Sawers, L. and Tabb, W.K. (eds), *Sunbelt/snowbelt: urban development and regional restructuring*. Oxford: Oxford University Press, 48–96.

Hart, J.F. 1976: *The South*. New York: D. Van Nostrand Co.

Hart, J.F. (ed.) 1991: *Our changing cities*. Baltimore: Johns Hopkins University Press.

Hart, J.F. and Mayda, C. 1998: The industrialization of livestock production in the United States. *Southeastern Geographer* **38**, 79–87.

Harvey, T. 1996: Portland Oregon: regional city in a global economy. *Urban Geography* **17**(1), 95–114.

Hays, S.P. 1959: *Conservation and the gospel of efficiency: the Progresssive Conservation Movement*. Cambridge, MA: Harvard University Press.

Hayward, D.J. and Erickson, R.A. 1995: The North American trade of US states: a comparative analysis of industrial shipments, 1983–1991. *International Regional Science Review* **18**, 1–32.

Heath, S.B. 1992: Why no official language? In Crawford, J. (ed.) *Language loyalties: a source book on the official English controversy*. Chicago: University of Chicago Press, 20–31.

Helm, J. 1981: *Handbook of North American Indians, Vol. VI: Subarctic*. Washington, DC: Smithsonian Institution.

Henderson-Sellers, A., and Robinson, P.J. 1986: *Contemporary climatology*. Harlow: Longman.

Hibbard, J.P., van Staal, C.R. and Cawood, P.A. (eds) 1995: *Current perspectives in the Appalachian–Caledonian Orogen*. Geological Association of Canada Special Paper 41.

Higgins, B. 1986: *The rise and fall? of Montréal.* Moncton: Institut Canadien de Recherche sur la Développement Régional.

Highan, J. 1992: Crusade for Americanization. In Crawford, J. (ed.) *Language loyalties: a source book on the official English controversy.* Chicago: University of Chicago Press, 72–84.

Hilliard, S.B. 1987: A robust new nation, 1783–1820. In Mitchell, R.D. and Groves, P.A. (eds), *North America: the historical geography of a changing continent.* London: Hutchinson, 149–71.

Hilliard, S.B. 1984: *Atlas of antebellum southern agriculture.* Baton Rouge: Louisiana State University Press.

Hogan, W.T. 1987: *Minimills and integrated mills: a comparison of steelmaking in the United States.* Lexington: D.C. Heath.

Holmes, J. 1992: The continental integration of the North American automobile industry: from the Auto Pact to the FTA and beyond. *Environment and Planning A* **24**, 33–48 and 95–119.

Horkheimer, M. 1947: *Eclipse of reason.* New York: Oxford University Press.

Houghton, J.T., Meira Filho, L.G., Callander, B.A., Harris, N., Kattenberg, A. and Maskell, K. 1996: *Climate change 1995: the science of climate change. The contribution of Working Group 1.* Cambridge: Cambridge University Press.

Howell, D.G. 1995: *Principles of terrane analysis.* Topics in the Earth Sciences 8. 2nd edn, New York: Chapman & Hall.

Hudson, J.C. 1979: The plains country town. In Blouet, B.W. and Luebke, F.C. (eds), *The Great Plains: environment and culture.* Lincoln: University of Nebraska Press, 99–118.

Hudson, J.C. 1996: *The geographer's Great Plains.* Occasional Publications in Geography. Manhattan: Kansas State University.

Hunter, F. 1953: *Community power structure: a study of decision makers.* Chapel Hill: University of North Carolina Press.

Hunter, F. 1980: *Community power succession: Atlanta's policy makers revisited.* Chapel Hill: University of North Carolina Press.

IJC 1988: *Revised Great Lakes Water Agreement of 1978.* Windsor: International Joint Commission.

Ilbery, B., Chiotti, Q. and Rickard, T. (eds) 1997: *Agricultural restructuring and sustainability, a geographical perspective.* Wallingford: CAB International.

Illeris, S. 1996: *The service economy: a geographical approach.* Chichester: Wiley.

Imhoff, G. 1991: English in the United States. *New Language Planning Newsletter* **5**. Mysore: Central Institute of Indian Languages Press 5, 1–2.

Indian and Northern Affairs Canada 1996a: Canada's first diamond mine one step closer. *News Release,* 8 August. Ottawa.

Indian and Northern Affairs Canada 1996b: *Northern oil and gas annual report 1995.* Ottawa.

Indian and Northern Affairs Canada 1996c: *Basic departmental data.* Ottawa.

Ip, G. 1996: The borderless world. *The Globe and Mail,* 6 July, D1.

Jacobs, J. 1965: *The death and life of great American cities.* Harmondsworth: Penguin.

Jackson, K.J. 1985: *Crabgrass frontier: the suburbanization of the United States.* New York: Oxford University Press.

James, P. 1941: *Latin America.* London: Cassell.

Janelle, D.G. (ed.) 1992: *Geographical snapshots of North America: commemorating the 27th Congress of the International Geographical Union and Assembly.* New York: Guilford Press.

Johnson, H.B. 1976: *Order upon the land.* New York: Oxford University Press.

Johnson, M.L. 1997: To restructure or not to restructure: contemplations on postwar industrial geography in the US South. *Southeastern Geographer* **37**, 162–92.

Jones, K.G. and Simmons, J.W. 1990: *The retail environment.* London: Routledge.

Kearey, P. and Vine, F.J. 1990: *Global tectonics.* Oxford: Blackwell Scientific Publications.

Keith, R.F. 1995–96: Aboriginal communities and mining in northern Canada. *Northern Perspectives* **23**(3–4), 1–9.

Kelly, C. 1994: Midwifing the new regional order [publisher's note]. *The New Pacific* **10**(6).

Kerr, D., Holdsworth, D.W., Laskin, S.L. and Matthews, G.J. 1990: *Historical atlas of Canada, Vol. III: addressing the twentieth century, 1891–1961.* Toronto: University of Toronto Press.

Knox, P. 1994: *Urbanization: an introduction to urban geography*. Englewood Cliffs, NJ: Prentice Hall.

Knox, P. 1997: Globalization and urban economic change. *Annals of the American Academy of Political and Social Science* **551**, 17–27.

Kodras, J. 1997: The changing map of American poverty in an era of economic restructuring and political realignment. *Economic Geography* **73**, 67–93.

Köppen, W. 1936: Das Geographische System der Klimate. *Handbuch der Klimatologie* **1**, 1–44.

Kresl, K. (ed.) 1995: *North American cities and the global economy*. London: Sage Publications.

Krotz, L. 1991: Dammed and diverted. *Canadian Geographic* **111**(1), 36–44.

Kutscher, R.E. 1988: Growth of services employment in the United States. In Guile, B.R. and Quinn, J.B. (eds), *Technology in services*. Washington, DC: National Academy of Sciences, 47–75.

Labov, W. and Harris, W.A. 1986: *De facto segregation of black and white vernaculars*. In Sankoff, D. (ed.), *Current issues in linguistic theory 53: diversity and diachrony*. Amsterdam: John Benjamins Publishing, 1–24.

Lamont, L. 1994: *Breakup: the coming end of Canada and the stakes for America*. New York: W.W. Norton.

Langston, N. 1995: *Forest dreams, forest nightmares: the paradox of old growth in the inland West*. Seattle: University of Washington Press.

Laporte, P.E. 1984: Status language planning in Québec: an evaluation. In Bouris, R.Y. (ed.), *Conflict and language planning in Québec*. Clevedon, PA: Multilingual Matters, 53–80.

Lasher, D. 1998: Golden – and global – in California. *Los Angeles Times*, 8 January, A1, 22.

Lasley, P., Leistritz, L., Lobao, L.M. and Meyer, K. (eds) 1995: *Beyond the amber waves of grain: an examination of social and economic restructure in the heartland*. Boulder: Westview Press.

Lee, F.R. 1994: English in black and white. *The Globe and Mail*, 10 January, A13.

Lees, L. (ed.) 1998: Vancouver: a portfolio. *Urban Geography* **19**(4) (special issue).

Leibowicz, J. 1992: Official English: another Americanization campaign? In Crawford, J. (ed.), *Language loyalties: a source book on the official English controversy*. Chicago: University of Chicago Press, 101–11.

Lemon, J. 1996: *Liberal dreams and nature's limits: great American cities since 1600*. Toronto: Oxford University Press.

Levesque, R. 1968: *An option for Québec*. Toronto: McClelland & Stewart Publishers.

Levine, A.G. 1982: *Love Canal: science, politics and people*. Lexington, MA: Lexington Books.

Levine, M.V. 1990: *The reconquest of Montréal*. Philadelphia: Temple University Press.

Levine, M.V. 1997: *La reconquête de Montréal*. Montréal: VLB Editeur.

Lewis, P.F. 1979: Axioms for reading the landscape. In Meinig, D.W. (ed.), *The interpretation of ordinary landscapes*. New York: Oxford University Press.

Lewis, P. 1983: The galactic metropolis. In Platt, R.H. and Macinko, G. (eds), *Beyond the urban fringe: land use issues of non-metropolitan America*. Minneapolis: University of Minnesota Press, 23–49.

Ley, D. 1974: *The black inner city as frontier outpost*. Washington, DC: Association of American Geographers.

Ley, D. 1995: Between Europe and Asia: the case of the missing sequoias. *Ecumene* **2**(2), 185–210.

Limerick, P.N. 1994: The adventures of the frontier in the twentieth century. In Grossman, J.R. (ed.), *The frontier in American culture*. Berkeley: University of California Press, 66–102.

Litwack, L.F. 1998: *Trouble in mind: black southerners in the age of Jim Crow*. New York: Alfred A. Knopf.

Locke, J. [1688] 1965: *Two treatises of government*, Laslett, P. (ed.). New York: Signet.

Lonsdale, R.L. and Browning, C.E. 1971: Rural–urban locational preferences of southern manufacturers. *Annals of the Association of American Geographers* **61**, 255–68.

Lord, J.D. 1996: The new geography of cotton production in North Carolina. *Southeastern Geographer* **36**, 93–112.

Lumley, E. 1983: *Speaking note – the Hon. Ed*

Lumley to the House of Commons on the Industrial and Regional Development Program, 27 June. Ottawa: Department of Regional Industrial Expansion, 1–2.

Lynch, K. 1960: *The image of the city*. Harmondsworth: Penguin.

Mackenzie, E. 1994: *Privatopia: homeowner associations and the rise of residential private government*. New Haven, CT: Yale University Press.

Mackenzie, S. and Norcliffe, G. 1997: Restructuring in the Canadian newsprint industry. *The Canadian Geographer* **41**, 2–6.

Maclean's 1997: 21 July.

MacMillan Bloedel Ltd n.d.: *Future forests: a look at the new forests MacMillan Bloedel is growing*. Vancouver: MacMillan Bloedel Ltd.

Mair, A., Florida, R. and Kenney, M. 1988: The new geography of automobile production: Japanese transplants in North America. *Economic Geography* **64**, 352–73.

Markusen, A.R. 1996: Sticky places in slippery space: a typology of industrial districts. *Economic Geography* **72**(3), 293–313.

Marschner, F.J. 1959: *Land use and its patterns in the United States*. Agriculture Handbook 153. Washington, DC: US Department of Agriculture.

Marsh, G.P. [1864] 1965: *Man and nature*, Lowenthal, D. (ed.). Cambridge, MA: Belknap Press.

Marsh, J. 1988: Railway history. In Marsh, J.H. (ed.), *The Canadian encyclopaedia, Vol. III*. 2nd edn, Edmonton: Hurtig, 1821–3.

Martin, L. 1993: *Pledge of allegiance: the Americanization of Canada in the Mulroney years*. Toronto: McClelland & Stewart.

Martinez, O. 1988: *Troublesome border*. Tucson: University of Arizona Press.

Massey, D. and Denton, N. 1993: *American apartheid: segregation and the making of the underclass*. Cambridge: Harvard University Press.

McCann, L.D. (ed.) 1982: *A geography of Canada: heartland and hinterland*. Scarborough, Ontario: Prentice Hall Canada Inc.

McDonald, M., *et al.*, 1997: Voices from the Bay: traditional ecological knowledge of Inuit and Cree in the Hudson Bay bioregion. *Northern Perspectives* **25**(1), 1–15.

McKnight, T.L. 1997: *Regional geography of the United States and Canada*. 2nd edn, Upper Saddle River: Prentice Hall.

McMichael, P. (ed.) 1994: *The global restructuring of agro-food systems*. Ithaca: Cornell University Press.

McNaught, K. 1969: *The Pelican history of Canada*. London: Penguin Books.

Meade, R.H. (ed.) 1996: *Contaminants in the Mississippi River 1987–92*. United States Geological Survey Circular 1133.

Meinig, D.W. 1986: *The shaping of America: a geographical perspective on 500 years of history, Vol. I: Atlantic America, 1492–1800*. New Haven: Yale University Press.

Meinig, D.W. 1993: *The shaping of America: a geographical perspective on 500 years of history, Vol. II: continental America, 1800–1867*. New Haven: Yale University Press.

Melbin, M. 1987: *Night as frontier: colonizing the world after dark*. New York: The Free Press.

Merchant, C. 1988: *Ecological revolutions: nature, gender and science in New England*. Chapel Hill: University of North Carolina Press.

Meyer, D.R. 1987: The national integration of regional economies 1860–1920. In Mitchell, R.D. and Groves, P.A. (eds), *North America: the historical geography of a changing continent*. London: Hutchinson, 321–46.

Milliman, J.D. and Meade, R.H. 1983: Worldwide delivery of river sediment to the oceans. *Journal of Geology* **91**, 1–21.

Mills, E.S. and McDonald, J.F. (eds) 1994: *Sources of metropolitan growth*. New Brunswick, NJ: Rutgers University Press.

Mishel, L and Frankel, D.M. 1991: *The state of working America 1990–1991*. Armonk: M.E. Sharpe.

Mitchell, M. 1936: *Gone with the wind*. New York: Macmillan.

Mitchell, R.D. and Groves, P.A. (eds) 1987: *North America: the historical geography of a changing continent*. Totowa, NJ: Rowman & Littlefield.

Muller, P.O. 1981: *Contemporary suburban America*. Englewood Cliffs: Prentice Hall.

Nader, G.A. 1972: *Cities of Canada, Vol. II: profiles of fifteen metropolitan centres*. Toronto: Macmillan.

Nassichuk, W.W. 1987: Forty years of northern non-renewable resource development. *Arctic* **40**(4), 274–84.

National Geographic Society 1988: *Historical atlas of the United States*. Washington, DC: National Geographic Society.

National Library of Canada 1998: The territorial evolution of Canada (/maps/htm); Confederation 1867 (/confed.htm). Information found at *http://www.nlc-bnc.ca/confed*

Newman, K. 1993: *Declining fortunes: the withering of the American dream*. New York: Basic Books.

Newman, P.C. 1995: *The Canadian revolution, 1985–1995: from deference to defiance*. Toronto: Penguin Books Canada.

Norcliffe, G. 1996: Mapping deindustrialization: Brian Kipping's 'Landscapes of Toronto'. *The Canadian Geographer/Le Géographe Canadien* **41**, 266–72.

O'Connell, S. 1990: *Imagining Boston: a literary landscape*. Boston: Beacon Press.

O'Connor, J.E. 1993: *Hydrology, hydraulics, and geomorphology of the Bonneville flood*. Geological Society of America Special Paper 274.

O'Loughlin, J. 1997: Economic globalization and income inequality in the United States. In Staeheli, L. Kodras, J. and Flint, C. (eds), *State restructuring in America: implications for a diverse society*. Thousand Oaks: Sage, 21–40.

Opie, J. 1994: *The law of the land: two hundred years of American farmland policy*. Lincoln: University of Nebraska Press.

Organization for Economic Cooperation and Development 1990: *National Accounts, Vols I and II*. Paris: OECD.

OTA 1995: *The technological reshaping of metropolitan America*. Washington, DC: US Congress, Office of Technology Assessment.

Pacific Northwest Economic Region 1998: Information found at *http://www.pnwer.org/index.html*

Page, P. 1996: Across the great divide: agriculture and industrial geography. *Economic Geography* **72**, 376–97.

Pandit, K. 1997: The Southern migration turnaround and current patterns. *Southeastern Geographer* **37**, 238–50.

Paterson, J.H. 1960: *North America: a regional geography*. London: Oxford University Press.

Paterson, J.H. 1994: *North America: a geography of the United States and Canada*. 9th edn, Oxford: Oxford University Press.

Paul, A.H. 1992: The Popper proposals for the Great Plains: a view from the Canadian Prairies. *Great Plains Research* **2**, 199–222.

Paullin, C.O. and Wright, J.K. 1932: *Atlas of the historical geography of the United States*. Washington, DC: Carnegie Institution and the American Geographical Society.

Pelly, D.A. 1993: Dawn of Nunavut. *Canadian Geographic* **113**(2), 20–8.

Perl, A. and Pucher, J. 1995: Transit in trouble? The policy challenge posed by Canada's changing urban mobility. *Canadian Public Policy* **21**, 261–83.

Peters, E.J. 1992: Protecting the land under modern land claim agreements: the effectiveness of the environmental regime negotiated by the James Bay Cree in the James Bay and Northern Québec Agreement. *Applied Geography* **12**, 133–45.

Petroleum Communication Foundation 1996: *News Release*, 6 February. Calgary.

Phillips, K. 1991: *The politics of rich and poor*, New York: Random House.

Polèse, M. 1990: La thèse du déclin économique de Montréal, revue et corrigée. *L'actualité économique* **66**(2), 133–46.

Pollan, M. 1991: *Second nature: a gardener's education*. New York: Atlantic Monthly Press.

Pollard, J. and Storper, M. 1996: A tale of twelve cities: metropolitan employment change in dynamic industries in the 1980s. *Economic Geography* **72**(1), 1–22.

Popper, D.E. and Popper, F.J. 1987: Great Plains: from dust to dust. *Planning*, December, 12–18.

Popper, F.J. and Popper, D.E. 1994: Great Plains: checkered past, hopeful future. *Forum for Applied Research and Public Policy* **9**(4), 89–100.

Porter, M.E. 1990: *The competitive advantage of nations*. New York: The Free Press.

Porteus, J.D. 1984: Beyond the company town: resource frontier settlements in Canada. In Gentilcore, R.L. (ed.), *China in Canada: a dialogue on resources and development*. Hamilton: McMaster University, Department of Geography.

Prest, V.K., Grant, D.R. and Rampton, V.N.

1967: *Glacial map of Canada,* Map 1253A. Ottawa: Geological Survey of Canada.

Prunty, M.C. 1951: Recent quantitative changes in the cotton regions of the southern states. *Economic Geography* **27**, 189–208.

Prunty, M.C. 1977: Two American Souths: the past and the future. *Southeastern Geographer* **17**, 1–24.

Quinn, F. 1991: As long as the rivers run: the impacts of corporate water development on native communities in Canada. *Canadian Journal of Native Studies* **11**(1), 137–54.

Quinn, J.B. 1992: *Intelligent enterprise: a knowledge and service based paradigm for industry.* New York: The Free Press.

Rabin, J. 1996: *Badland.* London: Picador.

Randall, J.E. and Ironside, R.G. 1996: Communities on the edge: an economic geography of resource-dependent communities in Canada. *The Canadian Geographer/Le Géographe Canadien* **40**(1), 17–35.

Ray, B. 1992: *Immigrants in a 'multicultural' Toronto,* (unpubl.) Ph.D. dissertation. Kingston, Ontario: Queen's University, Department of Geography.

Reader's Digest 1997: *Reader's Digest Great World Atlas.* Pleasantville, NY: Reader's Digest Association Inc., p. 59.

Rees, R. 1988: *New and naked land.* Saskatoon: Western Producer Prairie Books.

Reich, R. 1992: *The work of nations.* New York: Random House.

Reinelt, E.R. (ed.) 1971: *Proceedings of the Peace–Athabasca Delta symposium.* Edmonton: University of Alberta, Water Resources Centre.

Reynolds, H. 1982: *The other side of the frontier: aboriginal resistance to the European invasion of Australia.* Harmondsworth: Pelican Books.

Richardson, B.C. 1992: *The Caribbean in the wider world, 1492–1992.* Cambridge: Cambridge University Press.

Riche, M.F. 1991: We're all minorities now. *American Demographics* **28**, 26–34.

Riebsame, W.E., and Robb, J. 1997: *Atlas of the new West : portrait of a changing region.* New York: W.W. Norton.

Rifkin, J. 1995: *The end of work: the decline of the global labor force and the dawn of the post-market era.* New York: G.P. Putnam's Sons.

Robinson, E.B. 1966: *History of North Dakota.* Lincoln: University of Nebraska Press.

Robinson, I.M. 1962: *New industrial towns on Canada's resource frontier.* Research Paper 73. Chicago: University of Chicago, Department of Geography.

Robinson, J.L. 1982:*The physical environment of Canada and the evolution of settlement patterns.* Vancouver: Talon Books.

Rodrik, D. 1997: *Has globalization gone too far?* Washington, DC: Institute for International Economics.

Rogers, J.C. and Rohli, R.V. 1991: Florida citrus freezes and polar anticyclones in the Great Plains. *Journal of Climate* **4**, 1103–13.

Rohmer, R. 1970: *The green North.* Toronto: Maclean-Hunter.

Rooney, J.F. Jr., Zelinsky, W. and Louder D.R. (eds) 1982: *This remarkable continent: an atlas of United States and Canadian society and culture.* College Station: Texas A & M University Press.

Roosevelt, T. 1905: Forestry and foresters. *Proceedings of the Society of American Foresters* **1**, 3–9.

Rothblatt, D.N. 1994: North American metropolitan planning: Canadian and US perspectives. *Journal of the American Planning Association* **60**, 501–20.

Rouse, W.R., *et al.* 1992: Damming James Bay: I: potential impacts on coastal climate and the water balance. *The Canadian Geographer/Le Géographe Canadien* **36**(1), 17–35.

Royal Commission on Maritime Claims 1926: *Report of the Royal Commission on Maritime Claims.* Ottawa: Royal Commission on Maritime Claims.

Royal Commission on the Economic Union and Development 1986: *Prospects for Canada.* Ottawa: Royal Commission on the Economic Union and Development.

Rubenstein, J.M. 1992: *The changing US auto industry.* New York: Routledge.

Rugg, D.S. and Rundquist, D.C. 1981: Urbanization in the Great Plains: trends and prospects. In Lawson, M.P. and Baker, M.E. (eds), *The Great Plains: perspectives and prospects.* Lincoln: University of Nebraska Press, 221–46.

Rutheiser, C. 1996: *Imagineering Atlanta: the politics of place in the city of dreams.* London: Verso.

Salisbury, R.F. 1986: *A homeland for the Cree:*

regional development in James Bay 1971–1981. Montréal and Kingston: McGill-Queen's University Press.

Sassen, S. 1991: *The global city*. Princeton, NJ: Princeton University Press.

Sassen, S. 1994: *Cities in the world economy*. Thousand Oaks, CA: Pine Forge Press.

Sauer, C.O. 1968: *Northern mists*. Berkeley: University of California Press.

Savoie, D. 1986: *Regional economic development: Canada's search for solutions*. Toronto: Toronto University Press.

Saxenian, A. 1996: *Regional advantage: culture and competition in Silicon Valley and Route 128*. Cambridge, MA: Harvard University Press.

Schell, P. and Hamer, J. 1993: *What is the future of Cascadia?* Seattle: Discovery Institute.

Schertz, L.P. (ed.) 1979: *Another revolution in US farming?* Washington, DC: US Department of Agriculture.

Schmidt, R.J. 1993: Language policy conflict in the United States. In Young, C. (ed.), *The rising tide of cultural pluralism: the nation state at bay?* Madison: University of Wisconsin Press, 73–115.

Schwantes, C.A. 1996: *The Pacific Northwest: an interpretative history*. Lincoln: University of Nebraska Press.

Schwartz, J. 1994: Air pollution and daily mortality: a review and meta analysis. *Environmental Research* **64**, 36–52.

Scott, A.J. 1993: *Technopolis: high-technology industry and regional development in southern California*. Berkeley: University of California Press.

Semple, R.K. 1997: Quaternary places in Canada. In Britton, J.N.H. (ed.), *The Canadian space economy*. Montréal and Kingston: McGill-Queen's University Press.

Sénécal, G. and Manzagol, C. 1993: Montréal ou la metamorphose des territoires. *Cahiers de Géographie du Québec* **37**(3), 351–70.

Shortridge, J.R. 1989: *The Middle West: its meaning in American culture*. Lawrence: University of Kansas Press.

Siegel, A. 1983: *Politics and the media in Canada*. Toronto: McGraw-Hill Ryerson.

Simeon, R. 1990: *Thinking about constitutional futures: a framework*. Paper presented at the Global Competition and Canadian Federalism Conference. Toronto: University of Toronto, mimeo.

Simpson-Housley, P. and Norcliffe, G. (eds) 1992: *A few acres of snow: literary and artistic images of Canada*. Toronto: Dundurn Press.

Smith, D.M. 1994: *Geography and social justice*. Oxford: Blackwell Publishers.

Smith, D.M. 1996: Atlanta: inequality and social justice in the Olympic city. *Geography Review* **18**(1), 2–5.

Smith, H.N. 1950: *Virgin land: the American West as symbol and myth*. Cambridge, MA: Harvard University Press.

Smith, P. 1975: *Brinco: the story of Churchill Falls*. Toronto: McClelland & Stewart.

Smith, R. 1990: Let's look at the facts. *Vancouver Sun*, 30 November, A, 15.

Smith, T.G., and Wright, H. 1989: Economic status and the role of hunters in a modern Inuit village. *Polar Record* **25**(153), 93–8.

Sorensen, A.A. and Greene, R.P. 1997: *Farming on the edge*. DeKalb: North Illinois University, American Farmland Trust and Center for Agriculture in the Environment.

Sparling, E., Wilken, K. and McKenzie, J. 1993: *Marketing fresh organic produce in Colorado supermarkets*. Davis: University of California, Sustainable Agriculture Research and Education Program.

Stabler, J.C., Tolley, G. and Howe, E.C. 1990: Fur trappers in the Northwest Territories: an economic analysis of the factors influencing participation. *Arctic* **43**(1), 1–8.

Stabler, J.C. and Olfert, M.R. 1996: *The changing role of rural communities in an urbanizing world: Saskatchewan – an update to 1995*. Regina: University of Regina, Canadian Plains Research Center.

Stabler, J.C., Olfert, M.R. and Fulton, M. 1992: *The changing role of rural communities in an urbanizing world: Saskatchewan 1961–1990*. Regina: University of Regina, Canadian Plains Research Center.

Statistics Canada 1989: *Census Metropolitan Areas: dimensions*. Ottawa: Minister of Regional Industrial Expansion and the Minister of State for Science and Technology, Canada Cat. 93–156.

Statistics Canada 1992a: *Census divisions and census subdivisions*. 1991 Census of Ottawa: Supply and Services Canada.

Statistics Canada 1992b: *Age, sex and marital status.* 1991 Census of Canada. Ottawa: Supply and Services Canada.

Statistics Canada 1993: *Selected income statistics.* 1991 Census of Canada. Ottawa: Industry, Science and Technology Canada.

Statistics Canada 1995: *Profile of Canada's aboriginal population.* 1991 Census of Canada. Ottawa: Industry, Science and Technology Canada.

Statistics Canada 1996: *Catalogue 88–001.* Ottawa: Statistics Canada.

Statistics Canada 1997a: *1996 Census of Agriculture.* Ottawa: Statistics Canada.

Statistics Canada 1997b: *Canada yearbook.* Ottawa: Statistics Canada.

Stevens, B. and Moore, C. 1980: A critical review of shift-share analysis. *Journal of Regional Science* **30**, 420–50.

Stewart, J.H. 1978: Basin-range structure in western North America: a review. In Smith, R.B. and Eaton, G.P. (eds), *Cenozoic tectonics and regional geophysics of the Western Cordillera.* Geological Society of America Memoir 152, 1–31.

Stoddard, E.R., Nostrand, R.L. and West, J.P. 1983: *Borderlands source book: a guide to the literature on northern Mexico and the American Southwest.* Norman, Oklahoma: University of Oklahoma Press.

Stöffler, D., Deutsch, A., Avermann, M. *et al.* 1994: The formation of the Sudbury structure, Canada: toward a unified impact model. In Dressler, B.O., Grieve, R.A.F. and Sharpton, V.L. (eds), *Large meteorite impacts and planetary evolution.* Geological Society of America Special Paper 293, 303–18.

Stone, C.N. 1976: *Economic growth and neighborhood discontent: system bias in the urban renewal program of Atlanta.* Chapel Hill: University of North Carolina Press.

Stone, C.N. 1989: *Regime politics: governing Atlanta, 1946–1988.* Lawrence: University of Kansas Press.

Stover, C.W. and Coffman, J.L. 1993: *Seismicity of the United States, 1568–1989* (revised). United States Geological Survey Professional Paper 1527.

Sunday Herald 1987: PM launches new agency for Atlantic Canada. 7 June, 1.

Supply and Services 1981: *Fiscal federalism in Canada.* Ottawa: Supply and Services, 157–76.

Suro, R. 1998: *Strangers among us: how Latino immigration is transforming America.* New York: Alfred A. Knopf.

Tapscott, D. 1996: *The digital economy: promise and peril in the age of networked intelligence.* New York: McGraw-Hill.

Termote, P. and Gauvreau, D. 1988: *La situation demolinguistique au Québec.* Québec: Conseil de la Langue Française.

Thernstrom, J.H. and Thernstrom, A. 1997: *America in black and white: one nation indivisible.* New York: Simon & Schuster.

Thomas, J.K., Howell, F.M., Wang, G. and Albrecht, D.E. 1996: Visualizing trends in the structure of US agriculture. *Rural Sociology* **61**, 349–74.

Thompson, J.H. 1992: Canada's quest for cultural sovereignty: protection, promotion and popular culture. In Randall, S.J., Konrad, H. and Silverman, S. (eds), *North America without borders? Integrating Canada, the United States and Mexico.* Calgary, Alberta: University of Calgary Press.

Thoreau, H.D. 1937: *The works of Thoreau,* Canby, H.S. (ed.). Boston: Houghton Mifflin.

Thorsell, W. 1989: Meech Lake: shadow-boxing on the Plains of Abraham. *The Globe and Mail,* 30 December, A14.

Thurow, L.C. 1996: *The future of capitalism.* New York: William Morrow & Co.

Timoney, K.P., La Roi, G.H., Zoltai, S.C. and Robinson, A.L. 1992: The high subarctic forest-tundra of northwestern Canada: position, width and vegetation gradient in relation to climate. *Arctic* **45**(1), 1–9.

Tobias, T.N. and Kay, J.J. 1994: The bush harvest in Pinehouse, Saskatchewan. *Arctic* **47**(3), 207–21.

Trachtenberg, A. 1982: *The incorporation of America: culture and society in the gilded age.* New York: Hill & Wang.

Trubowitz, P. 1998: *Defining the national interest: conflict and change in American foreign policy.* Chicago: University of Chicago Press.

Tschetter, J. 1987: Producer services industries: why are they growing so rapidly? *Monthly Labor Review,* December, 31–40.

Tucker, M. 1980: *Canadian foreign policy: con-*

temporary issues and themes. Toronto: McGraw-Hill Ryerson.

Turner, F.J. 1893: The significance of the frontier in American history. *Report of the American Historical Association*, 199–227.

Turner, F.J. 1896: The problem of the West. *The Atlantic Monthly* **78**, 289–97.

Tweto, O. 1975: Laramide (late Cretaceous-early Tertiary) orogeny in the southern Rocky Mountains. In Curtis, B.F. (ed.), *Cenozoic history of the southern Rocky Mountains*. Geological Society of America Memoir 144, 1–44.

Twitchell, K. 1991: The not-so-pristine Arctic. *Canadian Geographic* **111**(1), 53–60.

United States Bureau of the Census 1960: *Historical statistics of the United States: colonial times to 1957*. Washington, DC: US Bureau of the Census.

United States Department of Commerce, Bureau of the Census 1984: *1982 Census of Agriculture*. Washington, DC: US Government Printing Office.

United States Department of Commerce, Bureau of the Census 1994: *1992 Census of Agriculture*. Washington, DC: US Government Printing Office.

United States Environmental Protection Agency 1996: *National air quality emissions trend report 1995*. EPA 454/R-96–005. Research Triangle Park: Office of Air Quality Planning and Standards.

United States Immigration and Naturalization Service 1998: Country of origin (/illegalalien/index.html); immigration and emigration by decade: 1901–90 (/299.html); illegal alien resident population (/300/html). Tables found at *http://www.ins.usdoj.gov/stats*

Usher, P.J., Tough, F.J. and Galois, R.N. 1992: Reclaiming the land: aboriginal title, rights and land claims in Canada. *Applied Geography* **12**, 109–32.

Vance, J.E. 1986: *Capturing the horizon*. Baltimore: Johns Hopkins University Press.

Vancouver Courier 1997: Fashionably fit. 1 October, 1, 4–5.

Vergara, C.J. 1995: *The new American ghetto*. New Brunswick, NJ: Rutgers University Press.

Vincent, S., and Bowers, G. (eds) 1988: *Baie James et Nord québécois: dix ans après/ James Bay and northern Quebec: ten years after*. Montréal: Recherches amérindiennes au Québec.

Waddell, E. and Gunn, A. 1998: Québec: a place and a people. In McCann, L. and Gunn, A. (eds), *Heartland and hinterland: a regional geography of Canada*. Toronto: Prentice Hall Canada, 146–68.

Waldram, J.B. 1984: Hydroelectric development and the process of negotiation in northern Manitoba, 1960–1977. *Canadian Journal of Native Studies* **4**(2), 205–39.

Wallace, J.M., Zhang, Y. and Renwick, J.A. 1995: Dynamic contribution to hemispheric mean temperature trends. *Science* **270**, 780–3.

Ward, D. 1971: *Cities and immigrants*. New York: Oxford University Press.

Ward, D. (ed.) 1979: *Geographic perspectives on America's past*. New York: Oxford University Press.

Warf, B. and Holly, B. 1997: The rise and fall and rise of Cleveland. *Annals of the American Academy of Political and Social Science* **551**, 208–21. Webb, W.P. 1931: *The Great Plains*. Boston: Ginn.

Webb, W.P. 1953: *The great frontier*. London: Secker & Warburg.

Webster, G.R. and Samson, S.A. 1992: On defining the Alabama black belt: historical changes and variations. *Southeastern Geographer* **32**, 163–72.

Wein, E.E. and Freeman, M.M.R. 1995: Frequency of traditonal food use by three Yukon First Nations living in four communities. *Arctic* **48**(2), 161–71.

Whalley, J. and Trela, I. 1986: *Regional aspects of Confederation*. Ottawa: Royal Commission on the Economic Union and Development Prospects for Canada.

Whitaker, R. 1989: The overriding right. *Policy Options* **10**, 3–6.

White, S.E. 1994: Ogallala oases: water use, population redistribution, and policy implications in the High Plains of western Kansas, 1980–1990. *Annals of the Association of American Geographers* **84**, 29–45.

Whitford, H.N. and Craig, R.D. 1918: *Forests of British Columbia*. Ottawa: Commission of Conservation.

Whittaker, L.M. and Horn, L.H. 1984: Northern Hemisphere extratropical cyclone activity for four mid-season months. *Journal of Climatology* **4**, 297–310.

Williams, G.P. and Guy, H.P. 1973: *Erosional and depositional aspects of Hurricane Camille in Virginia, 1969.* United States Geological Survey Professional Paper 804.

Williams, M. 1989: *Americans and their forests: a historical geography.* New York: Cambridge University Press.

Williams, R. 1983: *Keywords: a vocabulary of culture and society.* London: Flamingo.

Williams, W.A. 1969: *The roots of the modern American empire: a study of the growth and shaping of social consciousness in a marketplace society.* New York: Random House.

Wilson, D. (ed.) 1997: Globalization and the changing US city. *Annals of the American Academy of Political and Social Science,* **551**, May (special issue).

Wimberley, R.C. 1987: Dimensions of US agriculture: 1969–1982. *Rural Sociology* **52**, 445–61.

Winsberg, M.D. 1997: The great southern agricultural transformation and its social consequences. *Southeastern Geographer* **37**, 193–213.

Wolman, W. and Colamosca, A. 1997: *The Judas economy: the triumph of capital and the betrayal of work.* Reading: Addison-Wesley.

Wonders, W.C. (ed.) 1971: *Canada's changing North.* Toronto: McClelland & Stewart.

Wonders, W.C. 1981: Northern resources development. In Mitchell, B. and Sewell, W.R.D. (eds), *Canadian resource policies: problems and prospects.* Toronto: Methuen, 56–83.

Wonders, W.C. 1984a: The Canadian North: its nature and prospects. *Journal of Geography* **83**(5), 226–33.

Wonders, W.C. 1984b: *Overlapping land use and occupancy of Dene, Métis, Inuvialuit and Inuit in the Northwest Territories.* Ottawa: Indian and Northern Affairs Canada.

Wonders, W.C. 1987: The changing role and significance of Native people in Canada's Northwest Territories. *Polar Record* **23**(147), 661–71.

Wonders, W.C. 1990: Tree-line and politics in Canada's Northwest Territories. *Scottish Geographical Magazine* **106**(1), 54–60.

Wonnacott, R.J. 1991: The Canada–US free trade agreement: its broad potential effects on the Canadian economy and on Canada's regions. In Remie, C.H.W. and Lacroix, J-M. (eds), *Canada on the threshold of the twenty-first century.* Amsterdam: John Benjamins Publishing Company.

Woollard, G.P. and Joesting, H.R. 1965: *Bouguer gravity anomaly map of the United States.* American Geophysical Union and United States Geological Survey.

Wynn, G. 1987: Forging a Canadian nation. In Mitchell, R.D. and Groves, P.A. (eds), *North America: the historical geography of a changing continent.* London: Hutchinson, 373–409.

Yeates, M. 1998a: The industrial heartland: its changing role and internal structure. In McCann, L. and Gunn, A. (eds), *Heartland and hinterland: a regional geography of Canada.* Toronto: Prentice Hall Canada.

Yeates, M. 1998b *The North American city.* New York: Addison Wesley Longman.

Zazlow, M. (ed.) 1981: *A century of Canada's Arctic Islands, 1880–1980.* Ottawa: Royal Society of Canada.

Zelinsky, W. 1973: *The cultural geography of the United States.* Englewood Cliffs, NJ: Prentice Hall.

Zelinsky, W. 1992: *The cultural geography of the United States: a revised edition.* Englewood Cliffs, NJ: Prentice Hall.

INDEX

Chrétien, Jean 254
Churchill, Manitoba 31
Churchill, Ward 65
Churchill, Winston S. 266
Cities 191–206
  see also named cities
  Central cities/inner cities 196, 199
  Economic transformations 192
  Government 204
  Housing stock 203–4
  Inequality 193, 198–200
  Social dimensions 196–7
  Socio-geographical polarization 199–200, 219
  Transport 204
  US–Canada contrasts 193, 196–7
City of Angels 214
Civil Rights Act (US) 115
Civil Rights Movement 66
Civil War
  see War
Civilian employment (federal) 163–4
Clayoquot Sound, British Columbia 242
Clean air legislation 34
Cleveland, Ohio 44, 269
Climate 28–36
  Air circulation 29
  Climate change 35
  Climatic regions 29
  Cyclones 29
  Great Plains and Prairies 258
  Heating 28
  Ocean currents, effects of 28
  Polar front 29
  Pressure systems 29
  Regional climates
    Boreal and Polar 32–3
    East of Rockies, US 30–2
    Western (including mountains) 33–4
  Solar radiation 28
  Temperature range 28
Clinton, Bill 276
CMA 174–90
CMSA 174–90
CNN 208
Coal 260, 261, 264
Coastal processes 23
Cobscook River, Maine 41
Coca Cola 298
Cochise 63
Coefficients of industrial concentration 159–60
Coefficients of regional specialization 161–2
Cold War 314, 307–9
Colonialism 58, 306
Colorado 30, 34, 35
Colorado Plateau 34
Colstrip, Montana 261
Columbia River 240
Columbus, Christopher 61
Competitive shift 164–7
Confederation of Canada 130, 222
Congress, US 313
Connecticut 44, 267
Contaminant transport 23
Constitution, US 304
Contact with Europeans 59–61

Continental accretion 4–5
Continental destiny 306
Contras 313
Cortez, Hernando 50
Cotton 268, 270
Counterfactual geography 4–6
Covington, Ohio 269
Craig, Roland 44, 45
Cronon, W. 7, 42, 53
Crosby, A. 4
Cuba 304
Cubans 272
Cultural geography 90–1
Culture regions 91–101
Cyclones 29

Dakotas 32
Dallas-Fort Worth, Texas 266, 272, 273, 274, 276
Davies, W.K.D. 198
Davis Straight 29
de Bres, K.J. 263
De-industrialization 142, 168
  and urban social polarization 199
Delaware 266, 271
Democratic Party 266, 313
Denver, Colorado 31, 32
Department of Agriculture, US 138
Department of Indian Affairs 66
Department of Regional Economic Expansion 253
Department of Regional Industrial Expansion 253
Depression, Great 307, 308
Depression, Long 306
Detroit, Michigan 74, 117, 182, 194, 272
  Automobile manufacturing 149
  Ghetto 195
  Industrial change 200
Diamond, J. 4
Disney World 273
Dominion of Canada 64
Dowd, M. 240
Doyle, D.H. 269
Durham, North Carolina 273
Dutch colonists 74–5, 127

Economic development in US 308
Ecotopia 100, 240, 243, 247
Edgington, D. 244, 246, 247
Edge cities 202, 234
Education
  Montréal schools 227
  Native peoples 64
Ekstrom, F.H. 42
Elberton, Georgia 268
Ellesmere Island 59
Employment
  Change 157
  Restructuring in Los Angeles 217
  Shift-share analysis 164–6
Employment Assurance 252
Endangered Species Act 53
England 42, 43
Environment
  Pacific Northwest environmentalism 242–3
  Protection of 278
  Spotted owl controversy 53, 242